Biscuits of Number Theory

Cover illustration by Gregory Nemec
Cover design by Carole Goodman, Freedom by Design

© *2009 by*
The Mathematical Association of America (Incorporated)
Library of Congress Catalog Card Number 2008939107
ISBN 978-0-88385-340-5
Printed in the United States of America
Current Printing (last digit)
10 9 8 7 6 5 4 3 2 1

The Dolciani Mathematical Expositions

NUMBER THIRTY-FOUR

Biscuits of Number Theory

Edited by
Arthur T. Benjamin
and
Ezra Brown

To Gareth,
May all of your Biscuits
be tasty!

Bud Brown
11/6/2010

Published and Distributed by
The Mathematical Association of America

DOLCIANI MATHEMATICAL EXPOSITIONS

Council on Publications
Paul Zorn, *Chair*

Dolciani Mathematical Expositions Editorial Board
Underwood Dudley, *Editor*
Jeremy S. Case
Tevian Dray
Robert L. Devaney
Jerrold W. Grossman
Virginia E. Knight
Mark A. Peterson
Jonathan Rogness
Joe Alyn Stickles
James S. Tanton

The DOLCIANI MATHEMATICAL EXPOSITIONS series of the Mathematical Association of America was established through a generous gift to the Association from Mary P. Dolciani, Professor of Mathematics at Hunter College of the City University of New York. In making the gift, Professor Dolciani, herself an exceptionally talented and successful expositor of mathematics, had the purpose of furthering the ideal of excellence in mathematical exposition.

The Association, for its part, was delighted to accept the gracious gesture initiating the revolving fund for this series from one who has served the Association with distinction, both as a member of the Committee on Publications and as a member of the Board of Governors. It was with genuine pleasure that the Board chose to name the series in her honor.

The books in the series are selected for their lucid expository style and stimulating mathematical content. Typically, they contain an ample supply of exercises, many with accompanying solutions. They are intended to be sufficiently elementary for the undergraduate and even the mathematically inclined high-school student to understand and enjoy, but also to be interesting and sometimes challenging to the more advanced mathematician.

1. *Mathematical Gems*, Ross Honsberger
2. *Mathematical Gems II*, Ross Honsberger
3. *Mathematical Morsels*, Ross Honsberger
4. *Mathematical Plums*, Ross Honsberger (ed.)
5. *Great Moments in Mathematics (Before 1650)*, Howard Eves
6. *Maxima and Minima without Calculus*, Ivan Niven
7. *Great Moments in Mathematics (After 1650)*, Howard Eves
8. *Map Coloring, Polyhedra, and the Four-Color Problem*, David Barnette
9. *Mathematical Gems III*, Ross Honsberger
10. *More Mathematical Morsels*, Ross Honsberger
11. *Old and New Unsolved Problems in Plane Geometry and Number Theory*, Victor Klee and Stan Wagon
12. *Problems for Mathematicians, Young and Old*, Paul R. Halmos
13. *Excursions in Calculus: An Interplay of the Continuous and the Discrete*, Robert M. Young
14. *The Wohascum County Problem Book*, George T. Gilbert, Mark Krusemeyer, and Loren C. Larson
15. *Lion Hunting and Other Mathematical Pursuits: A Collection of Mathematics, Verse, and Stories by Ralph P. Boas, Jr.*, edited by Gerald L. Alexanderson and Dale H. Mugler
16. *Linear Algebra Problem Book*, Paul R. Halmos
17. *From Erdös to Kiev: Problems of Olympiad Caliber*, Ross Honsberger
18. *Which Way Did the Bicycle Go? ... and Other Intriguing Mathematical Mysteries*, Joseph D. E. Konhauser, Dan Velleman and Stan Wagon
19. *In Pólya's Footsteps: Miscellaneous Problems and Essays*, Ross Honsberger
20. *Diophantus and Diophantine Equations*, I. G. Bashmakova (Updated by Joseph Silverman and translated by Abe Shenitzer)
21. *Logic as Algebra*, Paul Halmos and Steven Givant
22. *Euler: The Master of Us All*, William Dunham
23. *The Beginning and Evolution of Algebra*, I. G. Bashmakova and G. S. Smirnova (Translated by Abe Shenitzer)
24. *Mathematical Chesnuts from Around the World*, Ross Honsberger
25. *Counting on Frameworks: Mathematics to Aid the Design of Rigid Structures*, Jack E. Graver

26. *Mathematical Diamonds*, Ross Honsberger
27. *Proofs that Really Count: The Art of Combinatorial Proof*, Arthur T. Benjamin and Jennifer J. Quinn
28. *Mathematical Delights*, Ross Honsberger
29. *Conics*, Keith Kendig
30. *Hesiod's Anvil: falling and spinning through heaven and earth*, Andrew J. Simoson
31. *A Garden of Integrals*, Frank E. Burk
32. *A Guide to Complex Variables* (MAA Guides #1), Steven G. Krantz
33. *Sink or Float? Thought Problems in Math and Physics*, Keith Kendig
34. *Biscuits of Number Theory*, edited by Arthur T. Benjamin and Ezra Brown

MAA Service Center
P. O. Box 91112
Washington, DC 20090-1112
1-800-331-1MAA   FAX: 1-301-206-9789

Contents

Introduction xi

Part I: Arithmetic   1

  1. **A Dozen Questions About the Powers of Two** 3
     James Tanton.
     Math Horizons, vol. 8, no. 1 (September 2001), pp. 5–10; 2002 Trevor Evans Award.

  2. **From 30 to 60 is Not Twice as Hard** 13
     Michael Dalezman.
     Mathematics Magazine, vol. 73, no. 2 (April 2000), pp. 151–153.

  3. **Reducing the Sum of Two Fractions** 17
     Harris S. Shultz and Ray C. Shiflett.
     Mathematics Teacher, vol. 98, no. 7 (March 2005), pp. 486–490.

  4. **A Postmodern View of Fractions and Reciprocals of Fermat Primes** 23
     Rafe Jones and Jan Pearce.
     Mathematics Magazine, vol. 73, no. 2 (April 2000), pp. 83–97; 2001 Allendoerfer
     Award.

  5. **Visible Structures in Number Theory** 39
     Peter Borwein and Loki Jörgenson.
     American Mathematical Monthly, vol. 108, no. 10 (December 2001), pp. 897–910;
     2002 Lester Ford Award.

  6. **Visual Gems of Number Theory** 53
     Roger B. Nelsen.
     Math Horizons, vol. 15, no. 3 (February 2008), pp. 7–9, 31.

Part II: Primes   59

  7. **A New Proof of Euclid's Theorem** 61
     Filip Saidak.
     American Mathematical Monthly, vol. 113, no. 10 (December 2006), pp. 937–938.

  8. **On the Infinitude of the Primes** 63
     Harry Furstenberg.
     American Mathematical Monthly, vol. 62, no. 5 (May 1955), p. 353.

  9. **On the Series of Prime Reciprocals** 65
     James A. Clarkson.
     Proceedings of the AMS, vol. 17, no. 2 (April 1966), p. 541.

 10. **Applications of a Simple Counting Technique** 67
     Melvin Hausner.
     American Mathematical Monthly, vol. 90, no. 2 (February 1983), pp. 127–129.

 11. **On Weird and Pseudoperfect Numbers** 69
     S. J. Benkoski and P. Erdős.
     Mathematics of Computation, vol. 28, no. 126 (April 1974), pp. 617–623.

 12. **A Heuristic for the Prime Number Theorem** 77
     Hugh L. Montgomery and Stan Wagon.
     Mathematical Intelligencer, vol. 28, no. 3 (2006), pp. 6–9.

13. A Tale of Two Sieves 85
Carl Pomerance.
Notices of the AMS, (December 1996), pp. 1473–1485; 2001 Conant Prize.

Part III: Irrationality and Continued Fractions   105

14. Irrationality of the Square Root of Two—A Geometric Proof 107
Tom M. Apostol.
American Mathematical Monthly, vol. 107, no. 9 (November 2000), pp. 841–842.

15. Math Bite: Irrationality of $\sqrt{m}$ 109
Harley Flanders.
Mathematics Magazine, vol. 72, no. 3 (June 1999), p. 235.

16. A Simple Proof that π is Irrational 111
Ivan Niven.
Bulletin of the AMS, vol. 53 (1947), p. 509.

17. π, e and Other Irrational Numbers 113
Alan E. Parks.
American Mathematical Monthly, vol. 93, no. 9 (November 1986), pp. 722–723.

18. A Short Proof of the Simple Continued Fraction of e 115
Henry Cohn.
American Mathematical Monthly, vol. 113, no. 1 (January 2006), pp. 57–61.

19. Diophantine Olympics and World Champions: Polynomials and Primes Down Under 121
Edward B. Burger.
American Mathematical Monthly, vol. 107, no. 9 (November 2000), pp. 822–829;
2004 Chauvenet Prize.

20. An Elementary Proof of the Wallis Product Formula for Pi 129
Johan Wästlund.
American Mathematical Monthly, vol. 114, no. 10 (December 2007), pp. 914–917.

21. The Orchard Problem 133
Ross Honsberger.
Mathematical Gems, Chapter 4, pp. 43–53, Dolciani Mathematical Expositions, MAA, 1973.

Part IV: Sums of Squares and Polygonal Numbers   141

22. A One-Sentence Proof that every Prime $p \equiv 1 \pmod 4$ is a Sum of Two Squares 143
D. Zagier.
American Mathematical Monthly, vol. 97, no. 2 (February 1990), p. 144.

23. Sum of Squares II 145
Martin Gardner and Dan Kalman.
Proofs Without Words: Exercises in Visual Thinking, Classroom Resource Materials, MAA, p. 78.

24. Sums of Squares VIII 147
Roger B. Nelsen.
Proofs Without Words II: More Exercises in Visual Thinking, Classroom Resource Materials, MAA, p. 88.

25. **A Short Proof of Cauchy's Polygonal Number Theorem** 149
    Melvyn B. Nathanson.
    Proceedings of the AMS, vol. 99, no. 1 (January 1987), pp. 22–24.

26. **Genealogy of Pythagorean Triads** 153
    A. Hall.
    Mathematical Gazette, vol. 54, no. 390 (December 1970), pp. 377–379.

Part V: Fibonacci Numbers   155

27. **A Dozen Questions About Fibonacci Numbers** 157
    James Tanton.
    Math Horizons, vol. 12, no. 3 (February 2005), pp. 5–9.

28. **The Fibonacci Numbers—Exposed** 167
    Dan Kalman and Robert Mena.
    Mathematics Magazine, vol. 76, no. 3 (June 2003), pp. 167–181.

29. **The Fibonacci Numbers—Exposed More Discretely** 183
    Arthur T. Benjamin and Jennifer J. Quinn.
    Mathematics Magazine, vol. 76, no. 3 (June 2003), pp. 182–192.

Part VI: Number-Theoretic Functions   195

30. **Great Moments of the Riemann zeta Function** 199
    Jennifer Beineke and Chris Hughes.
    Original article.

31. **The Collatz Chameleon** 217
    Marc Chamberland.
    Math Horizons, vol. 14, no. 2 (November 2006), pp. 5–8.

32. **Bijecting Euler's Partition Recurrence** 223
    David M. Bressoud and Doron Zeilberger.
    American Mathematical Monthly, vol. 92, no. 1 (January 1985), pp. 54–55.

33. **Discovery of a Most Extraordinary Law of the Numbers Concerning the Sum of Their Divisors** 225
    Leonard Euler. Translated by George Pólya.
    Mathematics and Plausible Reasoning, Volume I, Princeton University Press, (1954), pp. 90–98.

34. **The Factorial Function and Generalizations** 233
    Manjul Bhargava.
    American Mathematical Monthly, vol. 107, no. 9 (November 2000), pp. 783–799; 2003 Hasse Prize.

35. **An Elementary Proof of the Quadratic Reciprocity Law** 251
    Sey Y. Kim.
    American Mathematical Monthly, vol. 111, no. 1 (January 2004), pp. 48–50.

Part VII: Elliptic Curves, Cubes and Fermat's Last Theorem   255

36. **Proof Without Words: Cubes and Squares** 257
    J. Barry Love.
    Mathematics Magazine, vol. 50, no. 2 (March 1977), p. 74.

37. Taxicabs and Sums of Two Cubes 259
Joseph H. Silverman.
American Mathematical Monthly, vol. 100, no. 4 (April 1993), pp. 331–340;
1994 Lester Ford Award.

38. Three Fermat Trails to Elliptic Curves 273
Ezra Brown.
The College Mathematics Journal, vol. 31, no. 3 (May 2000), pp. 162–172;
2001 Pólya Award.

39. Fermat's Last Theorem in Combinatorial Form 285
W. V. Quine.
American Mathematical Monthly, vol. 95, no. 7 (August-September 1988), p. 636.

40. "A Marvelous Proof" 287
Fernando Q. Gouvêa.
American Mathematical Monthly, vol. 101, no. 3 (March 1994), pp. 203–222; 1995
Lester Ford Award.

About the Editors 311

Introduction

"Powdermilk Biscuits: Heavens, they're tasty and expeditious! They're made from whole wheat, to give shy persons the strength to get up and do what needs to be done." — Garrison Keillor, *A Prairie Home Companion*

You are probably wondering, "What exactly are biscuits of number theory?" In this book, we have an assortment of articles and notes on number theory, where each item is not too big, easily digested, and makes you feel all warm and fuzzy when you're through. We hope they will whet your appetite for more! Overall, we felt that the biscuit analogy hit the spot (in addition, one of the editors bakes biscuits for his students).

In this collection, we have chosen articles that we felt were exceptionally well-written and that could be appreciated by anyone who has taken (or is taking) a first course in number theory. This book could be used as a textbook supplement for a number theory course, especially one that requires students to write papers or do outside reading. Here are some of the possibilities:

- Each piece in the collection is a fine starting point for classroom discussions. After telling the Ramanujan story about sums of two cubes, an instructor can ask the class what questions this story might prompt. Students can be given a list of cubes and asked to verify Ramanujan's claim. Some of the material in Silverman's Taxicabs article can show them how following up on an innocent little remark can lead to some quite wonderful mathematics.
- The articles can be used as follow-ups to classroom presentations. For example, after introducing the Fibonacci numbers, have some students read the Tanton paper and assign one or more of the "Taking it Further" questions. Have others read the two matched "Fibonacci numbers—exposed" papers and describe how they are the same and how they are different.
- Students who see the standard proofs of the infinitude of the primes or the Quadratic Reciprocity Law and who are curious about other proofs can be directed to the alternative proofs in this collection and encouraged to present them to the class. This could lead to students' searching for other proofs in the literature and writing them up in a paper.
- Students in most courses in number theory learn about Paul Erdős. Have them read Erdős' paper in this collection, work out the details, and present them to the class, along with a determination (with proof) of the instructor's Erdős number.
- Zagier's one-sentence proof of the Two-Squares Theorem gives students an opportunity to work out the details of a very short proof. In doing this, they learn the mathematics behind that proof. Another possibility is for students to present Jackson's one-line proof that every prime $p = 8n + 3$ is of the form $a^2 + 2b^2$, and perhaps create proofs of their own for similar results.
- Many of the papers contain material that might give students a start on an undergraduate research project.
- For students who wonder how to do research in number theory, have them read

Pólya's translation of Euler's paper on sums of divisors, so they can see how one of the masters did it.
- The Arithmetic section contains articles that can give future secondary-school teachers ideas on how to introduce their students to patterns hidden in powers of two and fractions and to pictures hidden in numbers.

We have just mentioned many of the biscuits by title, but all 40 of them have interesting mathematics for number theorists, young and not-so-young. After some of the articles, we provide "second helpings" where readers can learn about further developments and other topics related to the article.

This project began at the MAA Northeastern Section meeting in November, 2004. During a break between talks, the two editors chatted about a possible invited paper session for MathFest 2005 in Albuquerque. Ten minutes later, we had a topic and a format: four speakers, each with a half-hour slot, would present what we called "Gems of Number Theory."

A half-hour after that, we had a list of potential speakers who would exactly fill the bill. A few weeks later, we had our program. The large and appreciative audience at MathFest 2005 heard Marc Chamberland speak on the Collatz $3x + 1$ Problem, Ed Burger on Diophantine Approximation, Jennifer Beineke on Great Moments of the Riemann zeta Function, and Roger Nelsen on Visual Gems of Number Theory. The two of us decided that the session was a success.

The MAA thought so, too. Not long after MathFest 2005, we got a request from Don Van Osdol of the MAA encouraging us to put together a book based on the four session talks. The four speakers at our session thought this was a fine idea and agreed to contribute their work.

We then set about looking for other articles that would be suitable for an undergraduate audience. We began with a list of all of the recent MAA prize-winning papers on number theory. Then, we sent emails to current and recent editors of the MAA journals, some well-known number theorists, and a selection of recent MAA presidents, asking them for suggestions; the response was both overwhelming and gratifying. We read through many back issues of the MAA journals and the AMS Notices. At one point, the number of suggestions was up into seven bits!

Then the whittling-down process began, and it was not easy. For one thing, no sooner had we brought the list down to a manageable size than one of us would toss a few more into the mix—and we'd start over again. Finally, the list converged to the collection you have before you.

And what a collection it is. Prizes won by these papers include the Allendoerfer, Chauvenet, Conant, Trevor Evans, Lester Ford, Hasse and Pólya Awards. Among the authors are recipients of the MAA's Haimo and Alder Awards for Distinguished Teaching, as well as Martin Gardner, George Pólya, Paul Erdős, and Leonhard Euler. We have included many visual delights, and some of the proofs are very likely proofs from "The Book", Erdős' imaginary compendium of perfect proofs.

We have divided the collection into seven chapters: Arithmetic, Primes, Irrationality, Sums of Squares and Polygonal Numbers, Fibonacci Numbers, Number Theoretic Functions, and Elliptic Curves, Cubes, and Fermat's Last Theorem. Before each chapter, we provide the reader with a summary of the articles that appear there. (You might call this an "aroma.") As with any anthology, you don't have to read the Biscuits in order. Dip into them anywhere: pick something from the Contents that strikes your fancy, and have at it. If the end of an article leaves you wondering what happens next, then by all means dive in and do some research. You just might discover something new!

We would like to acknowledge the following people for their advice and assistance with this book: Don Albers, Jerry Alexanderson, Tom Apostol, Jennifer Beineke, Lowell Beineke, Ed Burger, Marc Chamberland, Karl Dilcher, Underwood Dudley, Frank Farris, Ron Graham, Richard Guy, Deanna Haunsperger, Roger Horn, Christopher Hughes, Dan Kalman, Steve Kennedy, Hugh Montgomery, Roger Nelsen, Ken Ono, Bruce Palka, Carl Pomerance, Peter Ross, Martha Siegel, Jim Tattersall, Don Van Osdol, Stan Wagon, and Paul Zorn.

Finally, we wish to thank our families for their support. It has been a pleasure working with MAA putting this collection together. We are particularly grateful to Elaine Pedreira, Beverly Ruedi, and especially Rebecca Elmo for their hard work, high standards, and dedication to this project.

— Arthur Benjamin

— Ezra Brown

August 8, 2008

Part I: Arithmetic

In Chapter XI of Lewis Carroll's *Alice's Adventures in Wonderland*, the Mock Turtle, in describing his schooling to Alice, referred to the four branches of arithmetic as Ambition, Distraction, Uglification and Derision. Our first six Biscuits will treat all of these and more.

We begin with Jim Tanton's "A dozen questions about the powers of two" (*Math Horizons*, vol. 8, no. 1 (September 2001), pp. 5-10), an Evans Award-winning exploration of the surprising amount of interesting mathematics to be found among the powers of two. In Tanton's inimitable style, a dozen questions about the powers of two lead to answers– and even more questions. He asks questions about weighing problems, sums of truncated triangular numbers, Egyptian multiplication, checkers in a circle, a variant of the game Survivor, folding fractions, leading digits, and more. Most answers include a challenge to "Take It Further." Read this most enjoyable paper with pencil and paper in hand, and then explain the following complete sequence: 4, 1, 5, 2, 9, 6, 46, 3, 53.

Early in number theory classes, we learn that 30 is the largest integer n such that every positive integer strictly between 1 and n and relatively prime to n is a prime. In 1907 H. Bonse gave a clever proof that hinges on the fact that the product of the first n prime numbers exceeds the square of the $(n + 1)^{st}$ prime. Very neat, but what comes next? Is there a largest number n such that every positive integer less than n and relatively prime to n is a prime power? Do some exploring and guess. Then read our next Biscuit, in which Michael Dalezman strengthens Bonse's Inequality and answers this and other similar questions in "From 30 to 60 is not twice as hard" (*Math. Magazine*, vol. 73, no. 2 (April 2000), pp. 151–153). At the end of this paper, Dalezman invites you to verify that 2730 is the largest integer n such that all positive integers less than n and prime to n have at most two prime divisors, multiplicities included. Replace "two" by "three" and "four", and the answers are 210210 and 29099070, respectively.

We add the fractions $\frac{a}{c}$ and $\frac{b}{d}$ by finding the least common denominator, adjusting the numerators, and adding. If $\frac{a}{c}$ and $\frac{b}{d}$ are reduced fractions then, depending on the values of c and d, the resulting fraction might never, sometimes, or always be further reducible. The answer is "never" for $\frac{a}{10}$ and $\frac{b}{12}$, "sometimes" for $\frac{a}{15}$ and $\frac{b}{21}$, and "always" for $\frac{a}{14}$ and $\frac{b}{38}$. Harris Shultz and Ray Shiflett were at a workshop where this came up, and they wondered what the pattern was. In their article "Reducing the sum of two fractions" (*Mathematics Teacher*, vol. 98, no. 7 (March 2005), pp. 486–490), they tell us what they found and gently lead us to the answers. (Again, make a guess, then read the article.) At the end, they encourage their readers to explore deeper waters: "It would be interesting to know how much of the facts about reducing numerical fractions carry over to the class of rational functions." It would, indeed!

Let n be a positive integer, and let f be a mapping on $\mathbb{Z}$ mod n, the integers mod n.

For $x \in \mathbb{Z}$ mod n, the forward orbit of x under f is the set $\{x_1 = x, x_2 = f(x_1), x_3 = f(x_2) \ldots\}$. Now, join the points (x_1, x_1), (x_1, x_2), (x_2, x_2), (x_2, x_3), …. The picture you get should look familiar, because this graphical iteration method generates pictures similar to those that appear in every book on dynamical systems. Rafe Jones and Jan Pearce use this method to tackle a modest goal in our next Biscuit—nothing less than bringing the beauty of fractions to the postmodern era of America's visually-oriented, quantitatively illiterate culture. They begin their Allendoerfer Award-winning paper, "A postmodern view of fractions and the reciprocals of Fermat primes" (*Math. Magazine*, vol.73, no. 2, (April 2000), pp. 83–97), by reminding us that the amazing fractal images of chaos theory took popular culture by storm. Wouldn't it be nice, they wonder, if this transformation of chaotic dynamical systems from differential equations, analysis and algebra into posters and tee shirts of the Mandelbrot set can be applied to, say, fractions?

They do not simply wonder. They apply to the study of certain fractions the graphical techniques that were used to transform the Mandelbrot set from algebra to cultural icon. The resulting paper is a picture-filled treatment of such topics as rotational graph pairs, rotational symmetry, fractions with prime denominators, and perfectly symmetric numbers. At the end, they apologize that they were not able to include all of their striking images in the paper—but they provide a pointer to their web site.

Mathematics, including number theory, is the science of patterns, and nothing helps us better in discovering new patterns than a good picture. In "Visible structures in number theory" (*Amer. Math. Monthly*, vol. 108, no. 10 (December 2001), pp. 897–910), Peter Borwein and Loki Jörgenson present pleasing periodic patterns provided by polynomials, Pascal's triangle, and π. This paper, which was awarded the MAA's Lester Ford Award, presents a number of open questions that are provoked by these pictures, including the meta-mathematical question "What does it mean to prove a theorem visually or experimentally?"

If you read MAA journals, you know that Roger Nelsen is the preeminent collector of Proofs Without Words (PWWs) anywhere. Our final arithmetical Biscuit is a selection of his "Visual gems of number theory" (*Math Horizons*, vol. 15, no. 3 (February 2008), pp. 7–9, 31). These visual treats include characterizations of primitive Pythagorean triples, Euclid's perfect number formula, relationships between triangular numbers and squares, a proof that $\sqrt{2}$ is irrational, and Larson's elegant wordless proof—surely one for Paul Erdős' mythical "Book" of perfect proofs —that every triangular number is a "choose two" binomial coefficient. Read them and marvel, then try your hand at making up some yourself. For example, the sum of the first m even squares $2^2 + 4^2 + \cdots + (2m)^2$ is a "choose three" binomial coefficient. Find a PWW of this.

A Dozen Questions About the Powers of Two

James Tanton

Everyone is familiar with the powers of two: 1, 2, 4, 8, 16, 32, 64, 128, and so on. They appear with surprising frequency throughout mathematics and computer science. For example, the number of subsets of a finite set is a power of two, as too is the sum of the entries of any row of Pascal's triangle. (Mathematically, these two statements are the same!) The largest prime number known today is one less than a power of two, a cube of tofu can be sliced into a maximum of 2^n pieces with n planar cuts, and every even perfect number is the sum of consecutive integers from 1 up to one less than a power of two!

Here I have put together a dozen curiosities all about the powers of two. These puzzles toy with results and ideas from classic number theory and geometry, game theory, and even popular TV culture (one problem is about a variation of the game *Survivor*)! I hope you enjoy thinking about them as much as I did.

1. A Weighty Problem

A woman possesses five stones, each weighing an integral number of kilograms. She claims, with the use of a simple see-saw balance, she can match the weight of any stone you give her and thereby determine its weight. She makes this claim under the proviso that your stone is of integral weight and weighs no more than 31 kilograms.

What are the weights of her five stones?

2. Multiplication without Multiplying

Here's an alternative method to long multiplication: Head two columns with the numbers you wish to multiply. Progressively halve the figures in the left-hand column (ignoring remainders) while doubling the numbers on the right. Continue this operation until the left-hand column is reduced to 1. Delete all rows with an even number in the left-hand column and add all the surviving numbers from the right-hand column. This sum is the desired product.

73	×	23
~~36~~		~~46~~
~~18~~		~~92~~
9		184
~~4~~		~~368~~
~~2~~		~~736~~
1		1472
		1679

$73 \times 23 = 1679.$

Why does this work?

3. Truncated Triangular Numbers

The numbers 5, 12, and 51, for example, can be written as a sum of two or more consecutive positive integers:

$$5 = 2 + 3$$
$$12 = 3 + 4 + 5$$
$$51 = 6 + 7 + 8 + 9 + 10 + 11.$$

Which numbers *cannot* be written as a sum of at least two consecutive positive integers?

4. Survivor

N people, numbered from 1 to N, are stranded on an island. They play the following variation of the TV game *Survivor*:

Members of the group vote whether person number N should survive or be escorted off the island. If 50% or more of the people agree to this person's survival then the game ends here and the N people all take an equal share of a $1,000,000 cash prize. If, on the other hand, the Nth person is voted off the island, the remaining $N-1$ people will take a second vote to determine the survival of the $(N-1)$th player (again with a quota of 50%). They do this down the line until a vote eventually passes and a person survives. The cash prize is then shared equally among all the folks remaining after this successful vote.

Assume that all players are greedy, but rational, thinkers; that they will always vote for their own survival, for example, and will vote for the demise of another player provided it does not lead to their own demise as a consequence.

The question here is simple: who survives?

5. Pascal Curiosity

Prove that all entries in the 2^nth row ($n \in \mathbf{N}$) of Pascal's triangle are odd.

6. Checkers in a Circle

Betty places a number of black and white checkers in arbitrary order on the circumference of a circle. (Say Betty lays down N checkers.) Charlie now places new checkers between the pairs of adjacent checkers in Betty's ring: he places a white checker between every two that are the same color, a black checker between every pair of opposite color. He then removes Betty's original checkers to leave a new ring of N checkers in a circle.

Betty then performs the same operation on Charlie's ring of checkers, following the same rules. The two players alternate performing this maneuver over and over again on the circle of checkers before them. Show that if N is a power of two, all the checkers will eventually be white, no matter the arrangement of colors Betty initially puts down.

7. Classic Number Theory

Is $2^{91} - 1$ prime? What about $2^{91} + 1$?

8. De Polignac's Remarkable Conjecture

In the mid-nineteenth century, the French mathematician A. de Polignac made a remarkable observation: It seems that every odd number larger than one can be written as a sum of a power of two and a prime.

$$3 = 2^0 + 2$$
$$5 = 2^1 + 3$$
$$7 = 2^2 + 3$$
$$\vdots$$
$$53 = 2^4 + 37$$
$$\vdots$$
$$241 = 2^7 + 113$$
$$\vdots$$
$$999999 = 2^{16} + 944463$$

He claimed to have checked this for all odd numbers up to three million! Can you prove de Polignac's conjecture?

9. Stacking Dilemma

Two line segments can sit in one-dimensional space touching in a zero-dimensional subspace, namely, a point:

It is possible to arrange four triangles in a plane so that each triangle intersects each other triangle along a one-dimensional line segment of positive length:

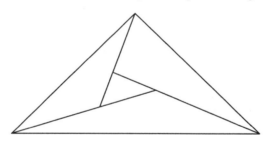

Is it possible to arrange eight tetrahedra in three-space so that each tetrahedron meets every other tetrahedron along a portion of surface of positive area?

10. The Game of 5-7-9

Here's a classic game for two players. It is played with three piles of coins, one with five coins, the second with seven and the third with nine coins. At each turn a player picks up as many or as few coins as she chooses from a single pile. The player who picks up the last coin wins.

What strategy could the first player employ to guarantee a win?

11. Folding Fractions

It is possible to place a crease mark exactly halfway along a strip of paper simply by folding the paper in half. Then, using this mark as a guide and folding the left and right portions of the paper, we can accurately place creases at positions $\frac{1}{4}$ and $\frac{3}{4}$ respectively.

Is it possible to accurately place a crease mark at position $\frac{5}{7}$ on the paper?

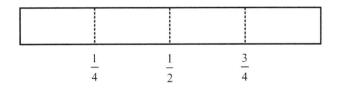

12. Where are the 7's?

Does any power of two begin with a seven? If so, does any power of two begin with 77?

Comments, Answers and Further Questions

1. Her stones weigh 1, 2, 4, 8 and 16 kilograms respectively. As every positive integer can be written uniquely as a sum of distinct powers of two, she can match any weight up to $1 + 2 + 4 + 8 + 16 = 31$ kilograms with these stones.

Taking It Further. Using a see-saw balance and a different set of five stones the woman can actually accomplish a much more impressive feat: she can determine the integral weight of any rock you give her weighing up to 243 kilograms! What are the weights of these five different stones?

Hint: The woman no longer claims she can match the weight of your stone, only that she can determine its integral value.

2. Removing the last digit of a number written in base two either divides the number in half or subtracts 1 and then divides by two, depending on whether the number is even or odd. (For example, the binary code for the number 13 is 1101. Deleting the last digit gives 110 which represents 6.) Thus one can determine the binary code for a number simply by repeatedly dividing by 2 (ignoring remainders), and keeping track of whether or not the result is even or odd along the way. We can thus read off the binary code of 73 from the left column of the table as 1001001. This means $73 = 1 + 2^3 + 2^6$, and so

$$73 \cdot 23 = (1 + 2^3 + 2^6) \cdot 23.$$

Expanding the brackets yields:

$$\begin{aligned} 73 \times 23 = & 1 \times 23 \\ \cancel{+} & \cancel{2 \times 23} \\ \cancel{+} & \cancel{2^2 \times 23} \\ + & 2^3 \times 23 \\ \cancel{+} & \cancel{2^4 \times 23} \\ \cancel{+} & \cancel{2^5 \times 23} \\ + & 2^6 \times 23. \end{aligned}$$

The desired product is precisely the sum of terms (resulting from doubling the number 23 multiple times) that correspond to the placement of ones in the binary code of the number 73, namely, the placement of odd terms in the left column. This method works for any pair of natural numbers you wish to multiply.

Comment. Computers perform multiplication in this way. The halving and doubling operations are trivial when all numbers are expressed in base 2.

3. All numbers *except* the powers of two can be written as a sum of at least two consecutive integers. If N is a number of the form
$$N = (n+1) + (n+2) + \cdots + (n+k)$$
for $n, k \in \mathbf{N}$ with $k \geq 2$, then
$$N = kn + \frac{k(k+1)}{2} = \frac{1}{2} \cdot k(2n+k+1).$$

If k is odd, then $(2n + k + 1)$ is even and it follows that k is an odd factor of N. If, on the other hand, k is even, then $(2n+k+1)$ is an odd factor of N. Either way, N possesses an odd factor and so cannot be a power of two.

Conversely, all numbers possessing an odd proper factor, can be written as a sum of two or more consecutive numbers. Suppose $N = ab$ with $a, b \in \mathbf{N}$, $b \geq 1$, $a \geq 3$, and a odd. Then
$$\begin{aligned} N &= b + b + b + b + \cdots + b \quad a \text{ times} \\ &= (b - (a-1)/2) + \cdots + (b-1) \\ &\quad + b + (b+1) + \cdots + (b + (a-1)/2). \end{aligned}$$

If $b - (a-1)/2 > 0$ we are done. If not, the first few terms of this sum are negative and will cancel the first few positive terms in the latter part of this sum. We need to show that, after cancellation, at least two positive terms survive. Simple algebra verifies that indeed $-(b - (a-1)/2) \leq b + (a-1)/2 - 2$.

Taking it Further. Classify those natural numbers that can be expressed as a sum of at least *three* consecutive positive integers.

4. One person on the island ($N = 1$) will certainly vote for himself and so survive. With two people on the island ($N = 2$ case), player 1 will not vote for player 2's survival (he'll survive without her) but player 2 will. Thus player 2 survives and both folks share the prize. With three players ($N = 3$ case), players 1 and 2 will vote for player 3's demise (they're fine without her) and even voting for herself, player 3 will not garner enough votes to survive. Players 1 and 2 will again share the prize.

Consider the game with four people on the island. Player 4 will certainly vote for his survival. So too will player 3, for without player 4, the previous analysis shows player 3 will not survive. Even though players 1 and 2 will vote against player 4, player 4 has enough votes to survive, and thus all four players stay on the island and share the prize.

For a game with $N < 8$ players, players 1 through 4 have no need to vote for the survival of higher numbered players. Players 5, 6, and 7 are therefore doomed to leave the island. We need the addition of an eighth player in the game to ensure their (and this eighth player's) survival.

In this way we see that the number of people who survive our game *Survivor* is the largest power of two less than or equal to the number of people initially on the island.

Taking it Further. How do the results of the game change if players must attain *strictly more* than 50% of the votes in order to survive?

5. We are asked to show that each entry of the 2^nth row of Pascal's triangle is congruent to 1 (mod 2). Regard the top row of Pascal's triangle as an infinite string of zeros except for a single "1" in the 'center.' Then every entry in the remaining rows of the triangle is the sum of the two nearest terms in the row above it. Working modulo 2, the first five rows of Pascal's triangle are thus:

$$
\begin{array}{ccccccccc}
\ldots & 0 & 0 & 0 & 1 & 0 & 0 & 0 & \ldots \\
\ldots & 0 & 0 & 1 & & 1 & 0 & 0 & \ldots \\
\ldots & 0 & 0 & 1 & 0 & 1 & 0 & 0 & \ldots \\
\ldots & 0 & 1 & 1 & 1 & 1 & 0 & & \ldots \\
\ldots & 0 & 1 & 0 & 0 & 0 & 1 & 0 & \ldots
\end{array}
$$

Certainly all the terms of the first, second, and fourth rows of the triangle are 1 (mod 2). Also, the two 1's on the fifth row are sufficiently spaced apart to generate their own copies of the first four rows of the triangle. We thus obtain a row of eight 1's in the eighth row of Pascal's triangle. The ninth row then consists of two single 1's which are sufficiently far apart to generate their own copies of the first eight rows of Pascal's triangle, ending with a sixteenth row which is nothing but 1's; and so on. An induction argument shows that all 2^n entries of the 2^nth row of Pascal's triangle are indeed congruent to 1 (mod 2).

Taking it Further. Prove that $\binom{2^n}{k}$ is even for $n, k \in \mathbf{N}$ with $1 < k < 2^n$. Is $\binom{2n}{n}$ ever odd for $n \geq 1$?

6. Break the circle and line the checkers in a row, noting that the first and last checkers are 'adjacent.' Represent this row of checkers as a string of 0's and 1's, where '0' represents a white checker, and '1' a black checker. The transformation described in the checker game creates a new string of 0's and 1's where each entry in the second row is the sum of the two nearest entries from the top row (with the appropriate interpretation for the end digits given the 'wrap around' effect for the string).

Suppose N is a power of 2. Consider a game starting with a single black checker. Due to the cyclic symmetry of the game, we may as well assume the black checker is placed at the beginning of the string, and thus the game can be represented by a string of the form:

$$1000\ldots0.$$

Note that $N - 1$ applications of the transformation generate the first N rows of Pascal's triangle modulo 2 (the 'wrap around' effect does not affect the formation of this initial portion of the triangle). By question 6, all entries of the final row are 1. Thus after $N - 1$ steps, all checkers in the checker game will be black. One more application of the transformation yields nothing but white checkers.

An arbitrary game can be thought of as a superposition of individual games involving single black checkers. For example, the game represented by 0110 is a superposition, in some sense, of the games 0100 and 0010. The checker game transformation commutes with 'exclusive or' binary addition (that is, binary addition *without* carrying 1's). As 0100 and 0010 both converge to 0000, it follows that 0110 converges to 0000 + 0000 = 0000. This argument shows that all checker games, involving a power of two number of checkers, yield nothing but white checkers in at most that many steps.

Taking It Further. Can anything be said for games with N checkers where N is not a power of two?

7. First note that $2^{91} = (2^7)^{13}$. The identities
$$x^{13} - 1 = (x - 1)(x^{12} + x^{11} + \cdots + x + 1)$$
and
$$x^{13} + 1 = (x + 1)(x^{12} - x^{11} + \cdots - x + 1)$$
show that $2^7 - 1 = 127$ and $2^7 + 1 = 129$ are factors of $2^{91} - 1$ and $2^{91} + 1$ respectively. (So too are 8191 and 8193.) Thus neither number is prime.

Comment. Numbers of the form $2^n - 1$ are called *Mersenne Numbers*. If n is composite, the above argument generalizes to show that $2^n - 1$ too must be composite. If n is prime, however, the situation is less clear. For example, $2^n - 1$ is prime for $n = 2, 3, 5, 7, 13, 17$ and 19, but not for $n = 11$. Nonetheless, Mersenne numbers have proven to be a rich source of large prime numbers, the largest known today being $2^{6972593} - 1$.

Taking It Further. Is $2^{91} - 3$ prime? What about $2^{91} - 5$ and $2^{91} - 7$?

8. Don't bother! It isn't true. De Polignac missed the number 127 which cannot be written as a sum of a power of two and a prime. (One only need check this for the powers 2^n with $n = 0, \ldots, 6$.) If only de Polignac noticed this slip, he could have saved himself literally months of very hard work!

9. To arrange eight tetrahedra in three-space in the appropriate way, use the following diagram of eight triangles in a plane. Connect the four shaded triangles to a point hovering below the plane of the drawing, and the four unshaded triangles to a point hovering above the plane of the drawing. This yields eight suitably touching tetrahedra.

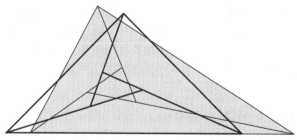

Taking It Further. Prove it is impossible to arrange *five* triangles in a plane that meet pairwise in a line segment of positive length. Is it possible to arrange *nine* tetrahedra in three-space in the appropriate way?

10. Express the numbers of coins in each pile in binary notation and write these numbers as rows of a table:

$$5 = 1\ 0\ 1$$
$$7 = 1\ 1\ 1$$
$$9 = 1\ 0\ 0\ 1$$

Notice there are an odd number of 1's appearing in the ones, twos and eights columns of the table. To guarantee a win, the first player should always move so as to produce an even number of 1's in each column. Her first turn is thus to convert the number 1001 into 0010, that is, to reduce the pile of nine coins to two coins. Her opponent will then be forced to create a table with an odd number of 1's in at least one column, and therefore a game with at least one remaining coin. Player 1 always operating this way thus offers her opponent no hope of ever winning.

Taking It Further. *Reverse 5-7-9* is played the same way except this time the person who picks up the last coin loses. Does either player have a best strategy in this variation of the game?

11. Placing a crease mark halfway between two previously constructed crease marks can only ever produce fractions of the form a/N, where N is a power of two. And conversely, every fraction of this form can be created via a finite sequence of such folding operations (see below). Thus there is no (finite) procedure for constructing the fraction $5/7$. However, as every number can be approximated arbitrarily closely by a fraction of the form indicated (just truncate its binary decimal expansion for example), it is possible to place a crease at any position we choose with any desired degree of precision.

Alternatively, one can use the following procedure for placing a crease arbitrarily close to the position $5/7$. Begin by making a guess as to where this fraction lies along the strip and place a crease at this location. Now fold the right portion of the strip in half to place a crease halfway between the guess and the right end of the paper. This produces a crease mark at position $6/7$ with the error reduced in half. Now fold from the left to produce a crease at position $3/7$ with error reduced in half yet again, and then, finally, fold from the right to produce a crease mark at position $5/7$ with one eighth the original error. Repeating the "RLR" sequence of folds multiple times yields a sequence of crease marks that rapidly converge to the true $5/7$ position.

Taking it Further 1. It is no coincidence that the sequence of folds "RLR" mimics the binary representation 101 of the number five. Show that if N is one less than a power of two, then the sequence of left and right folds that hones in on the fraction a/N is precisely the binary expansion of a read backwards with '1' representing 'right' and '0' representing 'left.'

Taking it Further 2. Consider a fraction of the form $a/2^n$. Write a as an n-digit binary number (you may need to place zeros at the beginning), and read this binary expansion backwards as a sequence of instructions with '1' and '0' representing 'right' and 'left' again. Show that if you follow these instructions through just once the final crease mark formed lies precisely at position $a/2^n$.

Taking it Further 3. Let q be *any* fraction in the interval [0, 1]. Show that, with a square sheet of paper, it *is* possible to produce, in a finite number of folds, a creased line segment precisely q units long.

12. We seek integers n and k such that

$$7 \cdot 10^k \leq 2^n < 8 \cdot 10^k,$$

that is,

$$\log_{10} 7 + k \leq n \log_{10} 2 < \log_{10} 8 + k.$$

If $\{x\}$ denotes the fractional part of the number x, we thus are being asked to find a value n for which

$$\log_{10} 7 \leq \{n \log_{10} 2\} < \log_{10} 8.$$

Working on the interval $[0, 1)$ with 'wrap around' effect (that is, working modulo 1), we hope to find a multiple of $\log_{10} 2$ that falls within the segment $[\log_{10} 7, \log_{10} 8)$, which is about 0.058 units long. It is easy to show that $\log_{10} 2$ is an irrational number (an equation of the form $2^b = 10^a$ can only hold if both exponents are zero) and consequently no two distinct multiples of $\log_{10} 2$ have the same fractional part. Thus, of the first 21 multiples of $\log_{10} 2$, at least two if them must lie within a distance of $\frac{1}{20} = 0.05$ from each other (considering the wrap around effect). Call these multiples $m \log_{10} 2$ and $(m + q) \log_{10} 2$. It then follows that the multiples $(m + q) \log_{10} 2$ and $(m + 2q) \log_{10} 2$ are also within this distance of each other, as too are $(m + 2q) \log_{10} 2$ and $(m + 3q) \log_{10} 2$, and so on. Creeping along this way, we must eventually hit upon a multiple of $\log_{10} 2$ that lies in the interval $[\log_{10} 7, \log_{10} 8)$. This shows that powers of two beginning with a seven do exist. (The diligent reader may have discovered that 2^{46} is the first power of two which begins with a 7.)

We can say more: The above argument shows that the multiples of $\log_{10} 2$ form a dense set in the interval $[0, 1)$ and so there are infinitely many multiples that lie in any given segment. Thus there are infinitely many powers of two that begin with seven and, in some sense, we can say that 7 occurs as a first digit 5.8% of the time! Similarly, $\log_{10} 78 - \log_{10} 77 \approx 0.56\%$ of the powers of two begin with '77' and there are infinitely many powers of two that begin with any prescribed (finite) set of digits! (For example, there are infinitely many powers of two that begin with the first billion digits of pi!)

Acknowledgments and Further Reading

Several of these puzzles explore classic ideas from number theory. The interested reader can look at Jay R. Goldman's beautiful text *The Queen of Mathematics, A Historically Motivated Guide to Number Theory* (A. K. Peters, Ltd., 1998), for example, to learn more about this fascinating and challenging subject. I first learned of the classification of truncated triangular numbers as a result discovered by eight year old Mit'ka Vaintrob of New Mexico. He followed a geometric approach, literally truncating triangular arrays of coins or buttons to create trapezoids with at least two rows. His analysis of those numbers which can be expressed as a sum of three consecutive integers is a remarkable achievement. Gengzhe Chang and Thomas Sederberg give a complete analysis of the circular checkers game in their wonderful book *Over and Over Again* (Mathematical Association of America, 1997).

Question 9, with its obvious extensions, is a very old and extremely hard question. It is known that it is always possible to arrange 2^n high-dimensional "simplices" in n-space that meet, pairwise, in regions of $(n-1)$-space of positive volume, but it is not at all clear whether one can do more. It has long been known that is impossible to arrange five triangles in a plane in this way, but it wasn't until 1991 that J. Zaks was able to prove the impossibility of arranging nine tetrahedra in three-space. Analysis of higher-dimensional spaces is still an area of open research. See Martin Aigner and Günter M. Ziegler's *Proofs from the Book* (Springer-Verlag, 1999), Chapter 13, for a discussion on this fascinating topic.

The 5-7-9 game, of course, is a specific instance of the famous game Nim. One can explore the subject of *nimbers* in John H. Conway and Richard K. Guy's *The Book of Numbers* (Copernicus, 1996). Changing the value of the quota in the game *Survivor* leads to some very interesting mathematics; my colleague Charles Adler and I are currently writing about some curious results from this game.

A Second Helping

This is one of my favourite "dozenal" articles. The weighty problem (question 1) caught me by surprise when I realized that ternary expansions using digits 1, 0, and −1 can be "stretched" to just the even numbers to yield the astounding result claimed in the solutions. Never forget: Even classic problems that have been bandied about for decades can yield new twists and delights! (See "Introducing binary and ternary expansions via weighings." *College Mathematics Journal*, 33 no. 4 (2002), 17–18, for details.)

The survivor problem (question 4) is closely related to the problem of iterating the approximate squaring function $f(x) = x[x]$, which caught the attention of J. C. Lagarias and N. J. A. Sloane. (See their paper "Approximate Squaring" *Experimental Mathematics*, vol 13 (2004) No 1; 113–128.)

I failed to mention in the article that counting the powers of two that begin with a seven (question 12) offers an explanation of an intriguing phenomenon known as Benford's Law.

It is said that, in 1881, American astronomer Simon Newcomb noticed that in public resource collections the first few pages of books of numerical tables—logarithms tables, for instance—were generally more worn and grimy than latter pages. These first pages contain data values that begin with "1." He thought this curious.

The same phenomenon was later independently observed in 1938 by physicist Frank Benford, who went further and examined large collections of data tables from a wide variety of sources: population growth, financial data, scientific activities. He observed that, with some consistency, about 30.1% of the entries in a data set begin with a "1", about 17.6% with a "2", about 12.5% with a "3," all the way down to about 4.6% with a "9". This observation has become known as Benford's Law. The IRS today uses Benford's law to quickly scan first digits for possibly falsified tax records.

Explanation: Any phenomenon that is based on some kind of exponential growth (population figures, scientific studies, interest rates, the powers of two) can be analyzed in exactly the same way we analyzed first digits of powers in the article!

Originally appeared as:
Tanton, James. "A Dozen Questions about the Powers of Two." *Math Horizons*. vol. 8 (September 2001): pp. 5–10.

From 30 to 60 is Not Twice as Hard

Michael Dalezman

Euclid's proof that there are infinitely many primes [3, p. 25] can be modified to yield a proof of a simple inequality: If $p_1, p_2, \ldots$ is the sequence of prime numbers and $n \geq 2$ then $p_1 p_2 \ldots p_n > p_{n+1}$. In 1907, Bonse [1] gave an elementary proof of a stronger inequality, now called Bonse's Inequality [4]: If $n \geq 4$ then $p_1 p_2 \ldots, p_n > p_{n+1}^2$. Bonse then used his inequality to prove that 30 is the largest integer m with the following property:

If $1 < k < m$ and $(k, m) = 1$, then k is a prime.

In this note we give an elementary proof of a stronger inequality and use it to prove that 60 is the largest integer m with the following property:

If $1 < k < m$ and $(k, m) = 1$, then k is a prime power.

We then indicate how the inequality can be strengthened and how the result can be generalized.

The following notations will be used:

- $\omega(k)$: the number of distinct prime divisors of k;
- $\Omega(k)$: the number of prime divisors of k, counting multiplicity;
- $\lfloor x \rfloor$: the largest integer not greater than x;
- $\pi(x)$: the number of primes not exceeding x;
- $\phi(k)$: the number of positive integers prime to k and not exceeding k.

THEOREM 1. *If $n \geq 4$, then $p_1 p_2 \cdots p_n > p_{n+1} p_{n+2}$.*

This inequality follows readily from Bertrand's postulate [3, p. 367] but we give here a proof that is self-contained.

Proof. The result can easily be verified for $n < 10$. Let $i = \lceil \frac{n}{2} \rceil$, and suppose

$$p_1 p_2 \cdots p_n \leq p_{n+1} p_{n+2} < p_{n+2}^2.$$

Then

$$(p_1 p_2 \cdots p_i)^2 < p_1 p_2 \cdots p_n < p_{n+2}^2 \quad \text{and} \quad p_1 p_2 \cdots p_i < p_{n+2}.$$

Let us consider the p_i integers $N_t = t p_1 p_2 \cdots p_{i-1} - 1$, $t = 1, 2, \ldots, p_i$. For all t, $N_t < p_1 p_2 \cdots p_i < p_{n+2}$ and is prime to $p_1, p_2, \ldots, p_{i-1}$. Thus if q_t is the smallest prime dividing N_t, then $p_i \leq q_t < p_{n+2}$. The q_t's are distinct, for if $q_t = q_{t'}$, with $t \neq t'$, then $q_t | N_t - N_{t'} = (t - t') p_1 p_2 \cdots p_{i-1}$, so $q_t | t - t'$; this is impossible since $1 \leq t, t' \leq p_i$. Hence the number of N_t's must be no greater than the number of primes q such that $p_i \leq q < p_{n+2}$. Therefore $p_i \leq n + 2 - i$. But $i = \lfloor \frac{n}{2} \rfloor$, so $n \leq 2i + 1$ and $p_i \leq i + 3$. The last inequality clearly fails for $i \geq 5$, so it fails for $n \geq 10$. ∎

DEFINITION. An integer m satisfies property $\mathscr{P}_s$ if

for all k such that $1 < k < m$ and $(k, m) = 1$, $\omega(k) \leq s$.

LEMMA. *If m satisfies $\mathscr{P}_1$ and $p_n p_{n+1} \leq m$, then $p_1 p_2 \cdots p_n \leq m$.*

Proof. Let m satisfy $\mathscr{P}_1$. Consider all the primes $p_1, p_2, \ldots, p_{n+1}$. If two of these primes, say p_ν and p_μ were both relatively prime to m, we would get $p_\nu p_\mu \leq m$ and $(p_\nu p_\mu, m) = 1$ contradicting the fact that m satisfies $\mathscr{P}_1$. Hence at most one of the primes $p_1, p_2, \ldots, p_{n+1}$ does not divide m, and the lemma follows. ∎

MAIN THEOREM. *60 is the largest integer satisfying $\mathscr{P}_1$.*

Proof. It is easy to verify that 60 satisfies $\mathscr{P}_1$. We need to prove that 60 is the largest such integer. Let $m > 60$ satisfy $\mathscr{P}_1$. If $m \geq 77 = 7 \cdot 11 = p_4 p_5$, we let n be the largest integer such that $p_1 p_2 \cdots p_n \leq m$. By the lemma, $n \geq 4$; by Theorem 1, $p_1 p_2 \cdots p_n > p_{n+1} p_{n+2}$. Hence, $p_{n+1} p_{n+2} < m$ and, by the lemma, $p_1 p_2 \cdots p_{n+1} \leq m$; this contradicts the maximality of n. Thus we must have $60 < m < 77$. Clearly $5 \cdot 7 < m$. By the argument in the proof of the lemma, m must be divisible by all of the primes 2, 3, 5, and 7 with at most one exception. Therefore, m is divisible by 105, by 70, by 42, or 30; the only possibility is 70. But 70 does not satisfy $\mathscr{P}_1$ because 33 is less than 70 and prime to 70, but not a prime power. Hence 60 is the largest integer satisfying $\mathscr{P}_1$. ∎

It is noteworthy that the Main Theorem implies Theorem 1. To see why, let $n \geq 4$ and let $a = p_1 p_2 \cdots p_n$. Since $a \geq 210$, a does not satisfy $\mathscr{P}_1$, so there exists an integer b such that $1 \leq b < a$, $(b, a) = 1$ and $\omega(b) \geq 2$. If p and q are 2 distinct primes that divide b, then $p_{n+1} p_{n+2} \leq pq \leq b < a = p_1 p_2 \cdots p_n$.

Generalizations Bonse went further and proved that if $n \geq 5$, then $p_1 p_2 \cdots p_n > p_{n+1}^3$. He used this result to show that 1260 is the largest integer with the property:

If $1 \leq k < m$ and $(k, m) = 1$, then $\Omega(k) \leq 2$.

Bonse also indicated that similar methods could be used to prove $p_1 p_2 \cdots p_n > p_{n+1}^4$ for sufficiently large n. Using this inequality, he wrote, he had found that 30,030 was the largest integer with the property:

If $1 \leq k < m$ and $(k, m) = 1$ then $\Omega(k) \leq 3$.

(Actually, Bonse erred; the correct number is 60,060.)

Landau [2] generalized Bonse's results, proving that for every integer $s \geq 1$ there exists an integer n_s such that

$$n \geq n_s \Rightarrow p_1 p_2 \cdots p_n > p_{n+1}^{s+1};$$

he concluded that there exists a largest integer m_s with the property:

If $1 \leq k < m_s$ and $(k, m_s) = 1$, then $\Omega(k) \leq s$.

Our results too, can be extended and generalized.

THEOREM 2. *For every integer $s \geq 1$, there exists an integer n_s such that*

$$n \geq n_s \Rightarrow p_1 p_2 \cdots p_n > p_{n+1} p_{n+2} \cdots p_{n+s+1}.$$

To prove this, we replace $i = \lfloor \frac{n}{2} \rfloor$ by $i = \lfloor \frac{n}{s+1} \rfloor$ in the proof of Theorem 1; we get

$$p_i \leq si + 2s + 1.$$

The reverse to this inequality will be a consequence of the following 3 lemmas:

LEMMA 1. *For all integers $a \geq 2$ and for all $x > 0$ $\pi(x) \leq \frac{x}{a}\phi(a) + (a-1)$.*

This follows from the fact that the interval from ka to $(k+1)a$ (for $k = 1, 2, \ldots \lfloor \frac{x}{a} \rfloor$) contains at most $\phi(a)$ primes.

LEMMA 2. *For all $i \geq 1$ $p_i > i\frac{a}{\phi(a)} - \frac{a^2}{\phi(a)}$.*

This is obtained as follows: For any $x > 0$, we let $i = \pi(x) + 1$, then $x < p_i$. We then substitute $i - 1$ for $\pi(x)$ and p_i for x in Lemma 1.

LEMMA 3. $\overline{\lim}_{a \to \infty} \frac{a}{\phi(a)} = +\infty$.

This follows from the fact that

$$\prod_{p \leq x} \frac{p}{p-1} = \prod_{p \leq x} \frac{1}{1 - \frac{1}{p}} = \prod_{p \leq x} \sum_{i=0}^{\infty} \frac{1}{p_i} > \sum_{n \leq x} \frac{1}{n}.$$

Theorem 2 can be used to prove that, for every s, there exists a largest integer m_s satisfying $\mathscr{P}_s$. As in the case $s = 1$, we have the following result:

LEMMA 4. *If m satisfies $\mathscr{P}_s$ and $p_n p_{n+1} \cdots p_{n+s} \leq m$, then $p_1 p_2 \cdots p_n \leq m$.*

To prove the existence of m_s, let m be an integer satisfying $\mathscr{P}_s$, and let us assume that, for some $l \geq n_s$, we have

$$p_1 p_2 \cdots p_{l+1} > m \geq p_1 p_2 \cdots p_l > p_{l+1} p_{l+2} \cdots p_{l+s+1}.$$

By Lemma 4, this implies $p_1 p_2 \cdots p_{l+1} \leq m$; this contradiction proves that $m < p_1 p_2 \cdots p_{n_s}$ and thus establishes the existence of m_s.

Bounds for m_s We will now show that

$$p_1 p_2 \cdots p_{n_s - 1} \leq m_s < p_{n_s} p_{n_s + 1} \cdots p_{n_s + s}.$$

Proof. We saw that $m_s < p_1 p_2 \cdots p_{n_s}$. If $p_{n_s} p_{n_s+1} \cdots p_{n_s+s} \leq m_s$ then Lemma 4 gives $p_1 p_2 \cdots p_{n_s} \leq m_s$. On the other hand, by definition of n_s we have $p_1 p_2 \cdots p_{n_s-1} < p_{n_s} p_{n_s+1} \cdots p_{n_s+s}$. This shows that $p_1 p_2 \cdots p_{n_s-1}$ satisfies $\mathscr{P}_s$ and gives the lower bound for m_s. ∎

The reader is invited to verify that $n_2 = 6$, $n_3 = 7$, $n_4 = 9$, $m_2 = 2730$, $m_3 = 210{,}210$ and $m_4 = 29{,}099{,}070$.

REFERENCES

1. H. Bonse, Über eine bekannte Eigenschaft der Zahl 30 und ihre Verallgemeinerung, *Archiv der Mathematik und Physik* (3) 12 (1907), 292–295.
2. Edmund Landau, *Handbuch der Lehre von der Verteilung der Primzahlen*, vol. I, Chelsea, New York, NY, 1953, pp. 229–234.
3. Ivan Niven, Herbert S. Zuckerman, and Hugh L. Montgomery, *An Introduction to the Theory of Numbers*, 5th Edition, Wiley, New York, NY, 1991.
4. J. V. Uspensky and M. A. Heaslet, *Elementary Number Theory*, McGraw Hill, New York, NY, 1939, p. 87.

Originally appeared as:
Dalezman, Michael. "From 30 to 60 is not Twice as Hard." *Mathematics Magazine*. vol. 73, no. 2 (April 2000): pp. 151–153.

Reducing the Sum of Two Fractions

Harris S. Shultz and
Ray C. Shiflett

Sometimes when we add fractions, the sum can be reduced even though we have used the least common denominator. Other times, the sum cannot be reduced. For example, if we add 1/3 + 1/6, the sum 3/6 can be reduced to 1/2. However, 2/5 + 1/6 = 17/30 cannot be reduced.

Let's look at another example:

$$\frac{4}{21}+\frac{7}{15}=\frac{20}{105}+\frac{49}{105}=\frac{69}{105}.$$

This can be reduced to 23/35. However, when we add 2/21 + 4/15, we obtain 10/105 + 28/105 = 38/105, which cannot be reduced.

There are pairs of denominators for which it seems we never can reduce the sum. For example,

$$\frac{7}{10}+\frac{11}{12}=\frac{42}{60}+\frac{55}{60}=\frac{97}{60},$$

which cannot be reduced. If we try 3/10 and 5/12, we obtain 18/60 + 25/60 = 43/60, which still cannot be reduced. By trying other numerator values, the reader will discover that whenever you add $a/10 + b/12$, where each of the two addends is in lowest terms, the resulting sum having denominator equal to 60 (the least common denominator) cannot be reduced.

On the other hand, there are pairs of denominators for which it seems we always can reduce the sum. Choose an a and b for which $a/20$ and $b/12$ are reduced and add them using the least common denominator 60. Every choice seems to produce an answer that can be reduced. You should now be wondering if this is true for all choices of a and b.

Several years ago, while discussing adding and reducing fractions in a workshop, we observed that certain sums of two fractions can be reduced while others cannot be reduced and wondered why. If there was a pattern, it was not immediately obvious. And searches of the literature did not provide us with any background information. This led us to pose the following general question: Given a pair of natural numbers, c and d, when does there exist a pair of natural numbers a and b for which when a/c and b/d are added to form a single fraction, that fraction can be reduced? It is assumed that the denominator of the sum is the least common multiple of c and d and that the addends a/c and b/d are in lowest terms.

As is common practice, we shall say that u and v are *relatively prime* if they have no common prime factor. For example, 15 and 14 are relatively prime since 15 = (3)(5) and 14 = (2)(7). If a/c is reduced to lowest terms, then a and c must be relatively prime.

In adding 7/60 to 11/36, we write $60 = (2^2)(5)(3)$ and $36 = (2^2)(3^2)$, find the least common multiple to be $(2^2)(5)(3^2)$, and multiply the numerator and denominator of the first fraction by $u = 3$ and the second by $v = 5$. Notice that u and v are relatively prime. This will always be true, as we shall see shortly. That is, if m is the least common multiple of numbers c and d, then there are relatively prime integers, u and v, where $m = cu = dv$.

To proceed from here, we need two useful facts:

> Fact 1: If r, s, t are given integers with $r + s = t$ and if the number n divides any two of the numbers r, s, and t, then n divides the third number.

For example, consider $18 + 49$. We know that 7 divides 49 but not 18. Therefore, by Fact 1, we know that 7 <u>cannot</u> divide $18 + 49$, without even doing the arithmetic.

> Fact 2: r and s are relatively prime when and only when there exist integers x and y for which $rx + sy = 1$.

For example, $15x + 14y = 1$ when $x = 1$ and $y = -1$. It should be noted that, in this result, x and y have to also be relatively prime, as are y and r and s and x. The proof of Fact 2 is based on the result in Number Theory stating that if r and s are any nonzero integers, then there exist x and y such that $rx + sy$ is equal to the greatest common factor of r and s. Proof of this can be found in Burton (1998).

The General Case

In number theory, we often need to know the exact power of a prime that divides an integer. This power for a prime p and an integer n is called the "p-order of n," written $\text{ord}_p(n)$. So, for example, since $150 = (2)(3)(5^2)$, $\text{ord}_5(150) = 2$, $\text{ord}_3(150) = 1$, and $\text{ord}_7(150) = 0$.

It turns out that the answers to our questions about whether or not the sum of two fractions can be reduced depends on the existence of a common prime factor of the denominators that shows up with exactly the same exponent in their prime factorizations. In other words, everything depends on the primes p with the property that $\text{ord}_p(c) = \text{ord}_p(d)$. As a shorthand, we call such a prime *balanced* for c and d. Primes for which $\text{ord}_p(c) \neq \text{ord}_p(d)$ will be called *unbalanced* for c and d.

Example

Suppose $\dfrac{a}{c} = \dfrac{a}{(2)(5^2)(7^5)(11^3)(13^2)}$ and $\dfrac{b}{d} = \dfrac{b}{(3^2)(5^3)(7^3)(11^3)(13^2)}$

are reduced. Adding these fractions, we obtain

$$\frac{au + bv}{(2)(3^2)(5^3)(7^5)(11^3)(13^2)}$$

where $u = (3^2)(5)$ and $v = (2)(7^2)$. Notice that 11 and 13 are balanced primes for the denominators while 2, 3, 5, and 7 are unbalanced prime factors of the denominators. We make the following observations in this example:
- u and v are relatively prime;
- the balanced primes 11 and 13 divide neither u nor v;

- the unbalanced prime factors 3 and 5 divide u but not b (because b/d is reduced);
- the unbalanced prime factors 2 and 7 divide v but not a (because a/c is reduced).

Generalizing from this example provides an understanding of Lemmas 1 and 2.

Throughout what follows, we will always use $m = cu = dv$ to mean the least common multiple of c and d.

Lemma 1. *u and v are relatively prime.*

Lemma 2. *Suppose that a/c and b/d are reduced. If p is an unbalanced prime factor of c or d, then p divides u and not b, or p divides v and not a. If p is a balanced prime for c and d, then it does not divide u, v, a or b.*

Lemma 3. *Suppose that a/c and b/d are reduced. If p is an unbalanced prime factor for c or d, then p does not divide $au + bv$.*

Proof. By Lemma 2, either p divides u and not b, or p divides v and not a. Without loss of generality, suppose p divides u and not b. Since u and v are relatively prime, p does not divide v. Therefore, p does not divide bv. Since p divides u, and therefore au, and since p does not divide bv, it follows from Fact 1 that p does not divide $au + bv$.

Theorem 1. *There exist natural numbers a and b for which a/c and b/d are reduced and $(au + bv)/m$ is not reduced if and only if c and d have a balanced prime factor.*

Proof. Assume that a/c and b/d are reduced and write

$$\frac{a}{c} + \frac{b}{d} = \frac{au+bv}{m}.$$

By Lemma 3, an unbalanced prime factor of c or d cannot be a divisor of $au + bv$. So, if there exist natural numbers a and b for which a/c and b/d are reduced and $(au + bv)/m$ can be reduced, then c and d must have a balanced prime factor. Notice that if $(au + bv)/m$ can be reduced, the only possible common prime factors of the numerator and denominator are the balanced prime factors of c and d.

Conversely, suppose c and d have a balanced prime factor. If P denotes the product of all the balanced primes for c and d, then P is a divisor of m. Since u and v are relatively prime, u^2 and v^2 are relatively prime. So, by Fact 2, there exist integers x and y such that

$$xu^2 + yv^2 = 1,$$

so

$$Pxu^2 + Pyv^2 = P$$

and

$$(Pxu - v + gPuv^2)u + (Pyv + u + gPvu^2)v = P(1 + 2gu^2v^2),$$

where g is a natural number large enough that $Pxu - v + gPuv^2$ and $Pyv + u + gPvu^2$ are both positive. If we define $a = Pxu - v + gPuv^2$ and $b = Pyv + u + gPvu^2$, then the fraction $(au + bv)/m$ can be reduced by P, since P is a divisor of $au + bv = P(1 + 2gu^2v^2)$ and of m.

Also, it turns out that a is relatively prime to c and b is relatively prime to d. Here's why:

To see that a and c are relatively prime, take any prime divisor q of c. If q is a balanced prime, then q divides P so it divides $Pxu + gPuv^2$ but not v so it does not divide a. If q is an unbalanced prime factor of c or d, then q divides u or v but not both. So q divides $-v + gPuv^2$ but not Pxu so q does not divide a, or q divides $Pxu + gPuv^2$ but not v so q does not divide a. So a and c are relatively prime. We may show that b and d are relatively prime in the same way.

Let us look at two of our earlier examples in the context of Theorem 1. We saw that when we add $4/21 + 7/15$, the sum $69/105$ can be reduced. Since the denominators 21 and 15 have a balanced prime factor, namely 3, the theorem assures us that this was no accident and that there do exist natural numbers a and b for which $a/21$ and $b/15$ are reduced and $(5a + 7b)/105$ is not reduced. It also seemed that each time we added $a/10 + b/12$, for any pair of integers a and b, where each of the two addends is in lowest terms, the resulting sum having denominator equal to 60 (the least common denominator) could not be reduced. Since the denominators 10 and 12 do not have a balanced prime factor, the theorem assures us that what seemed to be true is indeed true.

In the proof of Theorem 1, under the assumption that c and d have a balanced prime factor, we constructed a sum $a/c + b/d$ that could be reduced by the product P of all the balanced primes for c and d. This is summarized by the following.

Corollary. *If c and d have at least one balanced prime factor, then there exist natural numbers a and b for which a/c and b/d are reduced and $au + bv$ is divisible by every balanced prime factor of c and d.*

Theorem 2. *Let c and d be given natural numbers. The fraction*
$$\frac{a}{c} + \frac{b}{d} = \frac{au + bv}{m}$$
can be reduced for all natural numbers a and b where a/c and b/d are reduced if and only if 2 is a balanced prime factor for c and d.

Proof. If 2 is a balanced prime for c and d, then u and v are both odd. If a and b are natural numbers for which a/c and b/d are reduced, then a and b are odd. Therefore, $au + bv$ is even and,
$$\frac{a}{c} + \frac{b}{d} = \frac{au + bv}{m}$$
consequently, can be reduced.

Conversely, suppose 2 is not a balanced prime of c and d. We shall show that there exists a and b for which a/c, b/d, and $(au + bv)/m$ are all reduced. If there are no balanced prime factors of c and d, then by Theorem 1, $(au + bv)/m$ is always reduced when a/c and b/d are reduced and we are done. So assume c and d have at least one balanced prime. By the Corollary, there exist natural numbers, A and B, for which A/c and B/d are reduced and $Au + Bv$ is divisible by every balanced prime factor of c and d.

Let g be the maximum of all prime factors of c and d and define q to be the product of all odd primes less than or equal to g. Then $q + 2$ has no prime factors less than or equal to g. So, $c + q + 2$ is relatively prime to c.

Let $a = (c + q + 2)A$ and $b = B$ to get:

$$\frac{a}{c}+\frac{b}{d}=\frac{(c+q+2)A}{c}+\frac{B}{d}=\frac{(c+q+2)Au+Bv}{m}=\frac{(c+q+1)Au+(Au+Bv)}{m}$$

Suppose $(c + q + 1)Au + (Au + Bv)$ and m have a common prime factor p. By Lemma 3, p must be a balanced prime for c and d. Therefore, p is odd and p divides c. Also, since $Au + Bv$ is divisible by every balanced prime factor of c and d, p divides $Au + Bv$ as well. Recall that A and c, and u and c are relatively prime. Then p divides neither A nor u, since p divides c. Also, since p divides c and p divides q, p does not divide $c + q + 1$. Therefore, p does not divide $(c + q + 1)Au$. Consequently, by Fact 1, p does not divide $(c + q + 1)Au + (Au + Bv)$, contradicting our assumption. Therefore, a/c, b/d, and $(au + bv)/m$ are all reduced.

Let us look at an earlier example in the context of Theorem 2. We observed that for every choice of a and b, when you add $a/20 + b/12$ using the least common denominator 60, where each of the two addends is reduced, the resulting sum can be reduced and we wondered if this is always true. Since 2 is a balanced prime for the denominators 20 and 12, the theorem does in fact assure us that this is always true.

Rational Functions

The questions we have examined also arise in the addition of ratios of polynomials, called rational functions. For example, the sum

$$\frac{a}{x-3}+\frac{b}{x+2}=\frac{(a+b)x+(2a-3b)}{(x-3)(x+2)}$$

where a and b are nonzero numbers, cannot be reduced. To verify this, observe that there do not exist nonzero numbers a and b such that $(a + b)x + 2a - 3b$ is a constant multiple of either $x - 3$ or $x + 2$. Likewise, we can show that if

$$\frac{a}{(x-3)(x+1)}+\frac{b}{x(2x-5)(x+3)},$$

where a and b are nonzero numbers, is to be written as a single fraction having denominator $(x - 3)(x + 1)(x)(2x - 5)(x + 3)$, this rational expression cannot be reduced.

However, there do exist nonzero constants a and b such that the sum

$$\frac{a}{(x-2)(x+1)}+\frac{b}{(x-2)(x-3)}=\frac{(a+b)x+(-3a+b)}{(x-2)(x+1)(x-3)},$$

can be reduced. For example, let $a = 3$ and $b = 1$, then

$$\frac{3}{(x-2)(x+1)}+\frac{1}{(x-2)(x-3)}=\frac{4(x-2)}{(x-2)(x+1)(x-3)}=\frac{4}{(x+1)(x-3)}.$$

In this last example, the factor $x - 2$ appears with multiplicity 1 in each denominator. It appears to play a role analogous to that of a balanced prime factor in our previous work. There is much known about the way that polynomials factor and divide each other and even the way some of them behave like primes. It would be interesting to know how much of the facts about reducing numerical fractions carry over to the class of rational functions.

Reference

1. Burton, David M. *Elementary Number Theory*, Fourth Edition. New York: McGraw-Hill, 1998.

Editor's Notes: Harris Shultz and Ray Shiflett provide yet another example of "mathematics for teaching:" mathematical investigations connected to the teaching profession—the question "When can you reduce the sum of two fractions?" is certainly more likely to come up in the teachers' lounge than in an engineering firm. And, while Shultz and Shiflett carry out the investigation for its own sake, the results (Theorems 1 and 2) will be quite useful to teachers as they design activities and problem sets around rational number arithmetic. We suspect that more than a few areas of mathematics were inspired by mathematical problems teachers face as they design activities for their students. One that we recently encountered is how to devise two linear equations in two unknowns with small integer coefficients so that students will not be able to get exact coordinates for the intersection of the graphs simply by zooming with their calculators.

Shultz and Shiflett end the article with an intriguing question: How much of their result carries over to rational functions in, say, one variable with rational coefficients. This is again a question closely related to our profession, because two basic algebraic systems of precollege mathematics—the integers and polynomials in one variable—have deep structural similarities that allow many results (and proofs) in one system to be transported to the other with only minor modification. For example, Fact 2 (and the more general result about greatest common divisor) holds for polynomials, too, and for the same reasons. This is responsible for the fact that both systems have a "fundamental theorem of arithmetic:" the unique prime decomposition property that is used throughout this article.

Originally appeared as:
Shultz, Harris S. and Ray C. Shiflett. "Reducing the Sum of Two Fractions." *Mathematics Teacher*. vol. 98, no. 7 (March 2005): pp. 486–490.

A Postmodern View of Fractions and the Reciprocals of Fermat Primes

Rafe Jones and Jan Pearce

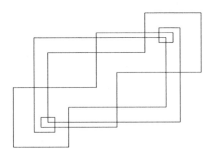

1/29 base 13

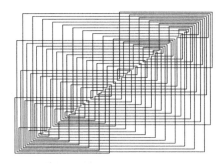

1/101 base 40

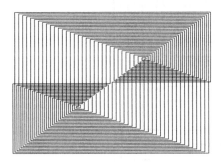

1/101 base 50

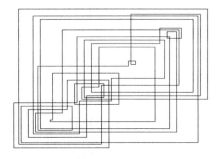

1/83 base 27

Foretaste

The story of this article began with a surprise visit of a high-school aged student and his mother to Dr. Jan Pearce's outer office at Berea college: Rafe Jones appeared with Dr. Libby Jones to inquire about Rafe doing a summer research project with Dr. Pearce. Caught a bit off guard by the apparition of a colleague and her son, and not immediately sold on doing research with so young a student, Dr. Pearce stalled for time, asking Rafe what he might like to work on. His quick reply: "Anything but number theory–I know nothing about number theory."

Dr. Pearce came round on doing the research project, and tried to accommodate Rafe's request. The work began with computer experiments in dynamical systems, but all roads seemed to lead to number theory. Soon the project required tools like the Euler totient function, and one of the main results involved Fermat primes. Steadily that old "queen of mathematics" (in the phrase of Gauss) wore down Rafe's defenses, and soon a hint of love was in the air. It blossomed into a Ph.D. in number theory at Brown University. Today Dr. Jones is an Assistant Professor at Holy Cross, working on both number theory and dynamics. Dr. Pearce continues to embrace unexpected opportunities at Berea.

Introduction and preliminaries

In America's visually-oriented, quantitatively illiterate culture, images have a great deal of power, so if a picture is today worth a thousand words, it must be worth at least a billion numbers. This power of the image is a hallmark of the postmodern era, in which the critical role of the observer has come to be recognized, and an understanding of the viewpoint has become inseparable from that of the object.

In some ways, the blossoming of chaos theory marked the arrival of mathematical postmodernism. Not so long ago, mathematical ideas were virtually unseen in American popular culture, and it took the enthralling fractal images of chaos theory to change that: the studies of chaos and fractals became some of the most widely discussed mathematical topics ever, and pictures of fractal images such as the

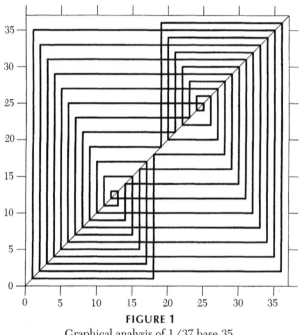

FIGURE 1
Graphical analysis of $1/37$ base 35.

Mandelbrot set began cropping up on T-shirts and posters selling in American malls. The power of an image is difficult to underestimate, particularly when it comes to creating interest in a topic widely regarded as bland. Perhaps we could fuel a greater excitement in traditionally underappreciated areas of mathematics if only we could present them in a flashier graphical fashion. Take fractions, for instance, which to many people appear to be merely seas of numbers; after all, infinitely many fractions have infinitely long strings of digits as their decimal expansions. Wouldn't it be nice if we could see complicated fractions, like $\frac{1}{37}$ base 35, as simple images? Wouldn't it be even nicer if, as for the Mandelbrot set, those graphical images exposed something about the inherent mathematical structure that the concise algebraic expression only implied?

In this paper, we apply to the study of certain fractions the same graphical techniques used to transform the Mandelbrot set from algebra to image. This will enable us to turn arcane algebraic objects into eye-catching designs, such as the one pictured in FIGURE 1. What's more, the mathematics behind this metamorphosis is not very hard to describe. We begin by describing a somewhat unusual method of representing a fraction, which will be useful for our purposes. Fractions can be viewed in a number of ways, many of which are base-dependent: reduced or unreduced, as pieces of a pie, expanded into decimal, binary, octal, etc. The method we adopt is quite base-dependent, and relies upon the remainders generated at each stage of the long-division process in base b. Consider $\frac{1}{7}$, which has a base 10 (decimal) expansion of $0.\overline{142857}$. We can calculate this using the usual long division process in base 10 as follows:

$$\begin{array}{r} 0.142857 \\ 7\overline{)1.000000} \\ \underline{7} \\ 30 \\ \underline{28} \\ 20 \\ \underline{14} \\ 60 \\ \underline{56} \\ 40 \\ \underline{35} \\ 50 \\ \underline{49} \\ 1 \end{array}$$

We can equivalently represent $\frac{1}{7}$ base 10 by writing the sequence of remainders produced in the above long division: $1 \to 3 \to 2 \to 6 \to 4 \to 5 \to 1$, a cycle that repeats infinitely. Note that what makes this a base ten long division is that we multiply the dividend by ten at every step; we could easily make it into a base b long division by multiplying by b at each step. This new long division would yield the sequence of remainders for $\frac{1}{7}$ base b; in fact, one can find the sequence of remainders for any fraction in any base simply by performing the appropriate long division. However, the laboriousness and iterative nature of long division make it desirable to have a simpler, more concise method of finding sequences of remainders. Happily, such a method exists, and it is simply the evaluation of the following function:

DEFINITION. Let a, b, and n be positive integers with $(n, a) = 1$ and $b > 1$. If r_i is the remainder produced at step i of the base b long division of $\frac{a}{n}$, the remainder produced at the $(i+1)$st step is given by $r_{i+1} = F_{b:n}(r_i) = b \times r_i \pmod{n}$. We call $F_{b:n}$ the *remainder function*, since if we begin with $r_0 = a$, iteration of $F_{b:n}$ yields the sequence of remainders of $\frac{a}{n}$ long divided in base b.

Note that a and n are relatively prime, so $\frac{a}{n}$ is a reduced fraction; we will assume throughout that all fractions are reduced. We can see the remainder function in action with the fraction used above, $\frac{1}{7}$ base 10. We begin with $r_0 = 1$. Next we have $r_1 = F_{10:7}(r_0) = 10 \times 1 \pmod{7} = 3$, followed by $r_2 = F_{10:7}(r_1) = 10 \times 3 \pmod{7} = 2$, $r_3 = F_{10:7}(r_2) = 10 \times 2 \pmod{7} = 6$, $r_4 = F_{10:7}(r_3) = 10 \times 6 \pmod{7} = 4$, $r_5 = F_{10:7}(r_4) = 10 \times 4 \pmod{7} = 5$, and $r_6 = F_{10:7}(r_5) = 10 \times 5 \pmod{7} = 1$.

Since $r_6 = r_0 = 1$, the sequence repeats. Note that each iteration of the remainder function simply multiplies by b and mods by n. Then, since r_0 is a, we can calculate the ith remainder directly using the formula $r_i = ab^i \pmod{n}$. This compact formula simplifies many arguments involving sequences of remainders, and you will see it often in the pages to come.

In the analysis above, our friend $\frac{1}{7}$ base 10 displays some surprising qualities. For example, $r_i + r_{i+3} = 7$ for all i. Moreover, if we let d_i represent the digit of the decimal expansion that is i places to the right of the decimal point, then in this example $d_i + d_{i+3} = 9$ for all i. These symmetries, as we shall see, have more than a numerical significance.

Before moving on to graphical topics, it will serve us well to discuss the three kinds of behavior a sequence of remainders (as well as the corresponding expansion) can exhibit. Each of these behaviors corresponds to a particular kind of graphical analysis graph, a concept we introduce in detail below. First, the sequence of remainders of $\frac{a}{n}$ in base b (as well as the corresponding base b expansion) may terminate; this happens if each remainder (and digit) is zero after some point, and such a fraction will have a graphical analysis graph that begins at some point and ends at some different point. This is the case if and only if every prime factor of n is also a prime factor of b. Second, the sequence may have a repeating cycle, but one that begins only after some initial string of remainders that never reappears. In this case, the graphical analysis graph will be an infinitely repeated figure, but with a tail created by the initial unrepeated string of remainders. This happens if and only if n has some factors that divide b and some that do not. Thirdly, the sequence may have only repeated cycles with no initial unrepeated string of remainders; this occurs if and only if n and b are relatively prime. This sort of fraction produces the neatest graphical analysis graph: a figure that retraces itself infinitely, with no unrepeated points.

The remainder function described above will allow us to work more easily with sequences of remainders. That it is a function also makes it a nice candidate for a graphical technique we will now introduce.

Graphical analysis

Graphical analysis or graphical iteration [2] gives us a visual way to explore function iteration. To graphically analyze a function $F(r)$, one does the following: Let r_0 be some number. Then, beginning with $i = 0$, draw a vertical line from (r_i, r_i) to the point $(r_i, F(r_i)) = (r_i, r_{i+1})$. (See FIGURE 2) From there, draw a horizontal line to $(F(r_i), F(r_i)) = (r_{i+1}, r_{i+1})$. Then increase i by one iteratively and repeat the preceding steps. Here, we will apply graphical analysis to our function $F_{b:n}(r)$. In order to avoid minor difficulties, we will say that if the remainder becomes zero at r_n, we stop the process at r_{n-1}. Although graphical analysis works only on functions, the remainder function associated with a given fraction is so closely tied to the fraction that we will refer to the graphical analysis of $F_{b:n}(r_i) = b \times r_i \pmod{n}$, with $r_0 = a$, as the graphical analysis of $\frac{a}{n}$ in base b.

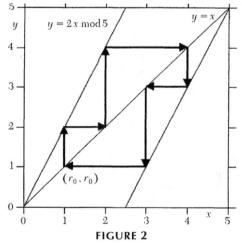

FIGURE 2

Graphical analysis of $\frac{1}{5}$ in base 2.

Note that the remainder function $F_{2,5}(x) = 2x$ (mod 5) plays a crucial role in FIGURE 2. However, you may have noticed that $F_{35:37}(x)$ does not appear in the graphical analysis graph of $\frac{1}{37}$ in base 35 (see FIGURE 1). The reason is that for so complex a picture, the slanted parallel lines of the remainder function become so dense as to obscure the image. Thus, despite their importance, for the sake of clarity we will omit them in the images to come.

Also, although the remainder function is theoretically important, one may graphically analyze a fraction without drawing the graph of the remainder function itself. In effect, the graphical analysis begins at the point (r_0, r_0), proceeds first vertically then horizontally to (r_1, r_1), then moves vertically then horizontally again to (r_2, r_2), and continues in this fashion. Hence in practice one can graphically analyze a fraction in a given base as follows: Compute the sequence of remainders; for each remainder, draw the appropriate dot on the line $y = x$; then connect the dots (following the order of the sequence of remainders), moving vertically then horizontally. Thus *the sequence of remainders entirely determines the graphical analysis graph of the fraction.* So when proving certain properties of graphical analysis graphs, such as various symmetries, we need not consider the entire image, but only the distribution of remainders.

Since the graphical analysis of a fraction varies from base to base, one might wonder how many distinct graphical pictures exist for a given fraction $\frac{a}{n}$. Bases zero and one are exempt from consideration. If b_1 and b_2 are bases such that $b_1 \equiv b_2$ (mod n), then $ab_1^m \equiv ab_2^m$ (mod n), so $\frac{a}{n}$ will generate identical sequences of remainders in both bases. Thus, we only have to consider for our bases a single representative from each congruence class modulo n. This means, of course, that at most n bases may produce distinct graphs. Further narrowing the field is the fact that if b is a base such that $b \equiv 0$ (mod n) or $b \equiv 1$ (mod n), the pictures are not very interesting: in the former case, all remainders save the first are zero, so the graphical analysis graph is merely a single point, since the analysis ends with the last nonzero remainder. In the latter case, if m is a positive integer, then $\frac{1}{n}$ written in base $mn + 1$ is $0.\overline{1}$, and the sequence of remainders is an infinite string of ones; again, the graphical analysis graph is a single point. We will exclude bases in the 0 congruence class in many later considerations. However, we will often be interested in all bases in which a fraction has a repeating expansion, and thus we will include bases in the 1 congruence class in spite of their graphical shortcomings.

The various graphs of a fraction in different bases often bear some relation to one another. The following definition will help us relate some of them to others.

Rotational graph pairs

DEFINITION 1. $\frac{a_1}{n_1}$ and $\frac{a_2}{n_2}$ are *rotational graph pairs* if the graphical analysis graph of $\frac{a_1}{n_1}$, when rotated 180° about the point $\left(\frac{n}{2}, \frac{n}{2}\right)$, produces the graphical analysis graph of $\frac{a_2}{n_2}$.

These pictures exemplify rotational graph pairs:

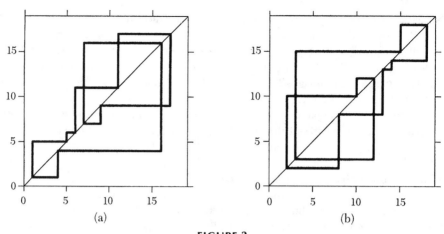

FIGURE 3
Graphical analysis of 17/19 base 5 vs. 2/19 base 5.

Since the graph of a fraction in base b depends entirely on its sequence of remainders, we can show that two fractions are rotational graph pairs simply by showing that "rotating" the sequence of remainders of one fraction about the point $\left(\frac{n}{2}, \frac{n}{2}\right)$ produces precisely the other sequence. In other words, the sequences must be zero in exactly the same places, and whenever the ith remainders of both sequences are nonzero, they must be equidistant from the point $\left(\frac{n}{2}, \frac{n}{2}\right)$. This is true if and only if the remainders in question sum to n. Thus we need only show that adding corresponding nonzero terms in the two sequences of remainders invariably yields n.

THEOREM 1. *In each base b, $\frac{a}{n}$ and $\frac{n-a}{n}$ are rotational graph pairs.*

Proof. First note that the only possible remainders at any stage of the long division of $\frac{a}{n}$ in base b belong to the set $\{0, 1, 2, \ldots, n-1\}$. Now, for any i, $ab^i \pmod{n} + (n-a)b^i \pmod{n} = (ab^i + (n-a)b^i) \pmod{n} = nb^i \pmod{n}$. Since $nb^i \equiv 0 \pmod{n}$ we have that the sum of the ith remainders of each sequence must be either 0 or n. Note that it is impossible for the ith remainder of one sequence to be zero and the ith remainder of the other nonzero: the nonzero remainder would make the sum necessarily greater than zero, and the zero remainder would make the sum necessarily less than n. Hence the sequences are zero in precisely the same places. Finally, if corresponding terms in the two sequences are nonzero, they cannot sum to zero, and so must sum to n. ∎

Part of the appeal of Theorem 1 lies in its breadth: it applies to any fraction in any base, regardless of the behavior of the fraction's sequence of remainders. However, in order to have breadth, one often must sacrifice depth. If we consider more restricted classes of fractions, we will be able to prove several stronger, more penetrating results.

We can extend Theorem 1 significantly if we restrict ourselves to fractions and bases that produce purely repeating sequences of remainders—that is, those satisfying $(b, n) = 1$. Since the graphs of these fractions consist of a single repeated figure, beginning with any point in the cycle will yield the same image. Thus if c_1 is a term in the sequence of remainders for $\frac{a}{n}$ base b, then the sequence of remainders of $\frac{c_1}{n}$ base b will go through exactly the same cycle, beginning at c_1 instead of a. Hence the two fractions $\frac{a}{n}$ and $\frac{c_1}{n}$ will produce identical graphs. Similarly, if c_2 is a term in the sequence of remainders of $\frac{n-a}{n}$, then $\frac{c_2}{n}$ and $\frac{n-a}{n}$ will produce identical graphs. This corollary then follows immediately from Theorem 1:

COROLLARY 2. *Suppose b and n are relatively prime. If $ab^i \equiv c_1$ (mod n) for some i and $(n-a)b^j \equiv c_2$ (mod n) for some j, then $\frac{c_1}{n}$ and $\frac{c_2}{n}$ are rotational graph pairs in base b.*

For example, $2 \times 100 \equiv 10$ (mod 19) and $17 \times 10 \equiv 18$ (mod 19), so $\frac{10}{19}$ base 10 and $\frac{18}{19}$ base 10 are rotational graph pairs.

Although we will return to this limited class of fractions later, in the next section we enlarge our consideration to include all sequences of remainders that do not terminate. The discussion hinges on a different sort of symmetry in the graphical analysis graph of a fraction: a rotational symmetry of a single graph, rather than of one graph to another.

Rotational symmetry

Consider the following two very different images in FIGURE 4:

The lovely rotational symmetry present in the graphical analysis graph of $\frac{1}{7}$ base 10 is strikingly absent in the graph of $\frac{1}{37}$. One might wonder why: after all, both 7 and 37

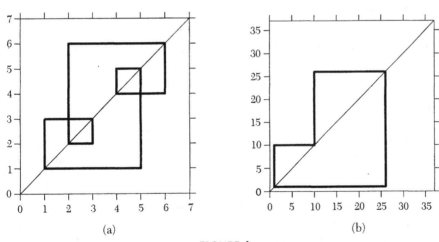

FIGURE 4
Graphical analysis of 1/7 base 10 vs. 1/37 base 10.

are not only relatively prime to 10, but also prime numbers. The following theorem will help to explain this difference.

THEOREM 3. *If $(n, a) = 1$ and n contains at least one prime factor that does not divide b then the following are equivalent:*

A. *$n - a$ appears in the sequence of remainders produced by the long division in base b of $\frac{a}{n}$ (i.e., $r_m = n - a$ for some m).*
B. *There exists an m, $0 < m < n$, such that for each natural number i, we have $r_i + r_{i+m} = n$.*

C. *The graphical analysis graph of the function $F_{b:n}$ beginning with $r_0 = a$ has $180°$ rotational symmetry about the point $\left(\frac{n}{2}, \frac{n}{2}\right)$.*

Proof. We will show A $\Rightarrow$ B by induction on i. By hypothesis, $r_0 + r_m = a + (n - a) = n$, so induction begins. Assuming that $r_i + r_{i+m} = n$, we must show that $r_{i+1} + r_{i+m+1} = n$. Using the remainder function, we have

$$r_{i+1} + r_{i+m+1} = F_{b:n}(r_i) + F_{b:n}(r_{i+m}) = b \times r_i \pmod{n} + b \times r_{i+m} \pmod{n}$$
$$= b \times (r_i + r_{i+m}) \pmod{n} = b \times n \pmod{n} = 0.$$

Thus $r_{i+1} + r_{i+m+1} \equiv 0 \pmod{n}$. Since n contains at least one prime factor that does not divide b, the sequence of remainders of $\frac{a}{n}$ base b does not terminate, so no remainder can be zero. Therefore $0 < r_{i+1} + r_{i+m+1} < 2n$, implying that $r_{i+1} + r_{i+m+1} = n$.

We now turn to B $\Rightarrow$ C. Condition B guarantees the existence of some positive integer m such that $r_i + r_{m+i} = n$ for each i. Let s be the smallest such integer. Since $r_s + r_{2s} = n = r_s + r_0$, it follows that $r_0 = r_{2s}$, and thus the length of the repeating cycle of the sequence of remainders is $2s$. Furthermore, the cycle is composed of the two halves $r_0, r_1, \ldots, r_{s-1}$ and $r_s, r_{s+1}, \ldots, r_{2s-1}$. Since $r_i + r_{s+i} = n$ for each i, these halves are essentially rotational graph pairs, and thus the whole graph is rotationally symmetric by itself.

Finally we address C $\Rightarrow$ A. Condition C means that our graph is rotationally symmetric about $\left(\frac{n}{2}, \frac{n}{2}\right)$, and since $r_0 = a$, (a, a) must be a point on the graph. Because of the graph's symmetry, $(n - a, n - a)$ must also be a point on the graph, implying that $n - a$ is a term in the sequence of remainders. Thus $r_m = n - a$ for some m. ∎

Remarks and observations

In the example given above, 36 is indeed nowhere to be found in the sequence of remainders for $\frac{1}{37}$ base 10, which is $1 \to 10 \to 26$, whereas 6 is the fourth number in the sequence for $\frac{1}{7}$ base 10. The equivalence of parts A and B thus predicts the visual discrepancy. In general, one need not go to the trouble of graphically analyzing a fraction to see if its graph is symmetric: it's enough to compute the sequence of remainders and examine it for a single number, $n - a$.

Interestingly, the symmetry among the remainders mentioned in part B of Theorem 3 is related to a similar symmetry among the digits. Suppose that the condition described in part B holds for a fraction $\frac{a}{n}$ in base b. The long division algorithm tells us that for each i, $b \times r_{i-1} = nd_i + r_i$ where d_i is the ith digit in the decimal expansion of $\frac{a}{n}$ in base b. Thus $nd_i + nd_{i+m} = b(r_{i-1} + r_{i+m-1}) - (r_i + r_{i+m}) = bn - n$. This implies that $d_i + d_{i+m} = b - 1$ for each i, a symmetry that we noted regarding $\frac{1}{7}$ base 10. A similar argument shows that the symmetry of remainders follows from the symmetry of digits, implying that the two are inseparable.

Symmetries in fractions with $(b, n) = 1$

Already the subject of Corollary 2, this class of fractions and its subclass of fractions with prime denominators will prove worthy of close scrutiny. Members of the larger class share one outstanding quality: in a given base b, rotational symmetry depends only on the denominator of the fraction in question (provided, of course, that the fraction is reduced). We make this precise in the next theorem.

THEOREM 4. *Let $\frac{a}{n}$ be a reduced fraction in base b, where $(b, n) = 1$. Then the graphical analysis graph of $\frac{a}{n}$ is rotationally symmetric in base b if and only if the graphical analysis graph of $\frac{1}{n}$ is rotationally symmetric in base b.*

Proof. Suppose that the graphical analysis graph of $\frac{1}{n}$ is rotationally symmetric in base b. The formula b^i (mod n) gives us the ith remainder of the long division of $\frac{1}{n}$ and ab^i (mod n) gives us the ith remainder of the long division of $\frac{a}{n}$. Since $\frac{1}{n}$ is rotationally symmetric, by Theorem 3 we have b^i (mod n) + b^{m+i} (mod n) = n for each i and for some m satisfying $0 < m < n$. Thus $b^i + b^{m+i} \equiv 0$ (mod n). Multiplying through by a yields $ab^i + ab^{m+i} = an \equiv 0$ (mod n), implying that ab^i (mod n) + ab^{m+i} (mod n) = 0 or n. Since $(b, n) = 1$, the sequence of remainders of $\frac{a}{n}$ base b does not terminate, and thus no remainders can be zero. We therefore conclude that ab^i (mod n) + ab^{m+i} (mod n) = n, proving the rotational symmetry of $\frac{a}{n}$ in base b.

The converse argument is quite similar. Supposing ab^i (mod n) + ab^{m+i} (mod n) = n for all i and for some m, we clearly have $ab^i + ab^{m+i} \equiv 0$ (mod n). We need only find a positive integer c such that $ca \equiv 1$ (mod n), and we will be able to

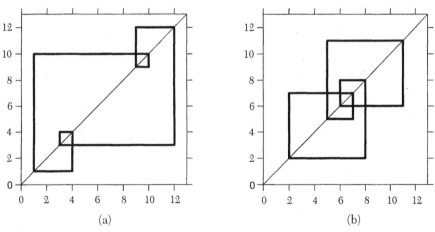

FIGURE 5
Graphical analysis of 1/13 base 10 vs. 5/13 base 10.

multiply through by c and complete the approach used above. Since our fraction is reduced, $(a, n) = 1$, so there exist positive integers c and d such that $ca + dn = 1$. This implies that $ca = 1 - dn \equiv 1$ (mod n), so the desired positive integer does indeed exist. ∎

This theorem guarantees that, for our limited class of fractions, if $\frac{1}{n}$ is rotationally symmetric in base b, then $\frac{a}{n}$ will be as well, provided $(a, n) = 1$. It often happens that $\frac{1}{n}$

and $\frac{a}{n}$ in fact produce identical graphs in base b; this is the case for $\frac{1}{7}$ and $\frac{a}{7}$ in base 10, where $(a, 7) = 1$. However, this need not happen, as FIGURE 5 shows.

Theorem 4 allows us to say that every reduced fraction with denominator n is either symmetric or not symmetric in any base b satisfying $(b, n) = 1$, since the value of the numerator plays no role. Thus for short, we will occasionally say simply that n is symmetric or not symmetric in base b.

The Euler totient function

We will be better able to understand symmetries in fractions with prime denominators with the aid of the *Euler totient function*. Denoted $\varphi(n)$, this function takes as input a positive integer n and produces as output the number of positive integers m that are less than or equal to n and satisfy $(m, n) = 1$. Some examples are $\varphi(4) = 2$, $\varphi(6) = 2$, $\varphi(12) = 4$, and, for any prime p, $\varphi(p) = p - 1$. The Euler totient function boasts two convenient properties which allow us to evaluate it easily for any small positive integer: First, if m and n are relatively prime positive integers, then $\varphi(mn) = \varphi(m)\varphi(n)$; and second, if p is prime and j is a positive integer, then $\varphi(p^j) = p^{j-1}(p-1)$ [3]. Thus, $\varphi(12) = \varphi(2^2 3) = \varphi(2^2)\varphi(3) = 2(2-1)2 = 4$. One of the better known theorems involving the Euler totient function is as follows:

LEMMA 5. (EULER'S FORMULA). *If b and m are positive integers and $(b, m) = 1$, then $b^{\varphi(m)} \equiv 1 \pmod{m}$.*

In particular, if p is prime and a not a multiple of p, we have $a^{p-1} \equiv 1 \pmod{p}$. Given a fraction with a purely repeating sequence of remainders, it's natural to be curious about the length of the repeating cycle (also known as the sequence's *period*). Euler's formula gives us some information about this period. Suppose $a < m$ and $\frac{a}{m}$ base b has a purely repeating sequence of remainders; we noted earlier that this is the case if and only if $(b, m) = 1$. The first remainder r_0 in the sequence is $ab^0 = a$, so the period of the sequence is the smallest nonzero k such that $r_k = a$. In other words, the period is the smallest nonzero k such that $ab^k \equiv a \pmod{m}$. Since $(b, m) = 1$, Euler's formula tells us that $b^{\varphi(m)} \equiv 1 \pmod{m}$, and thus $ab^{\varphi(m)} \equiv a \pmod{m}$. Because the period is the smallest nonzero k with $ab^k \equiv a \pmod{m}$, and $\varphi(m)$ satisfies this congruence, it follows that the period must divide $\varphi(m)$. In the special case where p is prime and b is not a multiple of p, we have the useful fact that the period of the sequence of remainders of $\frac{a}{p}$ in base b divides $p - 1$.

Fractions with prime denominators

Consider for a moment a reduced fraction with a prime denominator p in a base b that is not a multiple of p. Clearly $(b, p) = 1$, so Theorem 4 applies, showing that the value of the numerator does not affect the symmetry of the fraction's graph. Thus to determine if p is symmetric in base b, it is enough to examine the behavior of $\frac{1}{p}$ in base b. Although this is nice, we can use our restriction to fractions with prime denominators to get something even nicer: a convenient characterization of rotational symmetry.

Any reduced fraction with prime denominator p in a base b satisfying $(b, p) = 1$ must have a purely repeating sequence of remainders. The period of this sequence has everything to do with the rotational symmetry of the fraction: an even period means symmetry, an odd period no symmetry. We enshrine this convenient characterization in the following theorem:

THEOREM 6. *Let m be the smallest positive integer such that $b^m \equiv 1 \pmod{p}$, where p is an odd prime and $(b, p) = 1$. Then $\frac{1}{p}$ is rotationally symmetric in base b if and only if m is even.*

Proof. First note that because $(b, p) = 1$ and p is prime, it follows from Euler's formula that $b^{p-1} \equiv 1 \pmod{p}$, so there exists some positive integer satisfying $b^m \equiv 1 \pmod{p}$. Hence it makes sense to discuss the smallest such integer. Now suppose $\frac{1}{p}$ is rotationally symmetric in base b, and let c_1 be a term in the sequence of remainders of $\frac{1}{p}$ base b. Then for some i, $b^i \equiv c_1 \pmod{p}$. Since $0 < c_1 < p$, we have $(c_1, p) = 1$, so, by Theorem 6, $\frac{c_1}{p}$ must be rotationally symmetric in base b. Thus, by Theorem 3, $p - c_1$ must appear in the sequence of remainders of $\frac{c_1}{p}$ base b. Hence for some j, $c_1 b^j \equiv p - c_1 \pmod{p}$. Since $b^i \equiv c_1 \pmod{p}$, we have $b^{i+j} \equiv c_1 b^j \equiv p - c_1 \pmod{p}$, implying that $p - c_1$ is in the sequence of remainders of $\frac{1}{p}$ base b. Thus each remainder r in the repeating cycle of $\frac{1}{p}$ base b occurs together with $p - r$. Since p is odd, we cannot have $r = p - r$, so the elements of the cycle occur in distinct pairs. Hence the cycle length must be even. Given that the first remainder is $b^0 \pmod{p} = 1$, this means that the smallest positive integer satisfying $b^m \equiv 1 \pmod{p}$ is even.

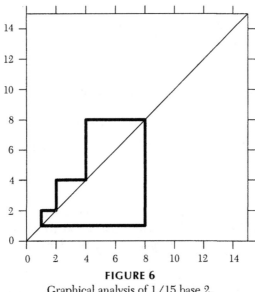

FIGURE 6
Graphical analysis of $1/15$ base 2.

To show the converse, let m be the smallest positive integer satisfying $b^m \equiv 1$ (mod p), and suppose m is even. Then $m = 2d$ for some positive integer d. Hence $b^m = b^{2d} = (b^d)^2 \equiv 1$ (mod p). Thus $(b^d + 1)(b^d - 1) \equiv 0$ (mod p) and since p is prime either $b^d \equiv 1$ (mod p) or $b^d \equiv -1$ (mod p). The first case is clearly impossible since $d = \frac{m}{2} < m$, and m was assumed to be the smallest positive integer such that $b^m \equiv 1$ (mod p). Thus we conclude that $b^d \equiv -1 \equiv p - 1$ (mod p). Therefore $p - 1$ is the dth remainder of $\frac{1}{p}$ base b, so by Theorem 3 $\frac{1}{p}$ is rotationally symmetric in base b. ■

Note that p must be prime for the above theorem to hold. Consider FIGURE 6, which shows that $\frac{1}{15}$ base 2 is not symmetric, though $2^m \equiv 1$ (mod 15) gives us a smallest m of 4.

Counting bases that produce symmetry

One might be tempted to guess that a prime number is symmetric in some randomly distributed number of bases; delightfully, this is not so. As we noted earlier, to find the ratio of bases in which a prime p is symmetric, we need only consider a single base b from each congruence class mod p. We will be interested here only in the bases in which $\frac{1}{p}$ has a repeating sequence of remainders. Therefore our considered bases will be all bases except those in the 0 congruence class.

For example, the reciprocal of 19 is symmetric in 9 of the 18 bases between 2 and 20, excluding 19; thus it is symmetric in half of the considered bases. The reciprocals of many other prime numbers are also symmetric in $\frac{1}{2}$ of the considered bases; some examples are 3, 7, 11, 23, 31, and 59. Other primes have reciprocals that are symmetric in $\frac{3}{4}$ of the considered bases; the first few are 5, 13, 29, 37, and 61. Still other primes, including 41, 73, and 89, have reciprocals symmetric in $\frac{7}{8}$ of the considered bases. In fact, one can find prime numbers that are symmetric in $\frac{(2^n - 1)}{2^n}$ of the considered bases for many positive integers n. We can explain this separation of the prime numbers into families, but to do so we will need a couple of number-theoretic results. [For details, see for example [3].]

LEMMA 7. *For any $n \geq 1$, we have $n = \sum_{d|n} \varphi(d)$ where the sum is taken over all divisors of n.*

LEMMA 8. *Let p be a prime number and d a positive divisor of $p - 1$. Then there are exactly $\varphi(d)$ numbers b that are incongruent (mod p) and have the property that d is the smallest positive integer satisfying $b^d \equiv 1$ (mod p).*

To illustrate Lemma 8, let $p = 7$ and choose $d = 3$. Lemma 8 tells us that there are $\varphi(3) = 2$ possible bases b which are not congruent (mod 7) and have the property that while b^1 and b^2 are not congruent to 1 (mod 7), b^3 is congruent to 1 (mod 7). In other words, were we to compute the sequence of remainders for $\frac{1}{7}$ in one base from each of the seven congruence classes mod 7, we would find that exactly two of them produce a sequence of period 3. If we want to know how many will yield a sequence with period 6, we simply have to calculate $\varphi(6)$, which is 2. The same holds for any other divisor of 6. To find the number of these bases in which the graph of $\frac{1}{7}$ is rotationally symmetric, Theorem 6 tells us we need only determine in how many of them $\frac{1}{7}$ has a sequence of remainders of even period. Since we ignore bases in the zero congruence class, all the bases we consider satisfy $(b, 7) = 1$. Since the period of $\frac{1}{7}$ in any of these bases must divide $\varphi(7) = 6$, the only possible even periods are 6 and 2. Thus our answer is $\varphi(6) + \varphi(2) = 3$, and we see that $\frac{1}{7}$ is symmetric in one half of the considered bases. This sort of analysis underlies the following proof.

THEOREM 9. *Suppose p is an odd prime number and n is the largest integer satisfying $2^n | p - 1$. Then, excluding bases $b \equiv 0 \pmod{p}$, $\frac{1}{p}$ is symmetric in $\frac{2^n - 1}{2^n}$ of the remaining bases.*

Proof. We need only consider a single representative base from each nonzero congruence class mod p. By Theorem 6, it suffices to find the number of bases in which $\frac{1}{p}$ produces a sequence of remainders of even period. The period of $\frac{1}{p}$ in any representative base must divide $\varphi(p) = p - 1$, so we want to find for each even divisor m of $p - 1$ the number of bases in which $\frac{1}{p}$ has a sequence of period m.

Lemma 8 tells us that for a divisor q of $p - 1$, $\frac{1}{p}$ will produce a sequence of period q in exactly $\varphi(q)$ of the representative bases. Hence we need only compute $\Sigma \varphi(m)$, where m varies over the even divisors of $p - 1$. We will call the value of this sum k.

Suppose 2 divides $p - 1$ exactly n times. Then the prime factorization of $p - 1$ is $2^n p_1^{\alpha_1} p_2^{\alpha_2} \cdots p_N^{\alpha_N}$, where each p_j is an odd prime. The largest odd divisor of $p - 1$ is thus $D = p_1^{\alpha_1} p_2^{\alpha_2} \cdots p_N^{\alpha_N}$. Now every even divisor m of $p - 1$ has the form $2^i t$, where $1 \le i \le n$ and t divides D. So we have

$$k = \sum_{i=1}^{n} \sum_{t | D} \varphi(2^i t).$$

Now since $t | D$ and D is odd, t must be odd. So $(2^i, t) = 1$ for any i, and by the first convenient property of the Euler function, we have $\varphi(2^i t) = \varphi(2^i) \varphi(t)$, so

$$k = \sum_{i=1}^{n} \sum_{t | D} \varphi(2^i) \varphi(t) = \sum_{i=1}^{n} \varphi(2^i) \sum_{t | D} \varphi(t).$$

By Lemma 7, $\sum_{t | D} \varphi(t) = D = p_1^{\alpha_1} p_2^{\alpha_2} \cdots p_N^{\alpha_N}$, so

$$k = p_1^{\alpha_1} p_2^{\alpha_2} \cdots p_N^{\alpha_N} \sum_{i=1}^{n} \varphi(2^i).$$

Now $\sum_{i=1}^{n} \varphi(2^i) = \varphi(2) + \varphi(2^2) + \cdots + \varphi(2^n)$. By the second convenient property of the Euler function, the right side is $2^0(1) + 2^1(1) + 2^2(1) + \cdots + 2^{n-1}(1) = 2^n - 1$, so we have $\sum_{i=1}^{n} \varphi(2^i) = 2^n - 1$. Finally we arrive at our value for k:

$$k = (2^n - 1) p_1^{\alpha_1} p_2^{\alpha_2} \cdots p_N^{\alpha_N}.$$

Since we are considering only a single base from each of the $p - 1$ nonzero congruence classes mod p, we have that $\frac{1}{p}$ is symmetric in

$$\frac{k}{p - 1} = \frac{(2^n - 1) p_1^{\alpha_1} p_2^{\alpha_2} \cdots p_N^{\alpha_N}}{2^n p_1^{\alpha_1} p_2^{\alpha_2} \cdots p_N^{\alpha_N}} = \frac{2^n - 1}{2^n}$$

of the considered bases. ∎

COROLLARY 10. *Suppose p is an odd prime and D is the largest odd divisor of $p - 1$. Then, excluding bases $b \equiv 0 \pmod{p}$, $\frac{1}{p}$ fails to be symmetric in $\frac{D}{p-1}$ of the remaining bases.*

Proof. Suppose 2 divides $p - 1$ exactly n times. Applying Theorem 9 we get that $\frac{1}{p}$ fails to be symmetric in $1 - \frac{2^n - 1}{2^n} = \frac{1}{2^n}$ of all the considered bases. The prime factorization of $p - 1$ is $2^n D$. Thus $\frac{1}{p}$ fails to be symmetric in $\frac{1}{2^n} = \frac{D}{2^n D} = \frac{D}{p-1}$ of the possible bases. ∎

We noted earlier that in a base of the form $ap + 1$, where a is a positive integer, $\frac{1}{p}$ will have a sequence of remainders that is simply an infinite string of ones. This leads to a graph consisting only of the fixed point $(1, 1)$. If $p = 2$, this graph is in fact symmetric about $(\frac{p}{2}, \frac{p}{2})$, but for any odd prime it fails to be symmetric. Thus an odd prime must fail to be symmetric in all bases belonging to the 1 congruence class mod p. However, there exist odd primes that are symmetric in *all* of the other considered bases, and thus are as symmetric as it is possible for an odd prime to be.

Perfectly symmetric numbers and Fermat primes

DEFINITION. *A positive integer $n > 1$ is perfectly symmetric if its reciprocal is symmetric in any base b provided $b \not\equiv 0 \pmod{n}$ and $b \not\equiv 1 \pmod{n}$.*

Clearly, 2 is trivially perfectly symmetric. This membership in the set of perfectly symmetric numbers makes 2 a spectacularly rare positive integer, joined only by widely-spaced comrades:

THEOREM 11. *The only perfectly symmetric numbers are 2 and the Fermat primes.*

Proof. Recall that a Fermat prime is a prime of the form $2^{2^m} + 1$, where m is a natural number. Suppose n is a perfectly symmetric number, and suppose also that n is composite. Then there is some base b that divides n and satisfies $1 < b < n$. Clearly b and n are not relatively prime. So the sequence of remainders of $\frac{1}{n}$ in base b cannot be purely repeating: either it terminates or has an initial string of unrepeated remainders. In the latter case, the string of unrepeated digits creates a tail in the graphical analysis graph of $\frac{1}{n}$ base b, and the tail ruins any symmetry. In the former case, $n - 1$ cannot appear in the sequence of remainders, for if it did, we would have $r_k = n - 1$ for some nonzero k, implying that $r_{2k} = 1$. But the sequence of remainders terminates, so this is not possible. In either case n is not symmetric in base b, contradicting our assumption.

Therefore n must be prime. We have already seen that 2 is perfectly symmetric. If n is an odd prime, then, by Corollary 10, $\frac{1}{n}$ will fail to be symmetric in bases belonging to D of the $n - 1$ nonzero congruence classes mod n, where D is the largest odd divisor of $n - 1$. Any odd prime must fail to be symmetric in at least one of these base congruence classes, but since n is perfectly symmetric, it cannot fail in any of the others; therefore $D = 1$. Thus no odd number greater than one can divide $n - 1$, implying that $n - 1$ is of the form 2^i for some i. Therefore $n = 2^i + 1$ and n is prime. If $i = uv$, where u is odd and $u > 1$, then $2^v + 1 | 2^i + 1$, so $2^i + 1$ fails to be prime. Thus in this case our i must be of the form 2^k, where $k \in \mathbb{N}$. Therefore our prime n is of the form $2^{2^k} + 1$, and is thus a Fermat prime.

Conversely, if n is either 2 or a Fermat prime, then clearly either $n = 2$, and is thus perfectly symmetric, or n is odd. In the latter case, by Corollary 10 we have that $\frac{1}{n}$ fails to be symmetric in bases belonging to D of the $n - 1$ nonzero congruence classes mod n, and must be symmetric in all the rest. Here $n - 1 = 2^i$, so $D = 1$. But the reciprocal of any number m must fail to be symmetric in bases belonging to the 1 congruence class mod m. Hence if $x > 1$ and $b \equiv x \pmod{n}$, $\frac{1}{n}$ is symmetric in base b. Therefore n is perfectly symmetric. ∎

Currently there are only five known Fermat primes: 3, 5, 17, 257, and 65537. Thus, only six known perfectly symmetric numbers lurk among all the positive integers greater than one, suggesting that perfect symmetry is among the more unusual properties a number can have. However, precisely how many perfectly symmetric numbers exist remains an open question.

Questions and conclusions

Our discussion of the number of symmetry-producing bases for various fractions raises two questions about certain kinds of prime numbers:

Question 1. *Does there exist, for each positive integer n, a natural number k such that $2^n(2k + 1) + 1$ is prime?*

If so, then for any positive integer n one can find a prime p such that 2 divides $p - 1$ exactly n times. This would mean that for any positive integer n, primes exist that are symmetric in $\frac{2^n - 1}{2^n}$ of the considered bases.

Question 2. *How many Fermat primes are there?*

No one has any idea; we know only that there are at least five. Pierre de Fermat thought that all numbers of the form $2^{2^k} + 1$ were prime, but history has proven otherwise: All the numbers generated using $k = 5, \ldots, 11$ have turned out to be composite, as well as selected others, including the monstrous $2^{2^{23471}} + 1$. There remain, however, infinitely many more as-yet-undetermined possibilities. An answer to this question would also tell how many perfectly symmetric numbers exist.

Thus ends our exploration of fractions and symmetry. Postmodernism has taught us that all ways of looking at a problem are not equivalent: different perspectives highlight different properties. Adopting our society's penchant for images led us to examine more closely the symmetries of certain fractions, and opened our eyes to unexpected visions.

Note on the computer program During the course of this project, we wrote a simple computer program that graphically analyzes any fraction in any base. We found many of the images quite striking and beautiful, and were sorry not to be able to include all of them in this article. For those of you who would like to generate some of these images yourselves, our program is in an electronic supplement at http://www.maa.org/pubs/mm_supplements/index.html.

Acknowledgments. We would like to thank Dr. Mark Hanisch for his help with MATLAB, Dr. James Lynch for his interest and encouragement, the Reverend Kent Gilbert for moral support and occasional entertainment, and Dr. Libby Jones for timely sustenance.

REFERENCES

1. K. H. Becker and M. Dörfler, *Dynamical Systems and Fractals*, Cambridge University Press, Cambridge, UK, 1989.
2. R. Devaney, *Chaos, Fractals, and Dynamics*, Addison-Wesley, Reading, MA, 1990.
3. J. Silverman, *A Friendly Introduction to Number Theory*, Prentice-Hall, Upper Saddle River, NJ, 1997.

Originally appeared as:
Jones, Rafe and Jan Pearce. "A Postmodern View of Fractions and Reciprocals of Fermat Primes." *Mathematics Magazine*. vol. 73, no. 2 (April 2000): pp. 83–97.

Visible Structures in Number Theory

Peter Borwein and
Loki Jörgenson

1. INTRODUCTION.

> I see a confused mass.   —Jacques Hadamard (1865–1963)

These are the words the great French mathematician used to describe his initial thoughts when he proved that there is a prime number greater than 11 [**11**, p. 76]. His final mental image he described as "...a place somehere between the confused mass and the first point". In commenting on this in his fascinating but quirky monograph, he asks "What may be the use of such a strange and cloudy imagery?".

Hadamard was of the opinion that mathematical thought is visual and that words only interfered. And when he inquired into the thought processes of his most distinguished mid-century colleagues, he discovered that most of them, in some measure, agreed (a notable exception being George Pólya).

For the non-professional, the idea that mathematicians "see" their ideas may be surprising. However the history of mathematics is marked by many notable developments grounded in the visual. Descartes' introduction of "cartesian" co–ordinates, for example, is arguably the most important advance in mathematics in the last millenium. It fundamentally reshaped the way mathematicians thought about mathematics, precisely because it allowed them to "see" better mathematically.

Indeed, mathematicians have long been aware of the significance of visualization and made great effort to exploit it. Carl Friedrich Gauss lamented, in a letter to Heinrich Christian Schumacher, how hard it was to draw the pictures required for making accurate conjectures. Gauss, whom many consider the greatest mathematician of all time, in reference to a diagram that accompanies his first proof of the fundamental theorem of algebra, wrote

> It still remains true that, with negative theorems such as this, transforming personal convictions into objective ones requires deterringly detailed work. To visualize the whole variety of cases, one would have to display a large number of equations by curves; each curve would have to be drawn by its points, and determining a single point alone requires lengthy computations. You do not see from Fig. 4 in my first paper of 1799, how much work was required for a proper drawing of that curve.   —Carl Friedrich Gauss (1777–1855)

The kind of pictures Gauss was looking for would now take seconds to generate on a computer screen.

Newer computational environments have greatly increased the scope for visualizing mathematics. Computer graphics offers magnitudes of improvement in resolution and speed over hand–drawn images and provides increased utility through color, animation, image processing, and user interactivity. And, to some degree, mathematics has evolved to exploit these new tools and techniques. We explore some subtle uses of interactive graphical tools that help us "see" the mathematics more clearly. In particular, we focus on cases where the right picture suggests the "right theorem", or where it indicates structure where none was expected, or where there is the possibility of "visual proof".

For all of our examples, we have developed Internet-accessible interfaces. They allow readers to interact and explore the mathematics and possibly even discover new results of their own—visit www.cecm.sfu.ca/projects/numbers/.

2. IN PURSUIT OF PATTERNS.

> Computers make it easier to do a lot of things, but most of the things they make it easier to do don't need to be done. —Andy Rooney

Mathematics can be described as the science of patterns, relationships, generalized descriptions, and recognizable structure in space, numbers, and other abstracted entities. This view is borne out in numerous examples such as [16] and [15]. Lynn Steen has observed [19]:

> Mathematical theories explain the relations among patterns; functions and maps, operators and morphisms bind one type of pattern to another to yield lasting mathematical structures. Application of mathematics use these patterns to "explain" and predict natural phenomena that fit the patterns. Patterns suggest other patterns, often yielding patterns of patterns.

This description conjures up images of cycloids, Sierpinski gaskets, "cowboy hat" surfaces, and multi-colored graphs. However it isn't immediately apparent that this patently visual reference to patterns applies throughout mathematics. Many of the higher order relationships in fields such as number theory defy pictorial representation or, at least, they don't immediately lend themselves intuitively to a graphic treatment. Much of what is "pattern" in the knowledge of mathematics is instead encoded in a linear textual format born out of the logical formalist practices that now dominate mathematics.

Within number theory, many problems offer large amounts of "data" that the human mind has difficulty assimilating directly. These include classes of numbers that satisfy certain criteria (e.g., primes), distributions of digits in expansions, finite and infinite series and summations, solutions to variable expressions (e.g., zeroes of polynomials), and other unmanageable masses of raw information. Typically, real insight into such problems has come directly from the mind of the mathematician who ferrets out their essence from formalized representations rather than from the data. Now computers make it possible to "enhance" the human perceptual/cognitive systems through many different kinds of visualization and patterns of a new sort emerge in the morass of numbers.

However the epistemological role of computational visualization in mathematics is still not clear, certainly not any clearer than the role of intuition where mental visualization takes place. However, it serves several useful functions in current practice. These include inspiration and discovery, informal communication and demonstration, and teaching and learning. Lately though, the area of experimental mathematics has expanded to include exploration and experimentation and, perhaps controversially, formal exposition and proof. Some carefully crafted questions have been posed about how experiment might contribute to mathematics [5]. Yet answers have been slow to come, due in part to general resistance and, in some cases, alarm [11] within the mathematical community. Moreover, experimental mathematics finds only conditional support from those who address the issues formally [7], [9].

The value of visualization hardly seems to be in question. The real issue seems to be what it can be used for. Can it contribute directly to the body of mathematical knowledge? Can an image act as a form of "visual proof"? Strong cases can be made to the affirmative [7], [3] (including in number theory), with examples typically in

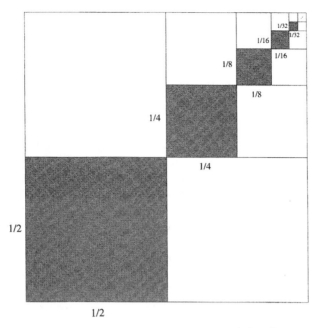

Figure 1. A simple "visual proof" of $\sum_{n=1}^{\infty}(\frac{1}{2})^{2n}=\frac{1}{3}$

the form of simplified, heuristic diagrams such as Figure 1. These carefully crafted examples call into question the epistemological criteria of an acceptable proof.

Establishing adequate criteria for mathematical proof is outside the scope of this paper; interested readers can visit [20], a repository for information related to reasoning with visual representations. Its authors suggest that three necessary, but perhaps not sufficient, conditions may be:

- *reliability*: the underlying means of arriving at the proof are reliable and the result is unvarying with each inspection
- *consistency*: the means and end of the proof are consistent with other known facts, beliefs, and proofs
- *repeatability*: the proof may be confirmed by or demonstrated to others

Each requirement is difficult to satisfy in a single, static visual representation. Most criticisms of images as mathematical knowledge or tools make this clear [8], [13].

Traditional exposition differs significantly from that of the visual. In the logical formal mode, proof is provided in linearly connected sentences composed of words that are carefully selected to convey unambiguous meaning. Each sentence follows the previous, specifying an unalterable path through the sequence of statements. Although error and misconception are still possible, the tolerances are extremely demanding and follow the strict conventions of deductivist presentation [12].

In graphical representations, the same facts and relationships are often presented in multiple modes and dimensions. For example, the path through the information is usually indeterminate, leaving the viewer to establish what is important (and what is not) and in what order the dependencies should be assessed. Further, unintended information and relationships may be perceived, either due to the unanticipated interaction of the complex array of details or due to the viewer's own perceptual and cognitive processes.

As a consequence, successful visual representations tend to be spartan in their detail. And the few examples of visual proof that withstand close inspection are limited in their scope and generalizability. The effort to bring images closer to conformity with the prevailing logical modes of proof has resulted in a loss of the richness that is intrinsic to the visual.

3. IN SUPPORT OF PROOF.

> Computers are useless. They can only give you answers. —Pablo Picasso (1881–1973)

In order to offer the reliability, consistency, and repeatability of the written word and still provide the potential inherent in the medium, visualization needs to offer more than just the static image. It too must guide, define, and relate the information presented. The logical formalist conventions for mathematics have evolved over many decades, resulting in a mode of discourse that is precise in its delivery. The order of presentation of ideas is critical, with definitions preceding their usage, proofs separated from the general flow of the argument for modularity, and references to foundational material listed at the end.

To do the same, visualization must include additional mechanisms or conventions beyond the base image. It isn't appropriate simply to ape the logical conventions and find some visual metaphor or mapping that works similarly (this approach is what limits existing successful visual proofs to very simple diagrams). Instead, an effective visualization needs to offer several key features

- *dynamic*: the representation should vary through some parameter(s) to demonstrate a range of behaviours (instead of the single instance of the static case)
- *guidance*: to lead the viewer through the appropriate steps in the correct order, the representation should offer a "path" through the information that builds the case for the proof
- *flexibility*: it should support the viewer's own exploration of the ideas presented, including the search for counterexamples or incompleteness
- *openness*: the underlying algorithms, libraries, and details of the programming languages and hardware should be available for inspection and confirmation

With these capabilities available in an interactive representation, the viewer could then follow the argument being made visually, explore all the ramifications, check for counterexamples, special cases, and incompleteness, and even confirm the correctness of the implementation. In fact, the viewer should be able to inspect a visual representation and a traditional logical formal proof with the same rigor.

Although current practice does not yet offer any conclusions as to how images and computational tools may impact mathematical methodologies or the underlying epistemology, it does indicate the direction that subsequent work may take. Examples offer some insight into how emerging technologies may eventually provide an unambiguous role for visualization in mathematics.

4. THE STRUCTURE OF NUMBERS.
Numbers may be generated by a myriad of means and techniques. Each offers a very small piece of an infinitely large puzzle. Number theory identifies patterns of relationship between numbers, sifting for the subtle suggestions of an underlying fundamental structure. The regularity of observable features belies the seeming abstractness of numbers.

Visible Structures in Number Theory

Figure 2. The first 1600 decimal digits of π mod 2.

Binary Expansions. In the 17th century, Gottfried Wilhelm Leibniz asked in a letter to one of the Bernoulli brothers if there might be a pattern in the binary expansion of π. Three hundred years later, his question remains unanswered. The numbers in the expansion appear to be completely random. In fact, the most that can now be said of any of the classical mathematical constants is that they are largely non-periodic.

With traditional analysis revealing no patterns of interest, generating images from the expansions offers intriguing alternatives. Figures 2 and 3 show 1600 decimal digits of π and 22/7 respectively, both taken mod 2. The light pixels are the even digits and the dark ones are the odd. The digits read from left to right, top to bottom, like words in a book.

What does one see? The even and odd digits of π in Figure 2 seem to be distributed randomly. And the fact that 22/7 (a widely used approximation for π) is rational appears clearly in Figure 3. Visually representing randomness is not a new idea; Pickover [18] and Voelcker [22] have previously examined the possibility of "seeing randomness". Rather, the intention here is to identify patterns where none has so far been seen, in this case in the expansions of irrational numbers.

These are simple examples but many numbers have structures that are hidden both from simple inspection of the digits and even from standard statistical analysis. Figure 4 shows the rational number 1/65537, this time as a binary expansion, with a period of 65536. Unless graphically represented with sufficient resolution, the presence of a regularity might otherwise be missed in the unending string of 0's and 1's.

Figures 5 a) and b) are based on similar calculations using 1600 terms of the simple continued fractions of π and e respectively. Continued fractions have the form

$$\cfrac{1}{a_1 + \cfrac{1}{a_2 + \cfrac{1}{a_3 + \cdots}}}$$

Figure 3. The first 1600 decimal digits of 22/7 mod 2.

Figure 4. The first million binary digits of 1/65537 reveal the subtle diagonal structure from the periodicity.

Visible Structures in Number Theory

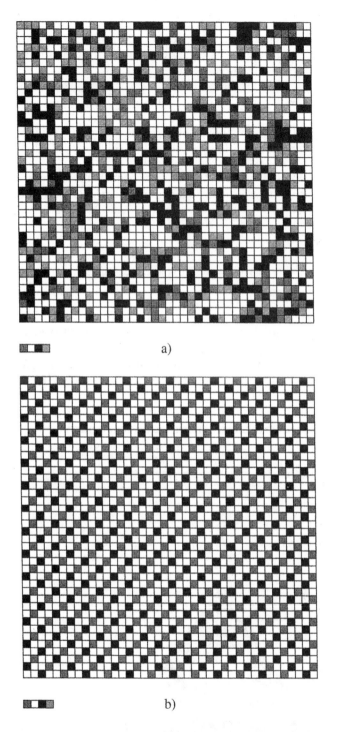

Figure 5. The first 1600 values of the continued fraction for a) π and b) e, both mod 4

In these images, the decimal values have been taken mod 4. Again the distribution of the a_i of π appears random though now, as one would expect, there are more odds than evens. However for e, the pattern appears highly structured. This is no surprise on closer examination, as the continued fraction for e is

$$[2, 1, 2, 1, 1, 4, 1, 1, 6, 1, 1, 8, 1, 1, 10, 1, 1, 12, \ldots]$$

and, if taken as a sequence of digits, is a rational number mod 4. It is apparent from the images that the natures of the various distributions are quite distinct and recognizable. In contrast no such simple pattern exists for $\exp(3)$ mod 4.

Presumably this particular visual representation offers a qualitative characterization of the numbers. It tags them in an instantly distinguishable fashion that would be almost impossible to do otherwise.

Sequences of Polynomials.

> Few things are harder to put up with than the annoyance of a good example.
> —Mark Twain (1835–1910)

In a similar vein, structures are found in the coefficients of sequences of polynomials. The first example in Figure 6 shows the binomial coefficients $\binom{n}{m}$ mod 3, or equivalently Pascal's Triangle mod 3. For the sake of what follows, it is convenient to think of the ith row as the coefficients of the polynomial $(1 + x)^i$ taken mod three. This apparently fractal pattern has been the object of much careful study [**10**].

Figure 7 shows the coefficients of the first eighty Chebyshev polynomials mod 3 laid out like the binomial coefficients of Figure 6. Recall that the nth Chebyshev polynomial T_n, defined by $T_n(x) := \cos(n \arccos x)$, has the explicit representation

$$T_n(x) = \frac{n}{2} \sum_{k=0}^{\lfloor n/2 \rfloor} (-1)^k \frac{(n-k-1)!}{k!(n-2k)!} (2x)^{n-2k},$$

and satisfies the recursion

$$T_n(x) = 2x T_{n-1}(x) - T_{n-2}(x), \quad n = 2, 3, \ldots.$$

Figure 6. Eighty rows of Pascal's Triangle mod 3

Visible Structures in Number Theory

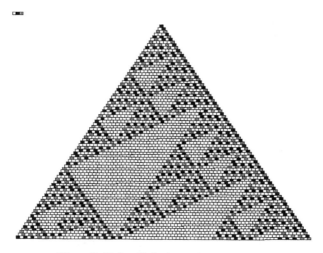

Figure 7. Eighty Chebyshev Polynomials mod 3

The expression for $T_n(x)$ resembles the $\binom{n}{m}$ form of the binomial coefficients and its recursion relation is similar to that for the Pascal's Triangle.

Figure 8 shows the Stirling numbers of the second kind mod 3, again organized as a triangle. They are defined by

$$S(n, m) := \frac{1}{m!} \sum_{k=0}^{m} \binom{m}{k}(-1)^{m-k} k^n$$

and give the number of ways of partitioning a set of n elements into m non-empty subsets. Once again the form of $\binom{n}{m}$ appears in its expression.

The well-known forms of the polynomials appear distinct. Yet it is apparent that the polynomials are graphically related to each other. In fact, the summations are variants of the binomial coefficient expression.

It is possible to find similar sorts of structure in virtually any sequence of polynomials: Legendre polynomials; Euler polynomials; sequences of Padé denominators to the exponential or to $(1 - x)^\alpha$ with α rational. Then, selecting any modulus, a distinct

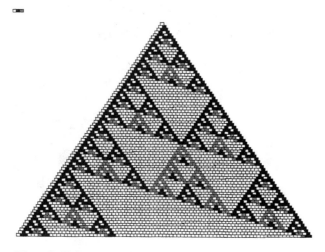

Figure 8. Eighty rows of Stirling Numbers of the second kind mod 3

pattern emerges. These images indicate an underlying structure within the polynomials themselves and demand some explanation. While conjectures exist for their origin, proofs for the theorems suggested by these pictures do not yet exist. And when there finally is a proof, might it be offered in some visual form?

Quasi-Rationals.

> For every problem, there is one solution which is simple, neat, and wrong.
> —H.L. Mencken (1880–1956)

Having established a visual character for irrationals and their expansions, it is interesting to note the existence of "quasi-rational" numbers. These are certain well-known irrational numbers whose images appear suspiciously rational. The sequences pictured in Figures 9 and 10 are $\{i\pi\}_{i=1}^{1600}$ mod 2 and $\{ie\}_{i=1}^{1600}$ mod 2, respectively. One way of thinking about these sequences is as binary expansions of the numbers

$$\sum_{n=1}^{\infty} \frac{\lfloor m\alpha \rfloor \bmod 2}{2^i},$$

where α is, respectively, π or e.

The resulting images are very regular. And yet these are transcendental numbers; having observed this phenomenon, we were subsequently able to explain this behavior rigorously from the study of

$$\sum_{n=1}^{\infty} \frac{\lfloor m\alpha \rfloor}{2^i},$$

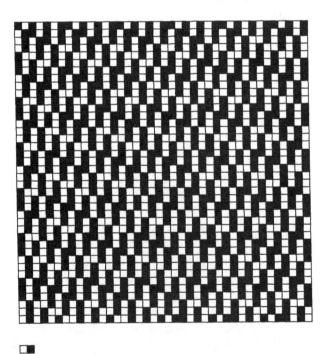

Figure 9. Integer part of $\{i\pi\}_{i=1}^{1600}$ mod 2; note the slight irregularities in the pseudo-periodic pattern.

Visible Structures in Number Theory

Figure 10. Integer part of $\{ie\}_{i=1}^{1600}$ mod 2; note the slight irregularities in the pseudo-periodic pattern.

which is transcendental for all irrational α. This follows from the remarkable continued fraction expansion of Böhmer [4]

$$\sum_{n=1}^{\infty}[m\alpha]z^n = \sum_{n=0}^{\infty} \frac{(-1)^{n+1}}{(1-z^{q_n})(1-z^{q_{n+1}})}.$$

Here (q_n) is the sequence of denominators in the simple continued fraction expansion of α.

Careful examination of Figures 9 and 10 show that they are only *pseudo*-periodic; slight irregularities appear in the pattern. Rational–like behaviour follows from the very good rational approximations evidenced by the expansions. Or put another way, there are very large terms in the continued fraction expansion. For example, the expansion of

$$\sum_{n=1}^{\infty} \frac{[m\pi]\bmod 2}{2^i}$$

is

[0, 1, 2, 42, 638816050508714029100700827905, 1, 126, ...],

with a similar phenomenon for e.

This behaviour makes it clear that there is subtlety in the nature of these numbers. Indeed, while we were able to establish these results rigorously, many related phenomena exist whose proofs are not yet in hand. For example, there is no proof or explanation for the visual representation of

$$\sum_{n=1}^{\infty} \frac{[m\pi] \bmod 2}{3^i}$$

Proofs for these graphic results might well offer further refinements to their representations, leading to yet another critical graphic characterization.

Complex Zeros. Polynomials with constrained coefficients have been much studied [2], [17], [6]. They relate to the Littlewood conjecture and many other problems. Littlewood notes that "these raise fascinating questions" [14].

Certain of these polynomials demonstrate suprising complexity when their zeros are plotted appropriately. Figure 11 shows the complex zeros of all polynomials

$$P_n(z) = a_0 + a_1 z + a_2 z^2 + \cdots + a_n z^n$$

of degree $n \leq 18$, where $a_i = \{-1, +1\}$. This image, reminiscent of pictures for polynomials with all coefficients in the set $\{0, +1\}$ [17], does raise many questions: Is the set fractal and what is its boundary? Are there holes at infinite degree? How do the holes vary with the degree? What is the relationship between these zeros and those of polynomials with real coefficients in the neighbourhood of $\{-1, +1\}$?

Some, but definitely not all, of these questions have found some analytic answer [17], [6]. Others have been shown to relate subtly to standing problems of some significance in number theory. For example, the nature of the holes involves a old problem known as *Lehmer's conjecture* [1]. It is not yet clear how these images contribute to a solution to such problems. However they are provoking mathematicians to look at numbers in new ways.

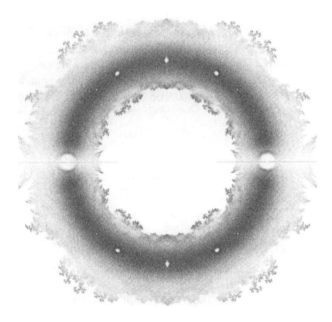

Figure 11. Roots of Littlewood Polynomials of degree at most 18 for coefficients ±1.

5. CONCLUSION. Visualization extends the natural capacity of the mathematician to envision his subject, to see the entities and objects that are part of his work with the aid of software and hardware. Since graphic representations are firmly rooted in verifiable algorithms and machines, the images and interfaces may also provide new forms of exposition and possibly even proof. Most important of all, like spacecraft, diving bells, and electron microscopes, visualization of mathematical structures takes the human mind to places it has never been and shows the mind's eye images from a realm previously unseen.

Readers are encouraged to review this paper in full color on-line [21].

ACKNOWLEDGEMENTS. This work has been supported in part by the NCE for Mathematics of Information Technology and Complex Systems (MITACS), and in part by research and equipment grants from the Natural Sciences and Engineering Research Council of Canada (NSERC). We thank the Centre for Experimental & Constructive Mathematics.

REFERENCES

1. F. Beaucoup, P. Borwein, D. Boyd, and C. Pinner, Multiple roots of $[-1, 1]$ power series. *London Math. Soc.* **57** (1998) 135–147.
2. A.T. Bharucha-Reid and M. Sambandham, *Random Polynomials*, Academic Press, Orlando, FL, 1986.
3. Alex Bogomolny, http://www.cut-the-knot.com/ctk/pww.html
4. J. Borwein and P. Borwein, On the generating function of the integer part: $[n\alpha + \gamma]$, *J. Number Theory* **43** (1993) 293–318.
5. J.M. Borwein, P.B. Borwein, R. Girgensohn, and S. Parnes, Making sense of experimental mathematics, www.cecm.sfu.ca/organics/vault/expmath/
6. P. Borwein and C. Pinner, Polynomials with $\{0, +1, -1\}$ coefficients and a root close to a given point, *Canadian J. Math.* **49** (1997) 887–915.
7. J.R. Brown, Proofs and pictures, *Brit. J. Phil. Sci.* **48** (1997) 161–180.
8. T. Eisenberg and T. Dreyfuss, On the reluctance to visualize in mathematics, in *Visualization in Teaching and Learning Mathematics*, Mathematical Association of America, Washington, DC, 1991, pp. 25–37.
9. Marcus Giaquinto, Epistemology of visual thinking in elementary real analysis, *Brit. J. Phil. Sci.* **45** (1994) 789–813.
10. A. Granville, The arithmetic properties of binomial coefficients, in *Proceedings of the Organic Mathematics Workshop, Dec. 12–14*, http://www.cecm.sfu.ca/organics/, 1995. IMpress, Simon Fraser University, Burnaby, BC.
11. John Horgan, The death of proof, *Scientific American* October, 1993, pp. 92–103.
12. Imre Lakatos, *Proofs and refutations: the logic of mathematical discovery*, Cambridge University Press, Cambridge, 1976.
13. Bruno Latour, Drawing things together, in *Representation in Scientific Practice*, Michael Lych and Steve Woolgar, eds., MIT Press, Cambridge, MA, 1990, pp. 25–37.
14. J.E. Littlewood, Some problems in real and complex analysis, in *Heath Mathematical Monographs*, D.C. Heath, Lexington, MA, 1968.
15. Roger A. Nelson, *Proofs Without Words II, More Exercises in Visual Thinking*, Mathematical Association of America, Washington, DC, 2000.
16. Roger A. Nelson, *Proofs Without Words: Exercises in Visual Thinking*, Mathematical Association of America, Washington, DC, 1993.
17. A. Odlyzko and B. Poonen, Zeros of polynomials with 0,1 coefficients, *Enseign. Math.* **39** (1993) 317–348.
18. C. Pickover, Picturing randomness on a graphics supercomputer, *IBM J. Res. Develop.* **35** (1991) 227–230.
19. Lynn Arthur Steen, The science of patterns, *Science* **240** (1988) 611–616.
20. http://www.hcrc.ed.ac.uk/gal/Diagrams/biblio.html
21. http://www.cecm.sfu.ca/~pborwein/
22. J. Voelcker, Picturing randomness, *IEEE Spectrum* **8** (1988) 13–16.

PETER BORWEIN is a Professor of Mathematics at Simon Fraser University, Vancouver, British Columbia. His Ph.D. is from the University of British Columbia under the supervision of David Boyd. After a postdoctoral year in Oxford and a dozen years at Dalhousie University in Halifax, Nova Scotia, he took up his current position. He has authored five books and over a hundred research articles. His research interests span diophantine and computational number theory, classical analysis, and symbolic computation. He is co-recipient of the Cauvenet Prize and the Hasse Prize, both for exposition in mathematics.
CECM, Simon Fraser University, Burnaby, BC, Canada V5A 1S6
pborwein@bb.cecm.sfu.ca

LOKI JÖRGENSON is an Adjunct Professor of Mathematics at Simon Fraser University, Vancouver, British Columbia. Previously the Research Manager for the Centre for Experimental and Constructive Mathematics, he is a senior scientist at Jaalam Research. He maintains his involvement in mathematics as the digital editor for the Canadian Mathematical Society. His Ph.D. is in computational physics from McGill University, and he has been active in visualization, simulation, and computation for over 15 years. His research has included forays into philosophy, graphics, educational technologies, high performance computing, statistical mechanics, high energy physics, logic, and number theory.
CECM, Simon Fraser University, Burnaby, BC, Canada V5A 1S6
loki@cecm.sfu.ca

Originally appeared as:
Borwein, Peter and Loki Jörgenson. "Visible Structures in Number Theory." *American Mathematical Monthly.* vol. 108, no. 10 (December 2001): pp. 897–910.

Visual Gems of Number Theory

Roger B. Nelsen

While looking through some elementary number theory texts recently, I was struck by how few illustrations most of them have. A number can represent the cardinality of a set, the length of a line segment, or the area of a plane region, and such representations naturally lead to a variety of visual arguments for topics in elementary number theory. Since a course in number theory usually begins with properties of the integers, specifically the positive integers, the texts should have more pictures. In the next few pages I'll present some "visual gems" illustrating one-to-one correspondence, congruence, Pythagorean triples, perfect numbers, and the irrationality of the square root of two.

We begin with *figurate numbers*, positive integers that can be represented by geometric patterns. The simplest examples are the triangular numbers t_n $(= 1 + 2 + 3 + \cdots + n)$ and square numbers n^2. In Figure 1(a) are the familiar patterns for t_4 and 4^2. The computational formula for t_n is $n(n + 1)/2$, which is easily derived from Figure 1(b) by seeing that $2t_n = n(n+1)$.

Figure 1. Triangular and Square Numbers.

You may have noticed that each triangular number is also a binomial coefficient:
$$1 + 2 + \cdots + n = \binom{n+1}{2}.$$

One explanation for this is that each is equal to $n(n + 1)/2$, but this answer sheds no light on why the relationship holds. A better explanation is the following: there exists a one to one correspondence between a set of $t_n = 1 + 2 + 3 + \cdots + n$ objects and the set of two-element subsets of a set with $n + 1$ objects. In Figure 2, we have a visual proof of this correspondence.

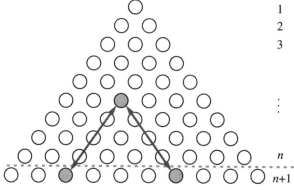

Figure 2. A visual proof that $1 + 2 + \cdots + n = \binom{n+1}{2}$.

There are many lovely relationships between triangular and square numbers. One is hidden in the following theorem, which appears in nearly every elementary number theory text as an example or exercise: if n is odd, then $n^2 \equiv 1 \pmod 8$. In Figure 3, we have a visual proof that every odd square is one more than eight times a triangular number, which proves the result. But it does more—it is an essential part of an induction proof that there are infinitely many numbers that are simultaneously square and triangular! First note that $t_1 = 1 = 1^2$. Then observe that

$$t_{8t_k} = \frac{8t_k(8t_k+1)}{2} = 4t_k(2k+1)^2,$$

so that if t_k is a square, so is t_{8t_k}. For example, $t_8 = 6^2$, $t_{288} = 204^2$, etc. But not all square triangular numbers are generated this way: $t_{49} = 35^2$, $t_{1681} = 1189^2$, etc.

Challenge: Can you find a visual argument to use in the proof for odd triangular numbers?

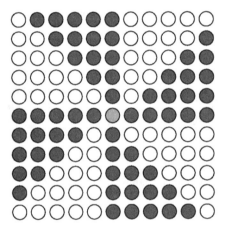

Figure 3. A visual proof that n odd implies $n^2 \equiv 1 \pmod 8$.

Congruence results involving triangular numbers can be illustrated in a similar manner. In Figure 4, we "see" that $t_{3n-1} \equiv 0 \pmod 3$.

Exercise: Find visual arguments to show that $t_{3n} \equiv 0 \pmod 3$ and $t_{3n+1} \equiv 1 \pmod 3$.

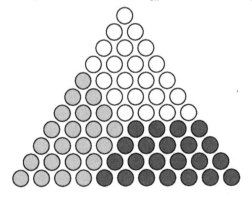

Figure 4. A visual proof that $t_{3n-1} \equiv 0 \pmod 3$.

Discussing triangles and squares in geometry always brings to mind the Pythagorean theorem. In number theory it brings to mind *Pythagorean triples*. A primitive Pythagorean triple (PPT) is a triple (a, b, c) of positive integers with no common factors satisfying $a^2 + b^2 = c^2$. Familiar examples are $(3, 4, 5)$, $(5, 12, 13)$ and $(8, 15, 17)$. Indeed, there are infinitely many PPTs, and in Figure 5 we have a visual proof that n^2 odd (i.e. $n^2 = 2k+1$) implies that $n^2 + k^2 = (k + 1)^2$, or equivalently, that $(n, k, k+1)$ is a PPT. In Figure 5, the $n^2 = 2k +1$ gray dots are first arranged into an "L" then k^2 white dots are drawn, for a total of $(k + 1)^2$ dots.

Exercise: In the proof, we've actually shown that there are infinitely many PPTs in which one leg and the hypotenuse are consecutive integers. Find a visual proof that there are infinitely many PPTs in which the hypotenuse and one leg differ by two.

Figure 5. There are infinitely many PPTs.

A classical result characterizing PPTs is the following theorem: (a, b, c) is a PPT if and only if there are relatively prime integers $m > n > 0$, one even and the other odd, such that (a, b, c) [or (b, a, c)] $= (2mn, m^2 - n^2, m^2 + n^2)$. Figure 6 uses the double angle formulas from trigonometry to illustrate this result.

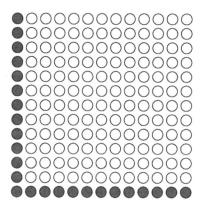

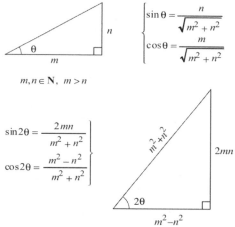

Figure 6. A characterization of PPTs.

Another less well-known characterization is the following: there exists a one-to-one correspondence between PPTs and factorizations of even squares of the form $n^2 = 2pq$

with p and q relatively prime. Figure 7 illustrates this result.

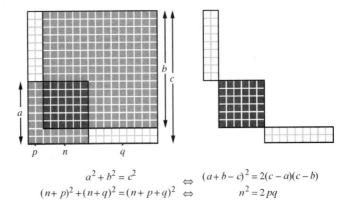

$$a^2 + b^2 = c^2$$
$$(n+p)^2 + (n+q)^2 = (n+p+q)^2 \Leftrightarrow$$
$$(a+b-c)^2 = 2(c-a)(c-b)$$
$$n^2 = 2pq$$

Figure 7. Another characterization of PPTs.

The PPT (3, 4, 5) is the only such triple where the sum of two consecutive squares is the next square. But what about sums of three consecutive squares? Four consecutive squares? Observe:

$$3^2 + 4^2 = 5^2$$
$$10^2 + 11^2 + 12^2 = 13^2 + 14^2$$
$$21^2 + 22^2 + 23^2 + 24^2 = 25^2 + 26^2 + 27^2$$

What's the pattern? A little reflection yields the observation that each square immediately to the left of the equals sign is the square of four times a triangular number, i.e., $4 = 4(1)$, $12 = 4(1 + 2)$, $24 = 4(1 + 2 + 3)$, etc., which leads to the identity

$$(4t_n - n)^2 + \cdots + (4t_n)^2 = (4t_n + 1)^2 + \cdots + (4t_n + n)^2.$$

While it is not difficult to verify this by induction, the illustration (for the $n = 3$ case, using $4t_3 = 4(1 + 2 + 3)$ of these "Pythagorean runs" in Figure 8 may better explain why the relationship holds.

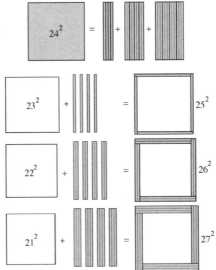

Figure 8. Pythagorean runs.

The ancient Greeks called positive integers like 6 and 28 *perfect*, because in each case the number is equal to the sum of its proper divisors (e.g., $28 = 1 + 2 + 4 + 7 + 14$). In Book IX of the *Elements*, Euclid gives a formula for generating even perfect numbers: If $p = 2^{n+1} - 1$ is a prime number, then $N = 2^n p$ is perfect. We can illustrate this result very nicely in Figure 9 by letting N be the area of a rectangle of dimensions $2^n \times p$, and noting that the rectangle can be partitioned into a union of smaller rectangles (and one square) whose areas are precisely the proper divisors of N.

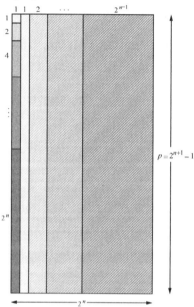

Figure 9. Euclid's formula for even perfect numbers.

We conclude with one more example from the Greeks. The Pythagoreans were probably the first to demonstrate that $\sqrt{2}$ is irrational, that is, to show that lengths of a side and the diagonal of a square are incommensurable. Here is a modern version of that classical Greek proof. From the Pythagorean theorem, an isosceles triangle of edge-length 1 has hypotenuse $\sqrt{2}$. If $\sqrt{2}$ is rational, then some multiple of this triangle must have three sides with integer lengths, and hence there must be a *smallest* isosceles right triangle with this property. But the difference of two segments of integer length is a segment of integer length and so, as illustrated in Figure 10, inside any isosceles right triangle whose three sides have integer lengths we can always construct a *smaller* one with the same property! Hence $\sqrt{2}$ must be irrational.

Exercise: Prove that $\sqrt{3}$ and $\sqrt{5}$ are irrational in a similar manner.

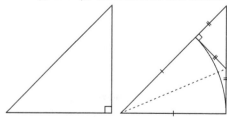

Figure 10. The irrationality of $\sqrt{2}$.

Further Reading

For more visual proofs with triangular numbers, see James Tanton's "Triangular Numbers" in the November 2005 issue of *Math Horizons*. For visual proofs in combinatorics, see Jennifer Quinn and Arthur Benjamin's *Proofs That Really Count: The Art of Combinatorial Proof* (MAA, 2003). For an introduction to creating visual proofs, see Claudi Alsina and Roger Nelsen's *Math Made Visual: Creating Images for Understanding Mathematics* (MAA, 2006). Some of the proofs in the article originally came from:

1. T. Apostol, "The irrationality of the square root of two—a geometric proof," *Amer. Math. Monthly*, 107 (2000), 841–842.

2. M. Boardman, "Proof without words: Pythagorean runs," *Math. Mag.*, 73 (2000), 59.

3. D. Goldberg, personal communication.

4. J. Gomez, "Proof without words: Pythagorean triples and factorizations of even squares," *Math. Mag.*, 78 (2005), 14.

5. L. Larson, "A discrete look at $1 + 2 + \cdots + n$," *College Math. J.*, 16 (1985), 369–382.

6. C. Vanden Eynden, *Elementary Number Theory*, Random House, 1987.

Originally appeared as:
Nelsen, Roger B. "Visual Gems of Number Theory." *Math Horizons*. vol. 15, no. 3 (February 2008): pp.7–9, 31.

Part II: Primes

The prime numbers are the building blocks of the integers, thanks to the Fundamental Theorem of Arithmetic that every natural number has a unique prime factorization. We open this chapter with "A new proof of Euclid's theorem," that there are infinitely many primes, given by Filip Saidak (*Amer. Math. Monthly*, vol. 113, no. 10 (December 2006), pp. 937–938). We won't say too much about it (since the proof is only four sentences long), but it is a constructive proof based on nothing more than the observation that consecutive numbers are relatively prime. The article references books that provide many other elegant proofs of Euclid's theorem. One of these proofs is "On the infinitude of primes" by Harry Furstenberg (*Amer. Math. Monthly*, vol. 62, no. 5 (May, 1955), p. 353). This paper is the only item in this book that requires more than a first course in number theory, since it is based on elementary ideas from topology. But the paper is a classic and is just one paragraph long, so we felt it was worth the space. In a similarly brief paper, "On the series of prime reciprocals" (*Proc. Amer. Math. Soc.*, vol. 17, no. 2 (April 1966), p. 541), James Clarkson proves a stronger theorem. Using little more than the divergence of the harmonic series, he proves that $\sum_{p \text{ prime}} \frac{1}{p}$ also diverges.

Any collection of articles on prime numbers ought to have at least one article by Paul Erdős, the most prolific mathematician of the twentieth century, and a wandering minstrel of mathematics. A perfect number is a number n whose proper divisors sum to n. We say n is pseudoperfect if some of its proper divisors sum to n. If the sum of the divisors of n exceeds n, but n is not pseudoperfect, then n is called weird. The number 70 is weird (in fact it's the smallest weird number) since its proper divisors 1, 2, 5, 7, 10, 14, 35 have a sum of 74, but there is no way to obtain a sum of 70 using some of these numbers at most once. In this paper "On weird and pseudoperfect numbers" (*Mathematics of Computation*, vol. 28, no. 126 (April, 1974), pp. 617–623), Stan Benkoski and Erdős prove and conjecture properties of weird, pseudoperfect, and related numbers. For instance weird numbers are not too rare—they have a positive density among the integers. Although maybe only the first half of the paper is suitable for undergraduates, we have kept the paper in its entirety. The proof of the first theorem is elementary and requires an integral best evaluated by using infinite series. After the proof, Erdős offers a $300 reward for the proof of a related conjecture. Although Erdős died in 1996, the problem remains open and the reward is still claimable.

Here are some famous theorems about primes that have names attached to them. Given a prime p, we have:
- Fermat's Little Theorem: *If a is an integer, then $a^p \equiv a \pmod{p}$.*
- Wilson's Theorem: $(p-1)! \equiv -1 \pmod{p}$.
- Cauchy's Theorem: *If G is a finite group whose order is a multiple of p, then G contains an element of order p.*

- Lucas's Theorem: *If n and r are written in base p, say $n = n_0 + n_1 p + \cdots + n_k p^k$ and $r = r_0 + r_1 p + \cdots + r_k p^k$, where $0 \leq n_i, r_i < p$, then*

$$\binom{n}{r} \equiv \binom{n_0}{r_0}\binom{n_1}{r_1}\cdots\binom{n_k}{r_k} \pmod{p}.$$

What do these theorems have in common? All can be obtained by "Applications of a simple counting technique," which is the title of an article by Melvin Hausner (*Amer. Math. Monthly*, vol. 90, no. 2 (February, 1983), pp. 127–129). Each is shown to be a consequence of a theorem about fixed points of functions. (Although the paper does not include Wilson's theorem, we leave that as an exercise, with this hint: For any circular arrangement of the numbers 0 through $p-1$, consider the operation of adding 1 to each element, modulo p.)

The Fundamental Theorem of Arithmetic says that every positive integer has a unique prime factorization. But given a large random number, how do you actually find its prime factors? In a paper dedicated to Paul Erdős, "A tale of two sieves" (*Notices of the AMS*, vol. 43 (December 1996), pp. 1473–1485), Carl Pomerance introduces the reader to sophisticated factoring algorithms, which are based on very elementary ideas. This paper was the first recipient of the Levi L. Conant Prize in 2001 for the best expository paper published in either the *Notices of the AMS* or the *Bulletin of the AMS* in the preceding five years. We are privileged to be able to include it in our collection.

Of course, there are many fundamental theorems about prime numbers, but only one of these theorems is called *the* Prime Number Theorem: The number of primes below x is asymptotic to $\frac{x}{\ln x}$. Although its proof is well beyond the scope of this book, Hugh Montgomery and Stan Wagon can prove something almost as good, using little more than elementary calculus. In "A heuristic for the Prime Number Theorem" (*Math Intelligencer*, vol. 28, no. 3 (2006), pp. 6–9), they show why the natural logarithm is the *natural* logarithm to use.

A New Proof of Euclid's Theorem

Filip Saidak

A prime number is an integer greater than 1 that is divisible only by 1 and itself. Mathematicians have been studying primes and their properties for over twenty-three centuries. One of the very first results concerning these numbers was presumably proved by Euclid of Alexandria, sometime before 300 B.C. In Book IX of his legendary *Elements* (see [2]) we find Proposition 20, which states:

Proposition. *There are infinitely many prime numbers.*

Euclid's proof (modernized). Assume to the contrary that the set P of all prime numbers is finite, say $P = \{p_1, p_2, ..., p_k\}$ for a positive integer k. If

$$Q := (p_1 p_2 \cdots p_k) + 1,$$

then $\gcd(Q, p_i) = 1$ for $i = 1, 2, ..., k$. Therefore Q has to have a prime factor different from all existing primes. That is a contradiction.

Today many proofs of Euclid's theorem are known. It may come as a surprise that the following almost trivial argument has not been given before:

New Proof. Let n be an arbitrary positive integer greater than 1. Since n and $n + 1$ are consecutive integers, they must be coprime. Hence the number $N_2 = n(n + 1)$ must have at least two different prime factors. Similarly, since the integers $n(n + 1)$ and $n(n + 1) + 1$ are consecutive, and therefore coprime, the number

$$N_3 = n(n + 1)[n(n + 1) + 1]$$

must have at least three different prime factors. This process can be continued indefinitely, so the number of primes must be infinite.

Analysis. The proof just given is conceptually even simpler than the original proof due to Euclid, since it does not use Eudoxus's method of "reductio ad absurdum," proof by contradiction. And unlike most other proofs of the theorem, it does not require Proposition 30 of *Elements* (sometimes called "Euclid's Lemma") that states: if p is a prime and $p \mid ab$, then either $p \mid a$ or $p \mid b$. Moreover, our proof is constructive, and it gives integers with an arbitrary number of prime factors.

Remarks. In Ribenboim [4, pp. 3–11] and Narkiewicz [3, pp. 1–10] one finds at least a dozen different proofs of the classical theorem of Euclid, and many other variations of the arguments listed in [1], [3], and [4] have been published over the years (in chronological order) by: Goldbach (1730), Euler (1737 and 1762), Kummer (1878), Perott (1881), Stieltjes (1890), Thue (1897), Brocard (1915), Pólya (1921), Erdős (1938), Bellman

(1947), Fürstenberg (1955), Barnes (1976), Washington (1980), and others. Goldbach's proof (see [4, p. 4]), which uses pairwise coprimality of Fermat numbers, seems to be closest in spirit to the argument we have presented.

Acknowledgements. Personal and virtual conversations with Professors Paulo Ribenboim (Queen's University) and Eduard Kostolansky (Bratislava) are gratefully acknowledged. I would also like to thank ProfessorWładyslaw Narkiewicz (Wroclaw) for bringing to my attention Hermite's very simple proof concerning $n! + 1$.

References

1. M. Aigner and G. M. & Ziegler, *Proofs from THE BOOK*, Springer-Verlag, Berlin, 1999.

2. T. L. Heath, *The Thirteen Books of Euclid's Elements*, vol. 2, University Press, Cambridge, 1908; 2nd ed. reprinted by Dover, New York, 1956.

3. W. Narkiewicz, *The Development of Prime Number Theory*, Springer-Verlag, New York, 2000.

4. P. Ribenboim, *The New Book of Prime Number Records*, Springer-Verlag, New York, 1996.

Originally appeared as:
Saidak, Filip. "A New Proof of Euclid's Theorem." *American Mathematical Monthly*. vol. 113, no. 10 (December 2006): pp. 937–938.

On the Infinitude of the Primes

Harry Furstenberg

In this note we would like to offer an elementary "topological" proof of the infinitude of the prime numbers. We introduce a topology into the space of integers S, by using the arithmetic progressions (from $-\infty$ to $+\infty$) as a basis. It is not difficult to verify that this actually yields a topological space. In fact, under this topology, S may be shown to be normal and hence metrizable. Each arithmetic progression is closed as well as open, since its complement is the union of other arithmetic progressions (having the same difference). As a result, the union of any finite number of arithmetic progressions is closed. Consider now the set $A = \cup A_p$, where A_p consists of all multiples of p, and p runs through the set of primes ≥ 2. The only numbers not belonging to A are -1 and 1, and since the set $\{-1, 1\}$ is clearly not an open set, A cannot be closed. Hence A is not a finite union of closed sets which proves that there are an infinity of primes.

Originally appeared as:
Furstenberg, Harry. "On the Infinitude of the Primes." *American Mathematical Monthly*. vol. 62, no. 5 (May 1955): p. 353.

On the Series of Prime Reciprocals

James A. Clarkson

Let p_n be the nth prime. We give another proof of the

Theorem. *The series $\sum_{n=1}^{\infty}(1/p_n)$ diverges.*

Proof. Assume the contrary, and fix k so that

(1) $$\sum_{n=k+1}^{\infty}(1/p_n) < 1/2.$$

Let $Q = p_1 p_2 \cdots p_k$.

We consider now the sum $S(r) = \sum_{i=1}^{r}[1/(1+iQ)]$, where r is any positive integer. Since $1+iQ$ is prime to Q, all the prime factors of all these denominators are from a finite segment of primes which we call $P(r)$:

$$P(r) = \{p_{k+1}, p_{k+2}, \ldots, p_{m(r)}\}.$$

Now let $S(r, j)$ stand for the sum of those terms in the sum $S(r)$ whose denominators $1+iQ$ have just j (not assumed distinct) prime factors. Each sum term has the form $1/q_1 q_2 \cdots q_j$, with each $q_i \in P(r)$. But every such term occurs at least once in the expansion of $\left[\sum_{n=k+1}^{m(r)}(1/p_n)\right]^j$, so by (1) $S(r, j) < 1/2^j$. Thus for each r,

$$S(r) = \sum_{j} S(r, j) < \sum_{j}(1/2^j) < 1.$$

So $\sum_{i=1}^{\infty}[1/(1+iQ)]$ converges, which in turn implies that the harmonic series does.

Originally appeared as:

Clarkson, James A. "On the Series of Prime Reciprocals." *Proceedings of the AMS.* vol. 17, no. 2 (April 1966): p. 541.

Applications of a Simple Counting Technique

Melvin Hausner

In this note, we show how the simple counting technique of Theorem 1 can be applied in several different settings. Throughout, X denotes a finite set, and p denotes a prime number. For $A \subseteq X$, let $|A|$ represent the number of elements in A.

1. THEOREM. *Let* $f: X \to X$ *with* $f^p = I$ (*the identity map*). *Let* X_0 *be the set of fixed points of* X. $X_0 = \{x \in X | f(x) = x\}$. *Then* $|X| \equiv |X_0| \bmod p$.

Proof. For any $x \in X$, we define the orbit $\bar{x}$ of x as the set $\{x, f(x), \ldots, f^{p-1}(x)\}$. It is an easy matter to show that the orbits partition X. Clearly, $|\bar{x}| = 1$ if and only if $x = f(x)$, that is, if $x \in X_0$. Now we claim that if $|\bar{x}| > 1$, then $|\bar{x}| = p$. For if we had any duplication in $\bar{x}$, then $f^i(x) = f^j(x)$ for some $i, j, 0 \leq i < j < p$, so that $f^{j-i}(x) = x$. Since $f^p(x) = x$ and $(j - i, p) = 1$, it follows that $f(x) = x$, and hence $|\bar{x}| = 1$.

Finally, since there are $|X_0|$ orbits of length 1, and all other orbits, say n of them, have length p, we have $|X| = |X_0| + np$. This yields the required congruence. ∎

Probably the most familiar use of Theorem 1 is for the solution of the following problem in group theory. "If G is a group of even order, prove it has an element $a \neq e$ (the identity) satisfying $a^2 = e$." (For example, see Herstein [1], p. 35, problem 11.) The familiar proof pairs off x and x^{-1} when they are distinct. The remaining even number of group elements satisfy $x = x^{-1}$ or $x^2 = e$. Since e is one such element, there are an odd number of elements $x \neq e$, with $x^2 = e$. In our notation, $X = G$, $p = 2$, $f(x) = x^{-1}$, so $f^2 = I$, and $X_0 = \{x | x = x^{-1}\} = \{x | x^2 = e\}$. By Theorem 1, $|X_0| \equiv |X| \equiv 0 \bmod 2$. McKay [3] gives a nice direct generalization of this proof which we put into our terminology.

2. EXAMPLE (CAUCHY'S THEOREM for Groups). *Let* G *be a finite group*, p *a prime dividing* $|G|$. *Then* G *has an element of order* p. *Indeed the number of such elements is congruent to* $-1 \bmod p$.

Proof. Let $X =$ the set of p-tuples $(x_1, \ldots, x_p)$ where $x_i \in G$, and $x_1 x_2 \cdots x_p = e$. Note that $|X| = |G|^{p-1} \equiv 0 \bmod p$. Since $x_1 x_2 \cdots x_p = e$ implies $x_2 \cdots x_p x_1 = e$, we may define $f(x_1, \ldots, x_p) = (x_2, \ldots, x_p, x_1)$ and we have $f: X \to X, f^p = I$. What is X_0? $f(x) = x$ if and only if $x = (g, g, \ldots, g)$ with $g^p = e$. Thus, $|X_0| =$ the number of $g \in X$ satisfying $g^p = e$. By Theorem 1, $|X_0| \equiv |X| \equiv 0 \bmod p$. Since $|X_0| = 1 +$ number of elements of order p, this is the result.

3. EXAMPLE (FERMAT'S THEOREM). $n^p \equiv n \bmod p$.

Proof. Consider the set X of all lattice points $(x_1, \ldots, x_p)$ with $1 \leq x_i \leq n$ and let $f(x_1, \ldots, x_p) = (x_2, \ldots, x_p, x_1)$. Clearly $f^p = I$ and $|X| = n^p$, $|X_0| = n$. Theorem 1 gives the result.

4. EXAMPLE (LUCAS' THEOREM). *Suppose*

$$n = n_0 + n_1 p + \cdots + n_k p^k; \quad r = r_0 + r_1 p + \cdots + r_k p^k$$

with $0 \leq n_i, r_i < p$. Then

$$\binom{n}{r} \equiv \binom{n_0}{r_0}\binom{n_1}{r_1} \cdots \binom{n_k}{r_k} \bmod p.$$

(*Note:* $\binom{a}{b} = 0$ if $b > a$.)

Proof. If we write $n = Np + n_0$, $r = Rp + r_0$, where $0 \leq n_0, r_0 < p$, it suffices to prove

$$\binom{n}{r} \equiv \binom{N}{R}\binom{n_0}{r_0} \bmod p.$$

We define $A_i = \{(i,1),\ldots,(i,N)\}$ for $i = 1,\ldots,p$ and $B = \{(0,1),\ldots,(0,n_0)\}$. Then, setting

$$A = A_1 \cup \cdots \cup A_p \cup B$$

we have $|A| = Np + n_0 = n$. We now define $f: A \to A$ by cyclically moving the A_i's and keeping B fixed:

$$f(i,x) = (i+1,x) \qquad 1 \leq i \leq p-1$$
$$f(p,x) = (1,x)$$
$$f(0,x) = (0,x).$$

Thus,

$$f(A_i) = A_{i+1}(1 \leq i \leq p-1), f(A_p) = A_1, f(B) = B.$$

Clearly $f^p = I$.

We now take X as the collection of subsets $C \subseteq A$ with $|C| = r$. Since $f: A \to A$, f acts naturally on subsets of A: $f(C) = \{f(x) | x \in C\}$. Since f is 1-1, $|f(C)| = |C|$, so $f: X \to X$, with $f^p = I$. Clearly,

$$|X| = \binom{n}{r}.$$

We now find $|X_0|$. Any subset C of A can be written uniquely as

$$C = C_1 \cup \cdots \cup C_p \cup C_0$$

where $C_i \subseteq A_i, C_0 \subseteq B$. Since f sends the A_i cyclically around, and keeps B fixed, we see that $f(C) = C$ if and only if

$$C_i = f^{i-1}(C_1) \qquad i = 1,\ldots,p.$$

For $C \in X$, we must have $|C| = r$, and if $C \in X_0$, we have $|C| = p|C_1| + |C_0| = r = Rp + r_0$. Note that $0 \leq |C_0|, r_0 < p$. Thus the cardinality restriction on C is satisfied if and only if $|C_1| = R$, $|C_0| = r_0$. But there are $\binom{N}{R}$ such choices for C_1 and $\binom{n_0}{r_0}$ independent choices for C_0. Thus,

$$|X_0| = \binom{N}{R}\binom{n_0}{r_0}.$$

Theorem 1 then provides the desired conclusion.

References

1. I. N. Herstein, Topics in Algebra, 2nd ed., Wiley New York, 1975.
2. E. Lucas, Bull. Soc. Math. France, 6 (1877–78) 52.
3. J. H. McKay, Another Proof of Cauchy's Group Theorem, this MONTHLY, 66 (1959) 119.

Originally appeared as:

Hausner, Melvin. "Applications of a Simple Counting Technique." *American Mathematical Monthly*. vol. 90, no. 2 (February 1983): pp. 127–129.

On Weird and Pseudoperfect Numbers

S. J. Benkoski and P. Erdős

Abstract. If n is a positive integer and $\sigma(n)$ denotes the sum of the divisors of n, then n is perfect if $\sigma(n) = 2n$, abundant if $\sigma(n) \geq 2n$ and deficient if $\sigma(n) < 2n$. n is called pseudoperfect if n is the sum of distinct proper divisors of n. If n is abundant but not pseudoperfect, then n is called weird. The smallest weird number is 70. We prove that the density of weird numbers is positive and discuss several related problems and results. A list of all weird numbers not exceeding 10^6 is given.

Let n be a positive integer. Denote by $\sigma(n)$ the sum of divisors of n. We call n perfect if $\sigma(n) = 2n$, abundant if $\sigma(n) \geq 2n$ and deficient if $\sigma(n) < 2n$. We further define n to be pseudoperfect if n is the distinct sum of some of the proper divisors of n, e.g., $20 = 1 + 4 + 5 + 10$ is pseudoperfect [6]. An integer is called primitive abundant, if it is abundant but all its proper divisors are deficient. It is primitive pseudoperfect if it is pseudoperfect but none of its proper divisors are pseudoperfect.

An integer n is called weird if n is abundant but not pseudoperfect. The smallest weird number is 70 and Table 1 is a list of all weird numbers not exceeding 10^6. The study of weird numbers leads to surprising and unexpected difficulties. In particular, we could not decide whether there are any odd weird numbers [1] nor whether $\sigma(n)/n$ could be arbitrarily large for weird n. We give an outline of the proof that the density of weird numbers is positive and discuss several related problems. Some of the proofs are only sketched, especially if they are similar to proofs which are already in the literature.

First, we consider the question of whether there are weird numbers n for which $\sigma(n)/n$ can take on arbitrarily large values. Tentatively, we would like to suggest that the answer is negative. We can decide a few related questions. Let n be an integer with $1 = d_1 < \cdots < d_k = n$ the divisors of n. We say that n has property P if all the 2^k sums $\sum_{i=1}^{k} \varepsilon_i d_i$, $\varepsilon_i = 0$ or 1 are distinct. P. Erdős proved that the density of integers having property P exists and is positive [2]. Clearly, 2^m has property P for every m. It is plausible to conjecture that if n has property P, then $\sigma(n)/n < 2$. The result is indeed true and follows from the next theorem. We conjectured this and the simple and ingenious proof is due to C. Ryavec.

Theorem 1. *Let $1 \leq a_1 < \cdots < a_n$ be a set of integers for which all the sums $\sum_{i=1}^{n} \varepsilon_i a_i$, $\varepsilon_i = 0$ or 1, are distinct. Then,*

$$\sum_{i=1}^{n} \frac{1}{a_i} < 2.$$

Proof. We have, for $0 < x < 1$,

Primitive	Nonprimitive
70	$70 \cdot p$ with $p > \sigma(70)$ and p a prime
836	$7192 \cdot 31$
4030	$836 \cdot 421$
5830	$836 \cdot 487$
7192	$836 \cdot 491$
7912	$836 \cdot p$ with $p \geq 557$ and p a prime
9272	
10792	
17272	
45356	
73616	
83312	
91388	
113072	
243892	
254012	
338572	
343876	
388076	
519712	
539744	
555616	
682592	
786208	

Table 1. Weird Numbers $\leq 10^6$

$$\prod_{i=1}^{n}(1+x^{a_i}) < \sum_{k=0}^{\infty} x^k = \frac{1}{1-x}$$

Thus, $\sum_{i=1}^{n} \log(1+x^{a_i}) < -\log(1-x)$ or

(1)
$$\sum_{i=1}^{n} \int_0^1 \frac{\log(1+x^{a_i})}{x} dx < -\int_0^1 \frac{\log(1-x)}{x} dx.$$

Now putting $x^{a_i} = y$, we obtain, from (1),

$$\sum_{i=1}^{n} \frac{1}{a_i} \int_0^1 \frac{\log(1+y)}{y} dy < -\int_0^1 \frac{\log(1-x)}{x} dx$$

i.e.,
$$\sum_{i=1}^{n} \frac{1}{a_i} \left(\frac{\pi^2}{12}\right) < \frac{\pi^2}{6}$$

Thus $\sum_{i=1}^{n} 1/a_i < 2$ and the theorem is proved.

The same argument can be used to show that if the sums $\sum_{i=1}^{n} \varepsilon_i a_i$ are all distinct, then

$$\sum_{i=1}^{n} \frac{1}{a_i} \leq 2 - \frac{1}{2^{n-1}}$$

and equality holds only if $a_i = 2^{i-1}$, $i = 1, 2, ..., n$.

Here, we call attention to an old conjecture of P. Erdős. If the sums $\sum_{i=1}^{n} \varepsilon_i a_i$,

---

*Editor's Note: By a weird coincidence, the table of weird numbers appears on page 70!

$\varepsilon_i = 0$ or 1, are all distinct, then, is it true that $a_n > 2^{n-C}$ for an absolute constant C? P. Erdős offered and still offers 300 dollars for a proof or disproof of this conjecture.

Consider next the property P'. An integer n is said to have property P' if no divisor of n is the distinct sum of other divisors of n. Here again, we can prove that there is an absolute constant C so that $\sigma(n)/n > C$ implies that n cannot have property P'. This is immediate from the following old result of P. Erdős [3].

Theorem 2. *Let $a_1 < a_2 < \cdots$ be a finite or infinite sequence of integers no term of which is the distinct sum of other terms; then $\sum_i 1/a_i < C$ where C is an absolute constant.*

Proof. In view of the fact that the proof appeared in Hungarian, we give the outline of the proof here.

Put $A(x) = \sum_{a_i \leq x} 1$. We split the positive integers into two classes. In the first class are the integers n for which

(2) $$A(2^{n+1}) - A(2^n) < 2^n/n^2.$$

Clearly, from (2),

(3) $$\sideset{}{'}\sum \frac{1}{a_j} < \sum_{n=1}^{\infty} \frac{1}{n^2} < 2$$

where $\sum'$ is over all j such that $2^n < a_j \leq 2^{n+1}$ for some n in the first class.

Let $n_1 < n_2 < \cdots$ be the integers belonging to the second class, i.e.,

(4) $$\sum_{2^{n_j} < a_i \leq 2^{n_j+1}} 1 \geq 2^{n_j}/n_j^2.$$

Observe that the integers

(5) $$a_1 + a_2 + \cdots + a_r + a_k \text{ with } 1 \leq r < k$$

are all distinct since if $a_1 + \cdots + a_{r_1} + a_{k_1} = a_1 + \cdots + a_{r_2} + a_{k_2}$, $r_2 > r_1$, then a_{k_1} would be a distinct sum of other a_i's.

Now, put $n_{[j/2]} = t$. Clearly,

(6) $$n_j \geq t + [j/2].$$

By (4),

(7) $$A(2^{t+1}) - A(2^t) \geq 2^t/t^2 > 5t^2 > j^2 \quad \text{for } j > 100.$$

Let $1 \leq a_1 < \cdots < a_{j^2}$ be the first j^2 of the a_i's. By (7), $a_{j^2} \leq 2^{t+1}$. Consider now the integers (5) for $1 \leq r \leq j^2$, $a_r < a_k \leq 2^{n_j+1}$. By (7), $a_r < 2^{t+1}$. Thus, by (6), the integers (5) are all less than

(8) $$2^{n_j+1} + j^2 2^{t+1} < 2^{n_j+2} \quad \text{for } j > 100.$$

Now, observe that there are at least

(9) $$j^2(A(2^{n_j+1}) - j^2)$$

integers of the form (5); they are all distinct and are all less than 2^{n_j+2}. Thus, from (8), (6), and (7),

(10) $$A\left(2^{n_j+1}\right) < j^2 + \frac{2^{n_j+2}}{j^2} < 10\frac{2^{n_j}}{j^2} \qquad \text{for } j > 100.$$

Now, (10) and (3) immediately imply the uniform boundedness of $\sum_{1}^{\infty} 1/a_i$. It is perhaps not quite easy to get the best possible value of C. It seems certain[i] that $C < 10$. Unfortunately, we obtain no information about pseudoperfect numbers by these methods.

It is known that the density of integers having property P exists and that the same holds for P' (see [2]). Denote by $u_1 < u_2 < \cdots$, respectively $v_1 < v_2 < \cdots$, the integers which do not have property P, respectively P', but all of whose proper divisors have property P, respectively P'. We expect that $\sum_i 1/u_i$ and $\sum_i 1/v_i$, both converge and, in fact, that

$$\sum_{u_i \leq x} 1 = O\left(\frac{x}{(\log x)^k}\right), \qquad \sum_{v_i \leq x} 1 = O\left(\frac{x}{(\log x)^k}\right)$$

for every k but have not been able to find a proof. For primitive abundant numbers, the analogous results and much more is true [4].

Now, consider weird and pseudoperfect numbers. An integer is primitive pseudoperfect if it is pseudoperfect but all its proper divisors are not pseudoperfect. It seems certain that the number of primitive pseudoperfect numbers not exceeding x is $O(x/(\log x)^k)$ and, hence, the sum of their reciprocals converges. This we could not prove, but the fact that the density of the pseudoperfect numbers exists follows by the methods of [2]. It is easy to prove that there are infinitely many primitive abundant numbers which are pseudoperfect and, therefore, primitive pseudoperfect. The integers $2^k p$, with p a prime such that $2^k < p < 2^{k+1}$, are easily seen to be primitive abundant and pseudoperfect. In fact, they are practical numbers of Grinivasan, i.e., every $m \leq \sigma(2^k p)$ is the distinct sum of divisors of $2^k p$. We leave the simple proof to the reader.

It is slightly less trivial to prove that there are infinitely many primitive abundant numbers all of whose prime factors are large and which are pseudoperfect. We only outline the proof.

For every k, let $f(k)$ be the smallest index for which ($p_1 < p_2 < \cdots$ are the consecutive prime numbers) $\sigma(p_k \cdots p_{f(k)}) \geq 2 p_k \cdots p_{f(k)}$.

Theorem 3. *There exists a positive integer k_0 such that, for $k > k_0$, the integers*

$$A_k = \prod_{k \leq l \leq f(k)} p_l \text{ and } B_k = (A_k / p_{f(k)}) p_{f(k)+1} p_{f(k)+2}$$

are both primitive pseudoperfect.

Note that $B_1 = 70$ which is not pseudoperfect. It appears that this is the only value of k for which Theorem 3 fails, but to prove this might be difficult and would certainly require long computations for B_k and perhaps a new idea for A_k.

We need two lemmas.

Lemma 1. *There is an absolute constant c such that every integer $m > c p_k$ is the distinct sum of primes not less than p_k.*

The lemma is probably well known and, in any case, easily follows by Brun's method.

Lemma 2. *There exists an integer k_0 such that, for every $k > k_0$,*

(11) $$cp_k < m < \sigma(A_k) - cp_k$$

implies that m is the distinct sum of divisors of A_k. The same result holds for B_k.

Lemma 2 follows easily from Lemma 1 and from the fact that, for $p_k \leq x < \frac{3}{2}x < A_k$, the interval $(x, \frac{3}{2}x)$ always contains a divisor of A_k and B_k. (To prove this last statement, we only need that, for $\varepsilon > 0$, there exists an integer $i_0(\varepsilon)$ such that $p_{i+1} < (1 + \varepsilon)p_i$ for $i > i_0(\varepsilon)$.)

Lemma 2 implies Theorem 3 if we can show

(12) $$\sigma(B_k) - 2B_k > cp_k \quad \text{and} \quad \sigma(A_k) - 2A_k > cp_k.$$

Statement (12) follows immediately for B_k by a very simple computation if we observe that there is an integer l_0 such that, for $l > l_0$,

$$(1 + 1/p_{l+1})(1 + 1/p_{l+2}) > 1 + 3/2p_l.$$

We do not have such a simple proof of (12) for A_k. Observe that

(13) $$\sigma(A_k) - 2A_k \geq (\sigma(A_k), A_k)$$

where (a, b) denotes the greatest common divisor of a and b.

Now, (13) implies (12) if we can show that $(\sigma(A_k), A_k)$ has, for $k > k_0$, at least two prime factors. In fact, we shall prove that ($\omega(n)$ denotes the number of prime factors of n)

(14) $$\lim_{k \to \infty} \omega((\sigma(A_k), A_k)) = \infty.$$

To prove (14), we first observe that, for $\varepsilon > 0$, there is an integer $k_0(\varepsilon)$ such that, for $k > k_0(\varepsilon)$,

(15) $$p_{f(k)} > p_k^{2-\varepsilon}$$

This is of course well known and follows from the theorems of Mertens. The following theorem now implies (14).

Theorem 4. *Denote by $g(x)$ the number of indices l_1 for which there is an l_2 satisfying*

$$x < p_{l_1} < p_{l_2} < x^2, \quad p_{l_2} \equiv -1 \pmod{p_{l_1}}.$$

We then have $\lim_{x \to \infty} g(x) = \infty$.

Theorem 4 follows easily from the proof of Motohashi's theorem [5]. It does not follow from the theorem of Motohashi but it is easy to deduce by the same proof. Motohashi uses some deep results of Bombieri. Thus, (14) and Theorem 3 are proved.

It seems likely that there are infinitely many primitive abundant numbers which are weird but this we cannot prove. We can, however, show that the density of weird numbers is positive.

It is clear that the weird numbers have a density since both the abundant numbers and the pseudoperfect numbers have a density. (A weird number is abundant and not pseudoperfect.) Hence, we need only show that the density of weird numbers cannot be 0. This follows from the following simple lemma.

Lemma. *If n is weird, then there is an $\varepsilon_n > 0$ such that nt is weird if*

$$\sum_{d|t} \frac{1}{d} < 1 + \varepsilon_n.$$

Proof. First define $(x)^+$ by

$$(x)^+ = x \quad \text{if } x \geq 0,$$
$$= \infty \quad \text{if } x < 0.$$

Put

(16) $$\delta_n = \min(1 - \sum{}' 1/d)^+$$

where, in $\sum'$, d runs over all subsets of the divisors greater than 1 of n. If n is deficient or weird, then $\delta_n > 0$.

If nt is not weird, then there is a set of divisors, greater than 1, of nt for which

(17) $$1 = \sum_1 1/d + \sum_2 1/d$$

where, in $\sum_1$, $d \mid n$ and, in $\sum_2$, $d \mid nt$ but $d \nmid n$.

From (16) and (17),

$$\frac{\sigma(t)}{t} \frac{\sigma(n)}{n} \geq \sum_2 \frac{1}{d} \geq \delta_n,$$

which proves the lemma for $\varepsilon_n = n\delta_n/\sigma(n)$.

Theorem 5. *The density of weird numbers is positive. Now, by the lemmas, if t is an integer and $\sigma(t)/t < 1 + \varepsilon_n$, then nt is weird. But the density of the integers t with $\sigma(t)/t < 1 + \varepsilon_n$ is positive for any $\varepsilon_n > 0$.*

Proof. If n is weird, then let ε_n be as in the proof of the lemma. Now, by the lemma, if t is an integer and $\sigma(t)/t < 1 + \varepsilon_n$ is positive for any $\varepsilon_n > 0$.

Actually, we proved a slightly stronger result. If n is weird, then the density of $\{m; n \mid m \text{ and } m \text{ is weird}\}$ is positive.

It is easy to see that if n is weird and p is a prime greater than $\sigma(n)$, then pn is also weird. More generally, the following result holds. Let n be an integer which is not pseudoperfect, i.e., n is deficient or weird. The integer pn is pseudoperfect if and only if there is a set A of proper divisors of n and a set B of divisors of n where no $b \in B$ is a multiple of p, such that

$$p\left(n - \sum_{d \in A} a\right) = \sum_{b \in B} b.$$

We leave that simple proof to the reader.

Finally, we state without proof the following result: Let $t \geq 0$ be an integer. The density of integers n for which $n + t$ is the distinct sum of proper divisors of n is positive. On the other hand, the density of the integers n, for which $n - t$ ($t > 0$) is the sum of distinct divisors of n, is 0.

References

1. S. J. Benkoski, "Elementary Problem and Solution E2308," *Amer. Math Monthly*, v. 79, 1972, p. 774.

2. P. Erdős, "Some Extremal Problems in Combinatorial Number Theory," *Mathematical Essays Dedicated to A. J. Macintyre*, Ohio Univ. Press, Athens, Ohio, 1970, pp. 123–133. MR 43 #1942.

3. P. Erdős, "Remarks on Number Theory. III," *Mat. Lapok*, v. 13, 1962, pp. 28–38. (Hungarian) MR 26 #2412.

4. P. Erdős, "On Primitive Abundant Numbers," *J. London Math. Soc.*, v. 10, 1935, pp. 49–58.

5. Y. Motohashi, "A Note on the Least Prime in an Arithmetic Progression with a Prime Difference," *Acta Arith.*, v. 17, 1970, pp. 283–285. MR 42 #3030.

6. W. Sierpinski, "Sur Les Nombres Pseudoparfaits," *Mat. Vesnik*, v. 2(17), 1965, pp. 212–213. MR 33 #7296.

7. A. & E. Zachariou, "Perfect, Semiperfect and Ore Numbers," *Bull. Soc. Math. Grèce*, v. 13. 1972, pp. 12–22.

Originally appeared as:
Benkoski, S. J. and P. Erdős. "On Weird and Pseudoperfect Numbers" *Mathematics of Computation*. vol. 28, no. 126 (April 1974): pp. 617–623.

Second Helping: My paper with Erdős

The story of how this article came to be is very interesting. I had a summer job in the summer of 1969 in San Diego as I prepared to go to graduate school at Penn State. I was hired to develop software to convert analog data into digital form. However, the data did not arrive until the day before I left for Penn State. (As you may have guessed, this was a government job.) Rather than just sit at my desk and read, I decided to try to get a head start on a possible thesis topic. I selected three possible problems. They were the Riemann Hypothesis, Fermat's last theorem, and perfect numbers. While just fooling around with perfect numbers and related ideas, I coined the name *weird* for a positive integer that was abundant but not pseudoperfect. (To this day, I am amazed that this was not defined centuries ago.) I investigated these weird numbers and used my programming skills and a government computer to find all weird numbers less than 100,000.

At the end of the summer, I left for Penn State and began my studies. The results for weird numbers were in a folder in my files. In late 1970, I came across that folder and decided weird numbers should see the light of day. So, I sent in a problem to the *American Mathematical Monthly*. It stated the definitions of abundant and pseudoperfect numbers and then posed two questions:

Are there any abundant numbers that are not pseudoperfect?

(*)Are there any odd abundant numbers that are not pseudoperfect?

A couple of months after I sent it in, I reread the problem and decided that it was a really bad problem. The first question can be answered with paper and pencil in about 15 minutes and I did not know the answer to the second part. So, I sent a letter to the *Monthly* withdrawing the problem. They wrote back and told me they were planning to use the problem and asked me not to withdraw it. The problem appeared in the summer of 1971 *without* the asterisk. (An asterisk indicates an open problem.)

In October, I received a letter from Erdős. (Of course, I still have that letter.) He said that he and Ernst Strauss had been working on the second question and wanted to know if I had found an odd weird number or could prove one did not exist. I was chagrined that the asterisk was left off because that implied that I had the answer. However, Erdős and Strauss missed the fact that 70 was the smallest weird number. I wrote back to Erdős and told him almost all I knew about weird numbers. We sent about 4 or 5 letters. Most of my letters were telling him I did not know the answers to the questions he sent. The correspondence ended in the Spring of 1972.

In the Fall of 1972, Erdős was scheduled to give a colloquium at Penn State. I was, of course, planning to attend but assumed he was not interested in weird numbers any more. When I arrived at the Math Department, my fellow students told me that the Department Chair was calling every 15 minutes. As soon as Erdős arrived, he asked to talk to me and kept asking the Chair where I was.

I was assigned to be his "escort" for the next morning. I was to arrive at his room by 8:00 and "entertain" him until 10:00. When I arrived, he was still in his pajamas. I helped him get dressed. We went to breakfast and talked for over an hour. Most of the time, he asked me math questions related to weird numbers that I could not answer. He also asked me about my family and other personal things. He seemed genuinely interested in me. Erdős asked me if I would like to write a joint paper. Easy question to answer. He also gave me a list of things to accomplish that I would do before we published.

I generated the first draft and sent it to him. I put his name as the first author, but he returned it with my name first. I submitted the paper to *Mathematics of Computation*. I have published other articles in my career, but this was the only time that the galley proofs were the first thing I received from the publisher.

—Stan Benkoski, Los Gatos, California

A Heuristic for the Prime Number Theorem

*Hugh L. Montgomery and
Stan Wagon*

Why e?

Why does e play such a central role in the distribution of prime numbers? Simply citing the Prime Number Theorem (PNT), which asserts that $\pi(x) \sim x/\ln x$, is not very illuminating. Here "$\sim$" means "is asymptotic to" and $\pi(x)$ is the number of primes less than or equal to x. So why do natural logs appear, as opposed to another flavor of logarithm?

The problem with an attempt at a heuristic explanation is that the sieve of Eratosthenes does not behave as one might guess from pure probabilistic considerations. One might think that sieving out the composites under x using primes up to $\sqrt{x}$ would lead to $x\Pi_{p<\sqrt{x}}\left(1-\frac{1}{p}\right)$ as an asymptotic estimate of the count of numbers remaining (the primes up to x; p always represents a prime). But this quantity turns out to be *not* asymptotic to $x/\ln x$. For F. Mertens proved in 1874 that the product is actually asymptotic to $2\,e^{-\gamma}/\ln x$, or about $1.12/\ln x$. Thus the sieve is 11% (from $1/1.12$) more efficient at eliminating composites than one might expect. Commenting on this phenomenon, which one might call the Mertens Paradox, Hardy and Wright [5, p. 372] said: "Considerations of this kind explain why the usual 'probability' arguments lead to the wrong asymptotic value for $\pi(x)$." For more on this theorem of Mertens and related results in prime counting see [3; 5; 6, exer. 8.27; 8].

Yet there ought to be a way to explain, using only elementary methods, why natural logarithms play a central role in the distribution of primes. A good starting place is two old theorems of Chebyshev (1849).

Chebyshev's First Theorem. *For any $x \geq 2$, $0.92x/\ln x < \pi(x) < 1.7x/\ln x$.*

Chebyshev's Second Theorem. *If $\pi(x) \sim x/\log_c x$, then $c = e$.*

A complete proof of Chebyshev's First Theorem (with slightly weaker constants) is not difficult and the reader is encouraged to read the beautiful article by Don Zagier [9] — the very first article ever published in this journal (see also [1, §4.1]). The first theorem tells us that $x/\log_c x$ is a reasonable rough approximation to the growth of $\pi(x)$, but it does not distinguish e from other bases.

The second theorem can be given a complete proof using only elementary calculus [8]. The result is certainly a partial heuristic for the centrality of e since it shows that, if any logarithm works, then the base must be e. Further, one can see the exact place in the proof where e arises ($\int 1/x\,dx = \ln x$). But the hypothesis for the second theorem is a strong one; here we will show, by a relatively simple proof, that the same conclusion follows from a much weaker hypothesis. Of course, the PNT eliminates the need for any hypothesis at all, but its proof requires either an understanding of complex analysis or the willingness

---

[1] Supported in part by NSF grant DMS--0244660.

to read the sophisticated "elementary proof." The first such was found by Erdos and Selberg; a modern approach appears in [7].

Our presentation here was inspired by a discussion in Courant and Robbins [2]. We show how their heuristic approach can be transformed into a proof of a strong result. So even though our original goal was just to motivate the PNT, we end up with a proved theorem that has a simple statement and quite a simple proof.

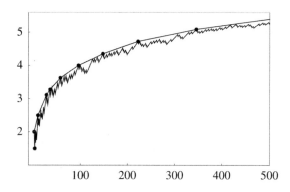

Figure 1: A graph of $x/\pi(x)$ shows that the function is not purely increasing. The upper convex hull of the graph is an increasing piecewise linear function that is a good approximation to $x/\pi(x)$.

Theorem. *If $x/\pi(x)$ is asymptotic to an increasing function, then $\pi(x) \sim x/\ln x$.*

Figure 1 shows that $x/\pi(x)$ is assuredly not increasing. Yet it does appear to be asymptotic to the piecewise linear function that is the upper part of the convex hull of the graph. Indeed, if we take the convex hull of the full infinite graph, then the piecewise linear function $L(x)$ corresponding to the part of the hull above the graph is increasing (see last section). If one could prove that $x/\pi(x) \sim L(x)$ then, by the theorem, the PNT would follow. In fact, using PNT it is not too hard to prove that $L(x)$ is indeed asymptotic to $x/\pi(x)$ (such proof is given at the end of this paper). In any case, the hypothesis of the theorem is certainly believable, if not so easy to prove, and so the theorem serves as a heuristic explanation of the PNT.

Nowadays we can look quite far into the prime realm. Zagier's article of 26 years ago was called *The First Fifty Million Prime Numbers*. Now we can look at the first 700 quintillion prime numbers. Not one at a time, perhaps, but the exact value of $\pi\left(10^i\right)$ is now known for i up to 22; the most recent value is due to Gourdon and Sabeh [4] and is $\pi\left(4 \cdot 10^{22}\right) = 783964159847056303858$. Figure 2 is a log-log plot that shows the error when these stratospheric π values are compared to $x/\ln x$ and also the much better logarithmic integral estimate $\text{li}(x)$ (which is $\int_0^x 1/\ln t\, dt$).

A Heuristic for the Prime Number Theorem

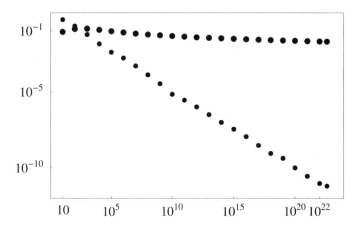

Figure 2: The large dots are the absolute value of the error when $x/\ln x$ is used to approximate $\pi(x)$, for $x = 10^i$. The smaller dots use $\mathrm{li}(x)$ as the approximant.

Two Lemmas

The proof requires two lemmas. The first is a consequence of Chebyshev's First Theorem, but can be given a short and elementary proof; it states that almost all numbers are composite.

Lemma 1. $\lim_{x \to \infty} \pi(x)/x = 0$.

Proof. First use an idea of Chebyshev to get $\pi(2n) - \pi(n) < 2n/\ln n$ for integers n. Take log-base-n of both sides of the following to get the needed inequality.

$$4^n = (1+1)^{2n} > \binom{2n}{n} \geq \prod_{n < p \leq 2n}^{p} > \prod_{n < p \leq 2n}^{n} = n^{\pi(2n)-\pi(n)}$$

This means that $\pi(2n) - \pi(n) \leq (\ln 4)n/\ln n$. Suppose n is a power of 2, say 2^k; then summing over $2 \leq k \leq K$, where K is chosen so that $2^K \leq n < 2^{K+1}$, gives

$$\pi(n) \leq 2 + \sum_{k=2}^{K} \frac{2^k \ln 4}{k}$$

Here each term in the sum is at most $3/4$ of the next term, so the entire sum is at most 4 times the last term. That is, $\pi(n) \leq cn/\ln n$, which implies $\pi(n)/n \to 0$. For any $x > 0$, we take n to be the first power of 2 past x, and then $\pi(x)/x \leq 2\pi(n)/n$, concluding the proof.

By keeping careful track of the constants, the preceding proof can be used to show that $\pi(x) \leq 8.2x/\ln x$, yielding one half of Chebyshev's first theorem, albeit with a weaker constant.

The second lemma is a type of Tauberian result, and the proof goes just slightly beyond elementary calculus. This lemma is where natural logs come up, well, naturally. For consider the hypothesis with $\log_c$ in place of $\ln$. Then the constant $\ln c$ will cancel, so the conclusion will be unchanged!

Lemma 2. *If $W(x)$ is decreasing and $\int_2^x W(t)\ln(t)/t\,dt \sim \ln x$, then $W(x) \sim 1/\ln x$.*

Proof. Let ϵ be small and positive; let $f(t) = \ln(t)/t$. The hypothesis implies $\int_x^{x^{1+\epsilon}} W(t)f(t)dt \sim \epsilon \ln x$ (to see this split the integral into two: from 2 to x and x to $x^{1+\epsilon}$). Since $W(x)$ is decreasing,

$$0009\epsilon \ln x \sim \int_x^{x^{1+\epsilon}} W(t)f(t)dt \leq W(x) \int_x^{x^{1+\epsilon}} \frac{\ln t}{t} dt = \epsilon \left(1 + \frac{\epsilon}{2}\right) W(x)(\ln x)^2$$

Thus $\liminf_{x \to \infty} W(x) \ln x \geq 1/\left(1 + \frac{\epsilon}{2}\right)$. A similar argument starting with $\int_{x^{1-\epsilon}}^{x} W(t)f(t)dt \sim \epsilon \ln x$ shows that $\limsup_{x \to \infty} W(x) \ln x \leq 1/\left(1 - \frac{\epsilon}{2}\right)$. Since ϵ can be arbitrarily small, we have $W(x) \ln x \sim 1$.

Proof of the Theorem

Theorem. *If $x/\pi(x)$ is asymptotic to an increasing function, then $\pi(x) \sim x/\ln x$.*

Proof. Let $L(x)$ be the hypothesized increasing function and let $W(x) = 1/L(x)$, a decreasing function. It suffices to show that the hypothesis of Lemma 2 holds, for then $L(x) \sim \ln x$. Let $f(t) = \ln(t)/t$. Note that if $\ln x \sim g(x) + h(x)$ where $h(x)/\ln x \to 0$, then $\ln x \sim g(x)$; we will use this several times in the following sequence, which reaches the desired conclusion by a chain of 11 relations. The notation $p^k \| n$ in the third expression means that k is the largest power of p that divides n; the equality that follows the $\|$ sum comes from considering each p^m for $1 \leq m \leq k$.

$$\ln x \sim \frac{1}{x} \sum_{n \leq x} \ln n \text{ (can be done by machine; note 1)}$$

$$= \frac{1}{x} \sum_{n \leq x} \sum_{p^k \| n} k \ln p$$

$$= \frac{1}{x} \sum_{n \leq x} \sum_{p^m | n, m \geq 1} \ln p$$

$$= \frac{1}{x} \sum_{p^m \leq x, 1 \leq m} \ln p \left\lfloor \frac{x}{p^m} \right\rfloor$$

$$\sim \frac{1}{x} \sum_{p^m \leq x, 1 \leq m} \ln p \frac{x}{p^m} \text{ (error is small; note 2)}$$

$$\sim \sum_{p \leq x} f(p) + \sum_{p^m \leq x, 2 \leq m} \frac{\ln p}{p^m}$$

$$\sim \sum_{p \leq x} f(p) \text{ (geometric series estimation; note 3)}$$

$$= \pi(\lfloor x \rfloor) f(x) - \int_2^x \pi(t) f'(t) dt \text{ (partial summation; note 4)}$$

$$\sim - \int_2^x \pi(t) f'(t) dt \text{ (because } \frac{\pi(\lfloor x \rfloor) f(x)}{\ln x} \leq \frac{\pi(x)}{x} \to 0 \text{ by Lemma 1)}$$

$$\sim - \int_2^x tW(t) \left(\frac{1}{t^2} - \frac{\ln t}{t^2}\right) dt \text{ (because } \pi(t) \sim tW(t))$$

$$\sim \int_2^x W(t)\frac{\ln t}{t}dt \text{ (l'Hôpital, note 5)}$$

Notes

1. It is easy to verify this relation using standard integral test ideas: start with the fact that the sum lies between $\int_1^x \ln t\, dt$ and $x \ln x$. But it is intriguing to see that *Mathematica* can resolve this using symbolic algebra. The sum is just $\ln(\lfloor x \rfloor !)$ and *Mathematica* quickly returns 1 when asked for the limit of $x \ln x / \ln(x!)$ as $x \to \infty$.

2. $\text{error}/\ln x \leq \frac{1}{x \ln x}\sum_{p^m \leq x} \ln p \leq \frac{1}{x \ln x}\sum_{p \leq x} \log_p x \, \ln p = \frac{1}{x \ln x}\sum_{p \leq x} \ln x = \pi(x)/x \to 0$ by Lemma 1

3. The second sum divided by $\ln x$ approaches 0 because:

$$\sum_{p^m \leq x, 2 \leq m} \frac{\ln p}{p^m} \leq \sum_{p \leq x} \ln p \sum_{m=2}^{\infty} \frac{1}{p^m}$$

$$= \sum_{p \leq x} \frac{\ln p}{p(p-1)} \leq \sum_{n=2}^{\infty} \frac{\ln n}{n(n-1)}$$

$$\leq \sum_{n=2}^{\infty} \frac{(n-1)^{1/2}}{(n-1)^2} < \infty$$

4. We use *partial summation*, a technique common in analytic number theory. Write the integral from 2 to x as a sum of integrals over $[n, n+1]$ together with one from $\lfloor x \rfloor$ to x and use the fact that $\pi(t)$ is constant on such intervals and jumps by 1 exactly at the primes. More precisely:

$$\int_2^x \pi(t)f'(t)dt = \int_{\lfloor x \rfloor}^x \pi(t)f'(t)dt + \sum_{n=2}^{\lfloor x \rfloor - 1} \int_n^{n+1} \pi(t)f'(t)dt$$

$$= \pi(\lfloor x \rfloor)(f(x) - f(\lfloor x \rfloor)) + \sum_{n=2}^{\lfloor x \rfloor - 1} \pi(n)(f(n+1) - f(n))$$

$$= \pi(\lfloor x \rfloor)f(x) - \sum_{p \leq x} f(p)$$

5. L'Hôpital's rule on $\frac{1}{\ln x}\int_2^x tW(t)\frac{1}{t^2}dt$ yields $\frac{W(x)/x}{1/x} = W(x) = 1/L(x)$, which approaches 0 by Lemma 1.

No line in the proof uses anything beyond elementary calculus except the call to Lemma 2. The result shows that if there is any nice function that characterizes the growth of $\pi(x)$ then that function must be asymptotic to $x/\ln x$. Of course, the PNT shows that this function does indeed do the job.

This proof works with no change if base-c logarithms are used throughout. But as noted, Lemma 2 will force the natural log to appear! The reason for this lies in the indefinite integration that takes places in the lemma's proof.

Conclusion

Might there be a chance of proving in a simple way that $x/\pi(x)$ is asymptotic to an increasing function, thus getting another proof of PNT? This is probably wishful thinking. However, there is a natural candidate for the increasing function. Let $L(x)$ be the upper convex hull of the full graph of $x/\pi(x)$ (precise definition to follow). The piecewise linear function $L(x)$ is increasing because $x/\pi(x) \to \infty$ as $x \to \infty$. Moreover, using PNT, we can give a proof that $L(x)$ is indeed asymptotic to $x/\pi(x)$. But the point of our work in this paper is that for someone who wishes to understand why the growth of primes is governed by natural logarithms, a reasonable approach is to convince oneself via computation that the convex hull just mentioned satisfies the hypothesis of our theorem, and then use the relatively simple proof to show that this hypothesis rigorously implies the prime number theorem.

We conclude with the convex hull definition and proof. Let B be the graph of $x/\pi(x)$: the set of all points $(x, x/\pi(x))$ where $x \geq 2$. Let C be the convex hull of B: the intersection of all convex sets containing B. The line segment from $(2,2)$ to any $(x, x/\pi(x))$ lies in C. As $x \to \infty$, the slope of this line segment tends to 0 (because $\pi(x) \to \infty$). Hence for any positive a and ϵ, the vertical line $x = a$ contains points in C of the form $(a, 2+\epsilon)$. Thus the intersection of the line $x = a$ with C is a set of points (a, y) where $2 < y \leq L(x)$. The function $L(x)$ is piecewise linear and $x/\pi(x) \leq L(x)$ for all x. This function is what we call the *upper convex hull* of $x/\pi(x)$.

Theorem. *The upper convex hull of $x/\pi(x)$ is asymptotic to $\pi(x)$.*

Proof. For given positive ϵ and x_0, define a convex set $A(x_0)$ whose boundary consists of the positive x-axis, the line segment from $(0,0)$ to $(0, (1+\epsilon)(\ln x_0))$. the line segment from that point to $(ex_0, (1+\epsilon)(1+\ln x_0))$, and finally the curve $(x, (1+\epsilon) \ln x)$ for $ex_0 \leq x$. The slopes match at ex_0, so this is indeed convex. The PNT implies that for any $\epsilon > 0$ there is an x_1 such that, beyond x_1, $(1-\epsilon) \ln x < x/\pi(x) < (1+\epsilon) \ln x$. Now choose $x_0 > x_1$ so that $\pi(x_1) < (1+\epsilon) \ln x_0$. It follows that $A(x_0)$ contains B: beyond x_0 this is because $x_0 > x_1$; below x_1 the straight part is high enough at $x = 0$ and only increases; and between x_1 and x_0 this is because the curved part is convex down and so the straight part is above where the curved part would be, and that dominates $\pi(x)$ by the choice of x_1. This means that the convex hull of the graph of $x/\pi(x)$ is contained in $A(x_0)$, because $A(x_0)$ is convex. That is, $L(x) \leq (1+\epsilon) \ln x$ for $x \geq x_0$. Indeed, $(1-\epsilon) \ln x \leq x/\pi(x) \leq L(x) \leq (1+\epsilon) \ln x$ for all sufficiently large x. Hence $x/\pi(x) \sim L(x)$.

References

1. D. Bressoud and S. Wagon, *A Course in Computational Number Theory*, Key College, San Francisco, 2000.

2. R. Courant and H. Robbins, *What is Mathematics?*, Oxford Univ. Press, London, 1941.

3. J. Friedlander, A. Granville, A. Hildebrand, and H. Maier, Oscillation theorems for primes in arithmetic progressions and for sifting functions, *J. Amer. Math. Soc.* **4** (1991) 25–86.

4. X. Gourdon and P. Sebah, The $\pi(x)$ project, http://numbers.computation.free.fr/Constants/constants.html.

5. G. H. Hardy and E. M. Wright, *An Introduction to the Theory of Numbers*, 4th ed., Oxford Univ. Pr., London, 1965.

6. I. Niven, H. S. Zuckerman, and H. L. Montgomery, *An Introduction to the Theory of Numbers*, 2nd ed., Wiley, New York, 1991.

7. G. Tenenbaum and M. Mendès France, *The Prime Numbers and their Distribution*, Amer. Math. Soc., Providence, 2000.

8. S. Wagon, It's only natural, *Math Horizons* **13**:1 (2005) 26–28.

9. D. Zagier, The first 50,000,000 prime numbers, *The Mathematical Intelligencer* **0** (1977) 7–19.

Originally appeared as:
Montgromery, Hugh L. and Stan Wagon. "A Heuristic for the Prime Number Theorem." *Mathematical Intelligencer*. vol. 28, no. 3 (2006): pp. 6–9.

A Tale of Two Sieves

Carl Pomerance

This paper is dedicated to the memory of my friend and teacher Paul Erdős

It is the best of times for the game of factoring large numbers into their prime factors. In 1970 it was barely possible to factor "hard" 20-digit numbers. In 1980, in the heyday of the Brillhart-Morrison *continued fraction factoring algorithm*, factoring of 50-digit numbers was becoming commonplace. In 1990 my own *quadratic sieve factoring algorithm* had doubled the length of the numbers that could be factored, the record having 116 digits.

By 1994 the quadratic sieve had factored the famous 129-digit RSA challenge number that had been estimated in Martin Gardner's 1976 *Scientific American* column to be safe for 40 quadrillion years (though other estimates around then were more modest). But the quadratic sieve is no longer the champion. It was replaced by Pollard's *number field sieve* in the spring of 1996, when that method successfully split a 130-digit RSA challenge number in about 15% of the time the quadratic sieve would have taken.

In this article we shall briefly meet these factorization algorithms—these two sieves—and some of the many people who helped to develop them.

In the middle part of this century, computational issues seemed to be out of fashion. In most books the problem of factoring big numbers was largely ignored, since it was considered trivial. After all, it was doable *in principle*, so what else was there to discuss? A few researchers ignored the fashions of the time and continued to try to find fast ways to factor. To these few it was a basic and fundamental problem, one that should not be shunted to the side.

But times change. In the last few decades we have seen the advent of accessible and fast computing power, and we have seen the rise of cryptographic systems that base their security on our supposed inability to factor quickly (and on other number theoretic problems). Today there are many people interested in factoring, recognizing it not only as a benchmark for the security of cryptographic systems, but for computing itself. In 1984 the Association for Computing Machinery presented a plaque to the Institute for Electrical and Electronics Engineers (IEEE) on the occasion of the IEEE centennial. It was inscribed with the prime factorization of the number $2^{251} - 1$, that was completed that year with the quadratic sieve. The president of the ACM made the following remarks:

> About 300 years ago the French mathematician Mersenne speculated that $2^{251} - 1$ was a composite, that is, a factorable number. About 100 years ago it was proved to be factorable, but even 20 years ago the computational load to factor the number was considered insurmountable. Indeed, using conventional machines and traditional search algorithms, the search time was estimated to be about 10^{20} years. The number was factored in February of this year at Sandia on a Cray computer in 32 hours, a world record. We've come a long way in computing, and to commemorate IEEE's contribution to computing we have inscribed the five factors of the Mersenne composite on a plaque. Happy Birthday, IEEE.

Factoring big numbers is a strange kind of mathematics that closely resembles the experimental sciences, where nature has the last and definitive word. If some method to factor n runs for awhile and ends with the statement "d is a factor of n," then this assertion may be easily checked; that is, the integers have the last and definitive word. One can thus get by quite nicely without proving a theorem that a method works in general. But, as with the experimental sciences, both rigorous and heuristic analyses can be valuable in understanding the subject and moving it forward. And, as with the experimental sciences, there is sometimes a tension between pure and applied practitioners. It is held by some that the theoretical study of factoring is a freeloader at the table (or as Hendrik Lenstra once colorfully put it, paraphrasing Siegel, "a pig in the rose garden"), enjoying undeserved attention by vapidly giving various algorithms labels, be they "polynomial," "exponential," "random," etc., and offering little or nothing in return to those hard workers who would seriously compute. There is an element of truth to this view. But as we shall see, theory played a significant role in the development of the title's two sieves.

A Contest Problem

But let us begin at the beginning, at least my beginning. When I give talks on factoring, I often repeat an incident that happened to me long ago in high school. I was involved in a math contest, and one of the problems was to factor the number 8051. A time limit of five minutes was given. It is not that we were not allowed to use pocket calculators; they did not exist in 1960, around when this event occurred! Well, I was fairly good at arithmetic, and I was sure I could trial divide up to the square root of 8051 (about 90) in the time allowed. But on any test, especially a contest, many students try to get into the mind of the person who made it up. Surely they would not give a problem where the only reasonable approach was to try possible divisors frantically until one was found. There must be a clever alternate route to the answer. So I spent a couple of minutes looking for the clever way, but grew worried that I was wasting too much time. I then belatedly started trial division, but I *had* wasted too much time, and I missed the problem.

So can you find the clever way? If you wish to think about this for a moment, delay reading the next paragraph.

Fermat and Kraitchik

The trick is to write 8051 as $8100 - 49$, which is $90^2 - 7^2$, so we may use algebra, namely, factoring a difference of squares, to factor 8051. It is 83×97. Does this always work? In fact, every odd composite can be factored as a difference of squares: just use the identity $ab = \left(\tfrac{1}{2}(a+b)\right)^2 - \left(\tfrac{1}{2}(a-b)\right)^2$. Trying to find a pair of squares which work is, in fact, a factorization method of Fermat. Just like trial division, which has some very easy cases (such as when there is a small prime factor), so too does the difference-of-squares method have easy cases. For example, if $n = ab$ where a and b are very close to $\sqrt{n}$, as in the case of $n = 8051$, it is easy to find the two squares. But in its worst cases, the difference-of-squares method can be far worse than trial division. It is worse in another way too. With trial division, most numbers fall into the easy case; namely, most numbers have a small factor. But with the difference-of-squares method, only a small fraction of numbers have a divisor near their square root, so the method works well on only a small fraction of possible inputs. (Though trial division allows one to *begin* a factorization for most inputs, finishing with a complete factorization is usually far more difficult. Most numbers resist this, even when a combination of trial division and difference-of-squares is used.)

In the 1920s Maurice Kraitchik came up with an interesting enhancement of Fermat's difference-of-squares technique, and it is this enhancement that is at the basis of most modern factoring algorithms. (The idea had roots in the work of Gauss and Seelhoff, but it was Kraitchik who brought it out of the shadows, introducing it to a new generation in a new century. For more on the early history of factoring, see [23].) Instead of trying to find integers u and v with $u^2 - v^2$ equal to n, Kraitchik reasoned that it might suffice to find u and v with $u^2 - v^2$ equal to a *multiple* of n, that is, $u^2 \equiv v^2$ mod n. Such a congruence can have interesting solutions, those where $u \not\equiv \pm v$ mod n, and uninteresting solutions, where $u \equiv \pm v$ mod n. In fact, if n is odd and divisible by at least two different primes, then at least half of the solutions to $u^2 \equiv v^2$ mod n, with uv coprime to n, are of the interesting variety. And for an interesting solution u, v, the greatest common factor of $u - v$ and n, denoted $(u - v, n)$, must be a nontrivial factor of n. Indeed, n divides $u^2 - v^2 = (u - v)(u + v)$ but divides neither factor. So n must be somehow split between $u - v$ and $u + v$.

As an aside, it should be remarked that finding the greatest common divisor (a, b) of two given numbers a and b is a very easy task. If $0 < a \le b$ and if a divides b, then $(a, b) = a$. If a does not divide b, with b leaving a remainder r when divided by a, then $(a, b) = (a, r)$. This neat idea of replacing a larger problem with a smaller one is over two thousand years old and is due to Euclid. It is very fast: it takes about as much time for a computer to find the greatest common divisor of a and b as it would take to multiply them together.

Let us see how Kraitchik might have factored $n = 2041$. The first square above n is $46^2 = 2116$. Consider the sequence of numbers $Q(x) = x^2 - n$ for $x = 46, 47, \ldots$. We get

$$75, 168, 263, 360, 459, 560, \ldots$$

So far no squares have appeared, so Fermat might still be searching. But Kraitchik has another option: namely, he tries to find several numbers x with the product of the corresponding numbers $Q(x)$ equal to a square. For if $Q(x_1) \cdots Q(x_k) = v^2$ and $x_1 \cdots x_k = u$, then

$$u^2 = x_1^2 \cdots x_k^2 \equiv \left(x_1^2 - n\right) \cdots \left(x_k^2 - n\right) = Q(x_1) \cdots Q(x_k) = v^2 \bmod n,$$

that is, we have a found a solution to $u^2 \equiv v^2$ mod n. But how to find the set $x_1, \ldots, x_k$? Kraitchik notices that some of the numbers $Q(x)$ factor very easily:

$$75 = 3 \times 5^2, \quad 168 = 2^3 \times 3 \times 7, \quad 360 = 2^3 \times 3^2 \times 5, \quad 560 = 2^4 \times 5 \times 7.$$

From these factorizations he can tell that the product of these four numbers is $2^{10} \times 3^4 \times 5^4 \times 7^2$, a square! Thus, he has $u^2 \equiv v^2$ mod n, where

$$u = 46 \cdot 47 \cdot 49 \cdot 51 \equiv 311 \bmod 2041,$$
$$v = 2^5 \cdot 3^2 \cdot 5^2 \cdot 7 \equiv 1416 \bmod 2041.$$

He is now nearly done, since $311 \not\equiv \pm 1416$ mod 2041. Using Euclid's algorithm to compute the greatest common factor $(1416 - 311, 2041)$, he finds that this is 13, and so $2041 = 13 \times 157$.

Continued Fractions

The essence of Kraitchik's method is to "play" with the sequence $x^2 - n$ as x runs through integers near $\sqrt{n}$ to find a subsequence with product a square. If the square root of this square is v and the product of the corresponding x values is u, then $u^2 \equiv v^2$ mod n, and there is now a hope that this congruence is "interesting," namely that $u \not\equiv \pm v$ mod n. In 1931, D. H. Lehmer and R. E. Powers suggested replacing Kraitchik's function $Q(x) = x^2 - n$ with another that is derived from the continued fraction expansion of $\sqrt{n}$.

If a_i/b_i is the i-th continued fraction convergent to $\sqrt{n}$, let $Q_i = a_i^2 - b_i^2 n$. Then $Q_i \equiv a_i^2 \mod n$. Thus, instead of playing with the numbers $Q(x)$, we may play with the numbers Q_i, since in both cases they are congruent modulo n to known squares. Although continued fractions can be ornery beasts as far as computation goes, the case for quadratic irrationals is quite pleasant. In fact there is a simple iterative procedure (see [16]) going back to Gauss, and perhaps earlier, for computing what's needed here, namely the sequence of integers Q_i and the residues a_i mod n.

But why mess up a perfectly simple quadratic polynomial with something as exotic as continued fractions? It is because of the inequality $|Q_i| < 2\sqrt{n}$. The numbers Q_i are smaller in absolute value than the numbers $Q(x)$. (As x moves away from $\sqrt{n}$, the numbers $Q(x)$ grow approximately linearly, with a slope of $2\sqrt{n}$.) If one wishes to "play" with numbers to find some of them with product a square, it is presumably easier to do this with smaller numbers than with larger numbers. So the continued fraction of Lehmer and Powers has an apparent advantage over the quadratic polynomial of Kraitchik.

How to "Play" with Numbers

It is certainly odd to have an instruction in an algorithm asking you to play with some numbers to find a subset with product a square. I am reminded of the cartoon with two white coated scientists standing at a blackboard filled with arcane notation, and one points to a particularly delicate juncture and says to the other that at this point a miracle occurs. Is it a miracle that we were able to find the numbers 75, 168, 360 and 560 in Kraitchik's sequence with product a square? Why should we expect to find such a subsequence and, if it exists, how can we find it efficiently?

A systematic strategy for finding a subsequence of a given sequence with product a square was found by John Brillhart and Michael Morrison, and surprisingly, it is at heart only linear algebra (see [16]). Every positive integer m has an *exponent vector* $v(m)$ that is based on the prime factorization of m. Let p_i denote the i-th prime, and say $m = \prod p_i^{v_i}$. (The product is over all primes, but only finitely many of the exponents v_i are nonzero.) Then $v(m)$ is the vector $(v_1, v_2, \ldots)$. For example, leaving off the infinite string of zeros after the fourth place, we have

$$v(75) = (0, 1, 2, 0),$$
$$v(168) = (3, 1, 0, 1),$$
$$v(360) = (3, 2, 1, 0),$$
$$v(560) = (4, 0, 1, 1).$$

For our purposes, the exponent vectors give *too much* information. We are only interested in squares, and since a positive integer m is a square if and only if every entry of $v(m)$ is even, we should be reducing exponents modulo 2. Since v takes products to sums, we are looking for numbers such that the sum of their exponent vectors is the zero vector mod 2. The mod 2 reductions of the above exponent vectors are

$$v(75) \equiv (0, 1, 0, 0) \bmod 2,$$
$$v(168) \equiv (1, 1, 0, 1) \bmod 2,$$
$$v(360) \equiv (1, 0, 1, 0) \bmod 2,$$
$$v(560) \equiv (0, 0, 1, 1) \bmod 2.$$

Note that their sum is the zero vector! Thus the product of 75, 168, 360 and 560 is a square.

To systematize this procedure, Brillhart and Morrison suggest that we choose some number B and only look at those numbers in the sequence that completely factor over the first B primes. So in the case above, we have $B = 4$. As soon as $B + 1$ such numbers have been assembled, we have $B + 1$ vectors in the B-dimensional vector space $\mathbb{F}_2^B$. By linear algebra, they must be linearly dependent. But what is a linear dependence relation over the field $\mathbb{F}_2$? Since the only scalars are 0 and 1, a linear dependence relation is simply a subset sum equaling the 0-vector. And we have many algorithms from linear algebra that can help us find this dependency.

Note that above we were a little lucky since we were able to find the dependency with just 4 vectors, rather than the 5 vectors needed in the worst case.

Brillhart and Morrison call the primes $p_1, p_2, \ldots, p_B$ the "factor base." (To be more precise, they discard those primes p_j for which n is not congruent to a square, since such primes will never divide a number Q_i in the continued fraction method, nor a number $Q(x)$ in Kraitchik's method.) How is B to be chosen? If it is chosen small, then we don't need to assemble too many numbers before we can stop. But if it is chosen too small, the likelihood of finding a number in the sequence that completely factors over the first B primes will be so minuscule, that it will be difficult to find even one number. Thus somewhere there is a happy balance, and with factoring 2041 via Kraitchik's method, the happy balance turned out to be $B = 4$.

Some of the auxiliary numbers may be negative. How do we handle their exponent vectors? Clearly we cannot ignore the sign, since squares are not negative. However, we can put an extra coordinate in each exponent vector, one that is 0 for positive numbers and 1 for negative numbers. (It is as if we are including the "prime" -1 in the factor base.) So allowing the auxiliary numbers to be negative just increases the dimension of the problem by 1.

For example, let us consider again the number 2041 and try to factor it via Kraitchik's polynomial, but now allowing negative values. So with $Q(x) = x^2 - 2041$ and the factor base 2, 3 and 5, we have

$$Q(43) = -192 = -2^6 \cdot 3 \leftrightarrow (1, 0, 1, 0)$$
$$Q(44) = -105$$
$$Q(45) = -16 = -2^4 \leftrightarrow (1, 0, 0, 0)$$
$$Q(46) = 75 = 3 \cdot 5^2 \leftrightarrow (0, 0, 1, 0)$$

where the first coordinates correspond to the exponent on -1. So, using the smaller factor base of 2, 3 and 5, but allowing also negatives, we are especially lucky, since the 3 vectors assembled so far are dependent. This leads to the congruence $(43 \cdot 45 \cdot 46)^2 \equiv (-192)(-16)(75) \bmod 2041$, or $1247^2 \equiv 480^2 \bmod 2041$. This again gives us the divisor 13 of 2041, since $(1247 - 480, 2041) = 13$.

Does the final greatest common divisor step always lead to a nontrivial factorization? No, it does not. The reader is invited to try still another assemblage of a square in connection with 2041. This one involves $Q(x)$ for $x = 41$, 45 and 49 and gives rise to the congruence $601^2 \equiv 1440^2$ mod 2041. In our earlier terminology, this congruence is uninteresting, since $601 \equiv -1440$ mod 2041. And sure enough, the greatest common divisor $(601 - 1440, 2041)$ is the quite uninteresting divisor 1.

Smooth Numbers and the Stirrings of Complexity Theory

With the advent of the RSA public key cryptosystem in the late 1970's, it became particularly important to try to predict just how hard factoring is. Not only should we know the state of the art at present, we would like to predict just what it would take to factor larger numbers beyond what is feasible now. In particular, it seems empirically that, dollar for dollar, computers double their speed and capacity every one and a half to two years. Assuming this and no new factoring algorithms, what will be the state of the art in ten years?

It is to this type of question that complexity theory is well-suited. So how then might one analyze factoring via Kraitchik's polynomial or the Lehmer-Powers continued fraction? Richard Schroeppel, in unpublished correspondence in the late 1970's, suggested a way. Essentially, he begins by thinking of the numbers Q_i in the continued fraction method, or the numbers $Q(x)$ in Kraitchik's method, as "random." If you are presented with a stream of truly random numbers below a particular bound X, how long should you expect to wait before you find some subset with product a square?

Call a number Y-smooth if it has no prime factor exceeding Y. (Thus a number which completely factors over the primes up to p_B is p_B-smooth.) What is the probability that a random positive integer up to X is Y-smooth? It is $\psi(X, Y)/[X] \approx \psi(X, Y)/X$, where $\psi(X, Y)$ is the number of Y-smooth numbers in the interval $[1, X]$. Thus the expected number of random numbers that must be examined to find just one that is Y-smooth is the reciprocal of this probability, namely $X/\psi(X, Y)$. But we must find about $\pi(Y)$ such Y-smooth numbers, where $\pi(Y)$ denotes the number of primes up to Y. So the expected number of auxiliary numbers that must be examined is about $\pi(Y)X/\psi(X, Y)$. And how much work does it take to examine an auxiliary number to see if it is Y-smooth? If one uses trial division for this task, it takes about $\pi(Y)$ steps. So the expected number of steps is $\pi(Y)^2 X/\psi(X, Y)$.

It is now a job for analytic number theory to choose Y as a function of X so as to minimize the expression $\pi(Y)^2 X/\psi(X, Y)$. In fact, in the late 1970's the tools did not yet quite exist to make this estimation accurately. This was remedied in a paper in 1983 (see [4]), though preprints of this paper were around for several years before then. So what's the minimum? It occurs when Y is about $\exp\left(\frac{1}{2}\sqrt{\log X \log \log X}\right)$ and the minimum value is about $\exp\left(2\sqrt{\log X \log \log X}\right)$. But what are "$X$" and "$Y$" anyway? The number X is an estimate for the typical auxiliary number the algorithm produces. In the continued fraction method, X can be taken as $2\sqrt{n}$. With Kraitchik's polynomial, X is a little larger, it is $n^{1/2+\varepsilon}$. And the number Y is an estimate for p_B, the largest prime in the factor base.

Thus, factoring n, either via the Lehmer-Powers continued fraction or via the Kraitchik polynomial, should take about $\exp\left(\sqrt{2\log n \log \log n}\right)$ steps. This is not a theorem; it is a conjecture. The conjecture is supported by the above heuristic argument which assumes that the auxiliary numbers generated by the continued fraction of $\sqrt{n}$ or by

Kraitchik's quadratic polynomial are "random" as far as the property of being Y-smooth goes. This has not been proved. In addition, getting many auxiliary numbers that are Y-smooth may not be sufficient for factoring n since each time we use linear algebra over $\mathbb{F}_2$ to assemble the congruent squares, we may be very unlucky and only come up with uninteresting solutions which don't help in the factorization. Again assuming randomness, we don't expect inordinately long strings of bad luck, and this heuristic again supports the conjecture.

As mentioned, this complexity argument was first made by Richard Schroeppel in unpublished work in the late 1970's. (He assumed the result mentioned above from [4], even though at that time it was not a theorem, or even really a conjecture.) Armed with the tools to study complexity, he used them during this time to come up with a new method that came to be known as the *linear sieve*. It was the forerunner of the quadratic sieve and also its inspiration.

Using Theory to Come Up with a Better Algorithm: The Quadratic Sieve

The above complexity sketch shows a place where we might gain some improvement. It is the time we are taking to recognize auxiliary numbers that factor completely with the primes up to $Y = p_B$, that is, the Y-smooth numbers. In the argument we assumed this is about $\pi(Y)$ steps, where $\pi(Y)$ is the number of primes up to Y. The probability that a number is Y-smooth is, according to the notation above, $\psi(X, Y)/X$. As you might expect, and as is easily checked in practice, when Y is a reasonable size and X is very large, this probability is very, very small. So one after the other, the auxiliary numbers pop up, and we have to invest all this time in each one, only to find out almost always that the number is not Y-smooth and is thus a number that we will discard.

It occurred to me early in 1981 that one might use something akin to the sieve of Eratosthenes to quickly recognize the smooth values of Kraitchik's quadratic polynomial $Q(x) = x^2 - n$. The sieve of Eratosthenes is the well-known device for finding all the primes in an initial interval of the natural numbers. One circles the first prime 2, and then crosses off every second number, namely 4, 6, 8, etc. The next unmarked number is 3. It is circled, and we then cross off every third number. And so on. After reaching the square root of the upper limit of the sieve, one can stop the procedure, and circle every remaining unmarked number. The circled numbers are the primes, the crossed-off numbers the composites.

It should be noted that the sieve of Eratosthenes does more than find primes. Some crossed-off numbers are crossed off many times. For example, 30 is crossed off 3 times, as is 42, since these numbers have 3 prime factors. Thus we can quickly scan the array looking for numbers that are crossed off a lot, and so quickly find the numbers which have many prime factors. And clearly there is a correlation between having many prime factors and having all small prime factors.

But we can do better than have a correlation. By dividing by the prime, instead of crossing off, numbers like 30 and 42 get transformed to the number 1 at the end of the sieve, since they are completely factored by the primes used in the sieve. So instead of sieving with the primes up to the square root of the upper bound of the sieve, say we only sieve with the primes up to Y. And instead of crossing a number off, we divide it by the prime. At the end of the sieve any number that has been changed to the number 1 is Y-smooth. But not every Y-smooth is caught in this sieve. For example, 60 gets divided by its prime factors, and is changed to the number 2. The problem is higher powers of the primes up to Y. We can rectify this by also sieving by these higher powers, and dividing hits by the underlying prime. Then the residual 1's at the end correspond exactly to the Y-smooth numbers in the interval.

The time for doing this is unbelievably fast compared with trial dividing each candidate number to see if it is Y-smooth. If the length of the interval is N, the number of steps is only about $N \log\log Y$, or about $\log\log Y$ steps on average per candidate.

So we can quickly recognize Y-smooth numbers in an initial interval. But can we use this idea to recognize Y-smooth values of the quadratic polynomial $Q(x) = x^2 - n$? What it takes for a sieve to work is that for each modulus m in the sieve, the multiples of the number m appear in regular places in the array. So take a prime p for example, and we ask: for which values of x do we have $Q(x)$ divisible by p? This is not a difficult problem. If n (the number being factored) is a nonzero square modulo p, then there are two residue classes a and b mod p such that $Q(x) \equiv 0 \mod p$ if and only if $x \equiv a$ or b mod p. If n is not a square modulo p, then $Q(x)$ is never divisible by p, and no further computations with p need be done.

So essentially the same idea can be used and we can recognize the Y-smooth values of $Q(x)$ in about $\log\log Y$ steps per candidate value.

What does the complexity argument give us? The time to factor n is now about

$$\exp\left(\sqrt{\log n \log\log n}\right);$$

namely the factor $\sqrt{2}$ in the exponent is missing. Is this a big deal? You bet. This lower complexity and other friendly features of the method allowed a two-fold increase in the length of the numbers that could be factored (compared with the continued fraction method discussed above). And so was born the quadratic sieve method, as a complexity argument and with no numerical experiments.

Implementations and Enhancements

In fact I was very lucky that the quadratic sieve turned out to be a competitive algorithm. More often than not, when one invents algorithms solely via complexity arguments and thought experiments, the result is likely to be too awkward to be a competitive method. In addition, even if the basic idea is sound, there well could be important enhancements waiting to be discovered by the people who actually try the thing out. This in fact happened with the quadratic sieve.

The first person to try out the quadratic sieve method on a big number was Joseph Gerver (see [9]). Using the task as an opportunity to learn programming, he successfully factored a 47-digit number from the Cunningham project. This project, begun early in this century by Lt.-Col. Allan J. Cunningham and H. J. Woodall, consists of factoring into primes the numbers $b^n \pm 1$ for b up to 12 (and not a power) and n up to high numbers (see [3]). Gerver's number was a factor of $3^{225} - 1$.

Actually I had a hard time getting people to try out the quadratic sieve. Many Cunningham project factorers seemed satisfied with the continued fraction method, and they thought that the larger values of Kraitchik's polynomial $Q(x)$, compared with the numbers Q_i in the continued fraction method, was too great a handicap for the fledgling quadratic sieve method. But at a conference in Winnipeg in the Fall of 1982, I convinced Gus Simmons and Tony Warnock of Sandia Laboratories to give it a try on their Cray computer.

Jim Davis and Diane Holdridge were assigned the task of coding up the quadratic sieve on the Sandia Cray. Not only did they succeed, but they quickly began setting records. And Davis found an important enhancement that mitigated the handicap mentioned above. He found a way of switching to other quadratic polynomials after values of the first one, $Q(x) = x^2 - n$, grew uncomfortably large. Though this idea did not substantially change the complexity estimate, it made the method much more practical. Their success not only made the cover of the *Mathematical Intelligencer* (the Volume 6, Number 3 cover in 1984 had on it a Cray computer and the factorization of the number consisting of 71 ones), but there was even a short article in *Time Magazine*, complete with a photo of Simmons.

It was ironic that shortly before the Sandia team began this project, another Sandia team had designed and manufactured an RSA chip for public key cryptography, whose security was based on our inability to factor numbers of about 100 digits. Clearly this was not safe enough, and the chip had to be scrapped.

Around this time, Peter Montgomery independently came up with another, slightly better way of changing polynomials, and we now use his method rather than that of Davis.

One great advantage of the quadratic sieve method over the continued fraction method is that with the quadratic sieve, it is especially easy to distribute the task of factoring to many computers. For example, using multiple polynomials, each computer can be given its own set of quadratic polynomials to sieve. At first, the greatest successes of the quadratic sieve came from supercomputers, such as the Cray XMP at Sandia Laboratories. But with the proliferation of low cost workstations and PC's and the natural way that the quadratic sieve can be distributed, the records passed on to those who organized distributed attacks on target numbers.

Robert Silverman was the first to factor a number using many computers. Later, Red Alford and I used over 100 very primitive, non-networked PC's to factor a couple of 100-digit numbers (see [2]). But we didn't set a record, because while we were tooling up, Arjen Lenstra and Mark Manasse [12] took the ultimate step in distributing the problem. They put the quadratic sieve on the Internet, soliciting computer time from people all over the world. It was via such a shared effort that the 129-digit RSA challenge number was eventually factored in 1994. This project, led by Derek Atkins, Michael Graff, Paul Leyland and Lenstra, took about 8 months of real time and involved over 10^{17} elementary steps.

The quadratic sieve is ultimately a very simple algorithm, and this is one of its strengths. Due to its simplicity one might think that it could be possible to design a special purpose computer solely dedicated to factoring big numbers. Jeff Smith and Sam Wagstaff at the University of Georgia had built a special purpose processor to implement the continued fraction method. Dubbed the "Georgia Cracker" it had some limited success, but was overshadowed by quadratic sieve factorizations on conventional computers. Smith, Randy Tuler and I (see [21]) thought we might build a special purpose quadratic sieve processor. "Quasimodo," for Quadratic Sieve Motor, was built, but never functioned properly. The point later became moot due to the exponential spread of low cost, high quality computers.

The Dawn of the Number Field Sieve

Taking his inspiration from a discrete logarithm algorithm of Don Coppersmith, Andrew Odlyzko and Richard Schroeppel [6] that used quadratic number fields, John Pollard in

Modern model of the Lehmer Bicycle Chain Sieve constructed by Robert Canepa and currently in storage at the Computer Museum, 1401 Shoreline Blvd., Mountain View, CA 94043.

1988 circulated a letter to several people outlining an idea of his for factoring certain big numbers via algebraic number fields. His original idea was not for any large composite, but for certain "pretty" composites that had the property that they were close to powers, and certain other nice properties as well. He illustrated the idea with a factorization of the number $2^{2^7} + 1$, the seventh Fermat number. It is interesting that this number was the first major success of the continued fraction factoring method, almost 20 years earlier.

I must admit that at first I was not too keen on Pollard's method since it seemed to be applicable to only a few numbers. However, some people were taking it seriously, one being Hendrik Lenstra. He improved some details in the algorithm, and along with his brother Arjen and Mark Manasse, set about using the method to factor several large numbers from the Cunningham project. After a few successes (most notably a 138-digit number), and after Brian LaMacchia and Andrew Odlyzko had made some inroads in dealing with the large, sparse matrices that come up in the method, the Lenstras and Manasse set their eyes on a real prize, $2^{2^9} + 1$, the ninth Fermat number. Clearly it was beyond the range of the quadratic sieve. Hendrik Lenstra's own elliptic curve method, which he discovered early in 1985 and which is especially good at splitting numbers which have a relatively small prime factor (say, "only" 30 or so digits), had so far not been of help in factoring it. The Lenstras and Manasse succeeded in getting the prime factorization in the spring of 1990. This sensational achievement announced to the world that Pollard's number field sieve had arrived.

But what of general numbers? In the summer of 1989 I was to give a talk at the meeting of the Canadian Number Theory Association in Vancouver. It was to be a survey talk on factoring, and I figured it would be a good idea to mention Pollard's new method. On the plane on the way to the meeting I did a complexity analysis of the method as to how it would work for general numbers, assuming myriad technical difficulties didn't exist and that it was possible to run it for general numbers. I was astounded. The complexity for this algorithm-that-did-not-yet-exist was of the shape $\exp(c(\log n)^{1/3}(\log\log n)^{2/3})$. The key difference over the complexity of the quadratic sieve was that the most important quantity in the exponent, the power of log n, had its exponent reduced from 1/2 to 1/3. If reducing the *constant* in the exponent had such a profound impact in passing from the continued fraction method to the quadratic sieve, think what reducing the *exponent* in the exponent might accomplish. Clearly this method deserved some serious thought!

I don't wish to give the impression that with this complexity analysis I had single-handedly found a way to apply the number field sieve to general composites. Far from it. I merely had a shrouded glimpse of exciting possibilities for the future. That these possibilities were ever realized was mostly due to Joe Buhler, Hendrik Lenstra and others. In addition, some months earlier, Lenstra had done a complexity analysis for Pollard's method applied to special numbers and he too arrived at the expression $\exp(c(\log n)^{1/3}(\log\log n)^{2/3})$. My own analysis was based on some optimistic algebraic assumptions and on arguments about what might be expected to hold, via averaging arguments, for a general number.

The starting point of Pollard's method to factor n is to come up with a monic polynomial $f(x)$ over the integers that is irreducible, and an integer m such that $f(m) \equiv 0 \bmod n$. The polynomial should have "moderate" degree d, meaning that if n has between 100 and 200 digits, then d should be 5 or 6. For a number such as the ninth Fermat number, $n = 2^{2^9} + 1$, it is easy to come up with such a polynomial. Note that $8n = 2^{515} + 8$. So, let $f(x) = x^5 + 8$ and let $m = 2^{103}$.

Of what possible use could such a polynomial be? Let α be a complex root of $f(x)$, and consider the ring $\mathbb{Z}[\alpha]$ consisting of all polynomial expressions in α with integer coefficients. Since $f(\alpha) = 0$ and $f(m) \equiv 0 \bmod n$, by substituting the residue $m \bmod n$ for each occurrence of α we have a natural map ϕ from $\mathbb{Z}[\alpha]$ to $\mathbb{Z}/(n\mathbb{Z})$. Our conditions on f, α and m ensure that ϕ is well-defined. And not only this, ϕ is a ring homomorphism.

Suppose now that S is a finite set of coprime integer pairs $\langle a,b \rangle$ with two properties. The first is that the product of the algebraic integers $a - \alpha b$ for all pairs $\langle a,b \rangle$ in S is a square in $\mathbb{Z}[\alpha]$, say γ^2. The second property for S is that the product of all the numbers $a - mb$ for pairs $\langle a,b \rangle$ in S is a square in $\mathbb{Z}$, say v^2. Since γ may be written as a polynomial expression in α, we may replace each occurrence of α with the integer m, coming up with an integer u with $\phi(\gamma) \equiv u \bmod n$. Then

$$u^2 \equiv \phi(\gamma)^2 = \phi(\gamma^2) = \phi\left(\prod_{\langle a,b \rangle \in S}(a - \alpha b)\right) = \prod_{\langle a,b \rangle \in S} \phi(a - \alpha b) \equiv \prod_{\langle a,b \rangle \in S}(a - mb) = v^2 \bmod n.$$

And we know what to do with u and v. Just as Kraitchik showed us 70 years ago, we hope that we have an interesting congruence, that is $u \not\equiv \pm v \bmod n$, and if so, we take the greatest common divisor $(u - v, n)$ to get a nontrivial factor of n.

Where is the set S of pairs $\langle a,b \rangle$ supposed to come from? For at least the second property S is supposed to have, namely that the product of the numbers $a - mb$ is a square, it is clear we might again use exponent vectors and a sieve. Here there are two variables a and b, instead of just the one variable in $Q(x)$ in the quadratic sieve. So we view this as a parametrized family of linear polynomials. We can fix b, and let a run over an interval then change to the next b, and repeat.

But S is to have a second property too: for the same pairs $\langle a,b \rangle$, the product of $a - \alpha b$ is a square in $\mathbb{Z}[\alpha]$. It was Pollard's thought that if we were in the nice situation that $\mathbb{Z}[\alpha]$ is the full ring of algebraic integers in $\mathbb{Q}(\alpha)$, if the ring is a unique factorization domain, and if we know a basis for the units, then we could equally well create exponent vectors for the algebraic integers $a - \alpha b$ and essentially repeat the same algorithm. To arrange for both properties of S to hold simultaneously, well, this would just involve longer exponent vectors, having coordinates for all the small prime numbers, for the sign of $a - \alpha b$, for all the "small" primes in $\mathbb{Z}[\alpha]$, and for each unit in the unit basis.

But how are we supposed to do this for a general number n? In fact, how do we even achieve the first step of finding the polynomial $f(x)$ and the integer m with $f(m) \equiv 0 \bmod n$? And if we could find it, why should we expect that $\mathbb{Z}[\alpha]$ has all of the nice properties to make Pollard's plan work?

The Number Field Sieve Evolves

There is at the least a very simple device to get started, that is, to find $f(x)$ and m. The trick is to find $f(x)$ *last*. First, one decides on the degree d of f. Next, one lets m be the integer part of $n^{1/d}$. Now write n in the base m, so that $n = m^d + c_{d-1}m^{d-1} + \cdots + c_0$, where the base m "digits" c_i satisfy $0 \le c_i < m$. (If $n > (2d)^d$, then the leading "digit" c_d is 1.) The polynomial $f(x)$ is now staring us in the face; it is $x^d + c_{d-1}x^{d-1} + \cdots + c_0$. So we have a monic polynomial $f(x)$, but is it irreducible?

There are many strategies for factoring primitive polynomials over $\mathbb{Z}$ into their irreducible factors. In fact, we have the celebrated polynomial-time algorithm of Arjen Lenstra, Hendrik Lenstra and László Lovász for factoring primitive polynomials in $\mathbb{Z}[x]$ (the running time is bounded by a fixed power of the sum of the degree and the number of digits in the coefficients). So suppose we are unlucky and the above procedure leads to a reducible polynomial $f(x)$, say $f(x) = g(x)h(x)$. Then $n = f(m) = g(m)h(m)$, and from a result of John Brillhart, Michael Filaseta and Andrew Odlyzko, this factorization of n is nontrivial. But our goal is to find a nontrivial factorization of n, so this is hardly unlucky at all! Since almost all polynomials are irreducible, it is much more likely that the construction will let us get started with the number field sieve, and we will not be able to factor n immediately.

There was still the main problem of how one might get around the fact that there is no reason to expect the ring $\mathbb{Z}[\alpha]$ to have any nice properties at all. By 1990, Joe Buhler, Hendrik Lenstra and I had worked out the remaining difficulties and, incorporating a very practical idea of Len Adleman [1], which simplified some of our constructions, published a description of the *general number field sieve* in [11].

Here is a brief summary of what we did. The norm $N(a - \alpha b)$ (over $\mathbb{Q}$) of $a - \alpha b$ is easily worked out to be $b^d f(a/b)$. This is the homogenized version of f. We define $a - \alpha b$ to be Y-smooth if $N(a - \alpha b)$ is Y-smooth. Since the norm is multiplicative, it follows that if the product of various algebraic integers $a - \alpha b$ is a square of an algebraic integer, then so too is the corresponding product of norms a square of an integer. Note too that we know how to find a set of pairs $\langle a,b \rangle$ with the product of $N(a - \alpha b)$ a square. This could be done by using a sieve to discover Y-smooth values of $N(a - \alpha b)$ and then combine them via exponent vector algebra over $\mathbb{F}_2$.

But having the product of the numbers $N(a - \alpha b)$ be a square, while a necessary condition for the product of the numbers $a - \alpha b$ to be a square, is far from sufficient. The principal reason for this is that the norm map takes various prime ideals to the same thing in $\mathbb{Z}$, and so the norm can easily be a square without the argument being a square. For example, the two degree-one primes in $\mathbb{Z}[i]$, $2 + i$ and $2 - i$, have norm 5. Their product is 5, which has norm $25 = 5^2$, but $(2 + i)(2 - i) = 5$ is not a square. We solve this as follows: For each prime p let R_p be the set of solutions to $f(x) \equiv 0 \bmod p$. When we come across a pair $\langle a,b \rangle$ with p dividing $N(a - \alpha b)$, then some prime ideal above p divides $a - \alpha b$. And we can tell which one, since a/b will be congruent modulo p to one of the members of R_p, and this will serve to distinguish the various prime ideals above p. Thus we can arrange for our exponent vectors to have $\#R_p$ coordinates for each prime p, and so keep track of the prime ideal factorization of $a - \alpha b$. Note that $\#R_p \le d$, the degree of $f(x)$.

So we have gotten over the principal hurdle, but there are still many obstructions. We are supposed to be working in the ring $\mathbb{Z}[\alpha]$ and this may not be the full ring of algebraic integers. In fact this ring may not be a Dedekind domain, so we may not even have factorization into prime ideals. And even if we have factorization into prime ideals, the above paragraph merely assures us that the principal ideal generated by the product of the algebraic integers $a - \alpha b$ is the square of some ideal, not necessarily the square of a principal ideal. And even if it is the square of a principal ideal, it may not be a square of an algebraic integer, because of units. (For example, the ideal generated by -9 is the square of an ideal in $\mathbb{Z}$, but -9 is not a square.) And even if the product of the numbers $a - \alpha b$ is a square of an algebraic integer, how do we know it is the square of an element of $\mathbb{Z}[\alpha]$?

The last obstruction is rather easily handled by using $f'(\alpha)^2$ as a multiplier, but the other obstructions seem difficult. However, there is a simple and ingenious idea of Len Adleman [1] that in one fell swoop overcomes them all. The point is that even though we are being faced with some nasty obstructions, they form, modulo squares, an $\mathbb{F}_2$-vector space of fairly small dimension. So the first thought might just be to ignore the problem. But the dimension is not *that* small. Adleman suggested randomly choosing some quadratic characters and using their values at the numbers $a - \alpha b$ to augment the exponent vectors. (There is one fixed choice of the random quadratic characters made at the start.) So we are arranging for a product of numbers to not only be a square up to the "obstruction space," but also highly likely to actually be a square. For example, consider the above problem with -9 not being a square. If somehow we cannot "see" the problem with the sign, but it sure looks like a square to us, because we know that for each prime p the exponent on p in the prime factorization of -9 is even, we might still detect the problem. Here is how: consider a quadratic character evaluated at -9, in this case the Legendre symbol $(-9/p)$, which is 1 if -9 is congruent to a square mod p and -1 if -9 is not congruent to a square mod p. Say we try this with $p = 7$. It is easy to compute this symbol, and it turns out to be -1. So -9 is not a square mod 7, and so it cannot be a square in $\mathbb{Z}$. If -9 is a square mod some prime p however, this does not guarantee it is a square in $\mathbb{Z}$. For example, if we had tried this with 5 instead of 7, then -9 would still be looking like a square. Adleman's idea is to evaluate smooth values of $a - \alpha b$ at the quadratic characters that were chosen, and use the linear algebra to create an element with two properties: its (un-augmented) exponent vector has all even entries and its value at each character is 1. This algebraic integer is highly likely, in a heuristic sense, to be a square. If it isn't a square, we can continue to use linear algebra over $\mathbb{F}_2$ to create another candidate.

To be sure, there are still difficulties. One of these is the "square root problem." If you have the prime factorizations of various rational integers, and their product is a square, you can easily find the square root of the square via its prime factorization. But in $\mathbb{Z}[\alpha]$ the problem does not seem so transparent. Nevertheless, there are devices for solving this too, though it still remains as a computationally interesting step in the algorithm. The interested reader should consult [15].

Perhaps it is not clear why the number field sieve is a good factoring algorithm. A key quantity in a factorization method such as the quadratic sieve or the number field sieve is what I was calling "X" earlier. It is an estimate for the size of the auxiliary numbers that we are hoping to combine into a square. Knowing X gives you the complexity; it is about $\exp\left(\sqrt{2\log X \log\log X}\right)$. In the quadratic sieve we have X about $n^{1/2+\varepsilon}$. But in the number field sieve, we may choose the polynomial $f(x)$ and the integer m in such a way that $N(a-\alpha b)(a-mb)$ (the numbers that we hope to find smooth) is bounded by a value of X of the form $\exp(c(\log n)^{2/3}(\log\log n)^{1/3})$. Thus the number of digits of the auxiliary numbers that we sieve over for smooth values is about the 2/3 *power* of the number of digits of n, as opposed to the quadratic sieve where the auxiliary numbers have more than half the number of digits of n. That is why the number field sieve is asymptotically so fast in comparison.

I mentioned earlier that the heuristic running time for the number field sieve to factor n is of the form $\exp(c(\log n)^{1/3}(\log\log n)^{2/3})$, but I did not reveal what "c" is. There are actually three values of c depending on which version of the number field sieve is used. The "special" number field sieve, more akin to Pollard's original method and well-suited to factoring numbers like $2^{2^9}+1$ which are near high powers, has $c = (32/9)^{1/3} \approx 1.523$. The "general" number field sieve is the method I sketched in this paper and is for use on any odd composite number that is not a power. It has $c = (64/9)^{1/3} \approx 1.923$. Finally, Don Coppersmith [5] proposed a version of the general number field sieve in which many polynomials are used. The value of "c" for this method is $\frac{1}{2}(92+26\sqrt{13})^{1/3} \approx 1.902$. This stands as the champion worst case factoring method asymptotically. It had been thought that Coppersmith's idea is completely impractical, but [8] considers whether the idea of using several polynomials may have some practical merit.

The State of the Art

In April, 1996, a large team (see [7]) finished the factorization of a 130-digit RSA challenge number, using the general number field sieve. Thus the gauntlet has finally been passed from the quadratic sieve, which had enjoyed champion status since 1983 for the largest "hard" number factored. Though the real time was about the same as with the quadratic sieve factorization of the 129-digit challenge number two years earlier, it was estimated that the new factorization took only about 15% of the computer time. This discrepancy was due to fewer computers being used on the project and some "down time" while code for the final stages of the algorithm was being written.

So where is the crossover between the quadratic sieve and the number field sieve? The answer to this depends somewhat on whom you talk to. One thing everyone agrees on: for smaller numbers, say less than 100 digits, the quadratic sieve is better, and for larger numbers, say more than 130 digits, the number field sieve is better. One reason a question like this does not have an easy answer is that the issue is highly dependent on fine points in the programming and on the kind of computers used. For example, as reported in [7], the performance of the number field sieve is sensitive to how much memory a computer has. The quadratic sieve is as well, but not to such a large degree.

There is much that was *not* said in this brief survey. An important omission is a discussion of the algorithms and complexity of the linear algebra part of the quadratic sieve and the number field sieve. At the beginning we used Gaussian elimination, as Brillhart and Morrison did with the continued fraction method. But the size of the problem has kept increasing. Nowadays a factor base of size one million is in the ballpark for record factorizations. Clearly, a linear algebra problem that is one million by one million is not a trifling matter. There is interesting new work on this that involves adapting iterative methods for dealing with sparse matrices over the real numbers to sparse matrices over $\mathbb{F}_2$. For a recent reference, see [14].

Several variations on the basic idea of the number field sieve show some promise. One can replace the linear expression $a - mb$ used in the number field sieve with $b^k g(a/b)$, where $g(x)$ is an irreducible polynomial over $\mathbb{Z}$ of degree k with $g(m) \equiv 0 \bmod n$. That is, we use two polynomials $f(x)$, $g(x)$ with a common root $m \bmod n$ (the original scenario has us take $g(x) = x - m$). It is a subject of current research to come up with good strategies for choosing polynomials. Another variation on the usual number field sieve is to replace the polynomial $f(x)$ with a family of polynomials along the lines suggested by Coppersmith. For a description of the number field sieve incorporating both of these ideas, see [8].

The discrete logarithm problem (given a cyclic group with generator g, and an element h in the group, find an integer x with $g^x = h$) is also of keen interest in cryptography. As mentioned, Pollard's original idea for the number field sieve was born out of a discrete logarithm algorithm. We have come full circle, since Dan Gordon, Oliver Schirokauer and Len Adleman have all given variations of the number field sieve that can be used to compute discrete logarithms in multiplicative groups of finite fields. For a recent survey, see [22].

I have said nothing on the subject of primality testing. It is generally much easier to recognize that a number is composite than to factor it. When we use complicated and time-consuming factorization methods on a number, we already know from other tests that it is an odd composite and it is not a power.

I have given scant mention of Hendrik Lenstra's elliptic curve factorization method. This algorithm is much superior to both the quadratic sieve and the number field sieve for all but a thin set of composites, the so-called "hard" numbers, for which we reserve the sieve methods.

There is also a rigorous side to factoring, where researchers try to dispense with heuristics and prove theorems about factorization algorithms. So far we have had much more success proving theorems about probabilistic methods than deterministic methods. We don't seem close to proving that various practical methods, such as the quadratic sieve and the number field sieve, actually work as advertised. It is fortunate that the numbers we are trying to factor have not been informed of this lack of proof!

For further reading I suggest several of the references already mentioned, and also [10, 13, 17, 18, 19, 20]. In addition, I am currently writing a book with Richard Crandall, *PRIMES: a computational perspective*, that should be out sometime in 1997.

I hope I have been able to communicate some of the ideas and excitement behind the development of the quadratic sieve and the number field sieve. This development saw an interplay between theoretical complexity estimates and good programming intuition. And neither could have gotten us to where we are now without the other.

Acknowledgements. This article is based on a lecture of the same title given as part of the Pitcher Lecture Series at Lehigh University, April 30, May 2, 1996. I gratefully acknowledge their support and encouragement for the writing of this article. I also thank the *Notices* editorial staff, especially Susan Landau, for their encouragement. I am grateful to the following individuals for their critical comments: Joe Buhler, Scott Contini, Richard Crandall, Bruce Dodson, Andrew Granville, Hendrik Lenstra, Kevin McCurley, Andrew Odlyzko, David Pomerance, Richard Schroeppel, John Selfridge and Hugh Williams.

Appendix: The First Twenty Fermat Numbers

m	known factorization of $F_m = 2^{2^m}+1$
0	3
1	5
2	17
3	257
4	65537
5	$641 \cdot P_7$
6	$274177 \cdot P_{14}$
7	$59649589127497217 \cdot P_{22}$
8	$1238926361552897 \cdot P_{62}$
9	$2424833 \cdot 7455602825647884208337395736200454918783366342657 \cdot P_{99}$
10	$45592577 \cdot 6487031809 \cdot 4659775785220018543264560743076778192897 \cdot P_{252}$
11	$319489 \cdot 974849 \cdot 167988556341760475137 \cdot 3560841906445833920513 \cdot P_{564}$
12	$114689 \cdot 26017793 \cdot 63766529 \cdot 190274191361 \cdot 1256132134125569 \cdot C_{1187}$
13	$2710954639361 \cdot 2663848877152141313 \cdot 3603109844542291969 \cdot 319546020820551643220672513 \cdot C_{2391}$
14	C_{4933}
15	$1214251009 \cdot 2327042503868417 \cdot C_{9840}$
16	$825753601 \cdot C_{19720}$
17	$31065037602817 \cdot C_{39444}$
18	$13631489 \cdot C_{78906}$
19	$70525124609 \cdot 646730219521 \cdot C_{157804}$

In the table, the notation P_k means a prime number of k decimal digits, while the notation C_k means a composite number of k decimal digits for which we know no nontrivial factorization.

The history of the factorization of Fermat numbers is a microcosm of the history of factoring. Fermat himself knew about F_0 through F_4, and he conjectured that all of the remaining numbers in the sequence $2^{2^m} + 1$ are prime. However, Euler found the factorization of F_5. It is not too hard to find this factorization, if one uses the result, essentially due to Fermat, that for p to be a prime factor of F_m it is necessary that $p \equiv 1 \bmod 2^{m+2}$, when m is at least 2. Thus the prime factors of F_5 are all 1 mod 128, and the first such prime, which is not a smaller Fermat number, is 641. It is via this idea that F_6 was factored (by Landry in 1880) and that "small" prime factors of many other Fermat numbers have been found, including more than 80 beyond this table.

The Fermat number F_7 was the first success of the Brillhart-Morrison continued fraction factoring method. Brent and Pollard used an adaptation of Pollard's "rho" method to factor F_8. As discussed in the main article, F_9 was factored by the number field sieve. The Fermat numbers F_{10} and F_{11} were factored by Brent using Lenstra's elliptic curve method.

We know that F_{14}, F_{20} and F_{22} are composite, but we do not know any prime factors of these numbers. That they are composite was discovered via *Pepin's criterion*: F_m is prime if and only if $3^{(F_m - 1)/2} \equiv -1 \mod F_m$. The smallest Fermat number for which we do not know if it is prime or composite is F_{24}. It is now thought by many number theorists that every Fermat number after F_4 is composite. Fermat numbers are connected with an ancient problem of Euclid: for which n is it possible to construct a regular n-gon with straightedge and compass? Gauss showed that a regular n-gon is constructible if $n \geq 3$ and the largest odd factor of n is a product of distinct, prime Fermat numbers. Wanke later proved the easy converse, so we have a complete classification of constructible regular n-gons up to a determination of prime Fermat numbers.

References

1. L. M. Adelman, *Factoring numbers using singular integers*, Proc. 23rd Annual ACM Symp. on Theory of Computing (STOC) (1991), 64–71.

2. W. R. Alford and C. Pomerance, *Implementing the self initializing quadratic sieve on a distributed network*: Number Theoretic and Algebraic Methods in Computer Science, Proc. Internat. Moscow Conference, June–July, 1993 (A. J. van der Poorten, I. Shparlinski, H. G. Zimmer, eds.), World Scientific, 1995, pp. 163–174.

3. J. Brillhart, D. H. Lehmer, J. L. Selfridge, B. Tuckerman and S. S. Wagstaff, Jr., *Factorizations of $b^n \pm 1$, $b = 2, 3, 5, 6, 7, 10, 11, 12$, up to high powers* (second edition), Vol. 22, Contemporary Mathematics, Amer. Math. Soc., Providence, RI, 1988.

4. E. R. Canfield, P. Erdős and C. Pomerance, *On a problem of Oppenheim concerning "Factorisatio Numerorum,"* J. Number Theory 17 (1983), 1–28.

5. D. Coppersmith, *Modifications to the number field sieve*, J. Cryptology 6 (1993), 169–180.

6. D. Coppersmith, A. M. Odlyzko and R. Schroeppel, *Discrete logarithms in GF(p)*, Algorithmica 1 (1986), 1–15.

7. J. Cowie, B. Dodson, R. Marije Elkenbracht-Huizing, A. K. Lenstra, P. L. Montgomery and J. Zayer, *A world wide number field sieve factoring record: On to 512 bits*, Advances in cryptology – Asiacrypt '96, to appear.

8. M. Elkenbracht-Huizing, *A multiple polynomial general number field sieve*: Algorithmic Number Theory, Second International Symposium, ANTS-II, to appear.

9. J. Gerver, *Factoring large numbers with a quadratic sieve*, Math. Comp. 41 (1983), 287–294.

10. A. K. Lenstra, *Integer factoring. Towards a quarter-century of public-key cryptography*, Des. Codes Cryptogr. 19 (2000), 101–128.

11. A. K. Lenstra and H. W. Lenstra, Jr. (eds.), The development of the number field sieve, Lecture Notes in Math. 1554, Springer-Verlag, Berlin, 1993.

12. A. K. Lenstra and M. S. Manasse, *Factoring by electronic mail*: Advances in cryptology—Eurocrypt '89 (J.-J. Quisquater and J. Vandewalle, eds.), Springer-Verlag, Berlin, Heidelberg, 1990, 355–371.

13. H. W. Lenstra, Jr., *Elliptic curves and number theoretic algorithms*, in: Proc. Internat. Cong. Math., Berkeley, CA, 1986, vol. 1 (A. M. Gleason, ed.), Amer. Math. Soc., Providence, RI, 1987, 99–120.

14. P. L. Montgomery, *A block Lanczos algorithm for finding dependencies over GF(2)*: Advances in cryptology—Eurocrypt '95 (L. C. Guillou and J.-J. Quisquater, eds.), Springer-Verlag, Berlin, Heidelberg, 1995, 106–120.

15. P. L. Montgomery, *Square roots of products of algebraic integers*: Mathematics of Computation 1943–1993, Fifty Years of Computational Mathematics (W. Gautschi, ed.), Proc. Symp. Appl. Math. 48, American Mathematical Society, Providence, 1994, pp. 567–571.

16. M. A. Morrison and J. Brillhart, *A method of factorization and the factorization of* F_7, Math. Comp. 29 (1975), 183–205.

17. A. M. Odlyzko, *The future of integer factorization*, CryptoBytes (the technical news letter of RSA Laboratories) 1, no. 2 (1995), 5–12.

18. C. Pomerance (ed.), Cryptology and computational number theory, Proc. Symp. Appl. Math. 42, Amer. Math. Soc., Providence, RI, 1990.

19. C. Pomerance, *The number field sieve*: Mathematics of Computation 1943–1993, Fifty Years of Computational Mathematics (W. Gautschi, ed.), Proc. Symp. Appl. Math. 48, American Mathematical Society, Providence, 1994, pp. 465–480.

20. C. Pomerance, *On the role of smooth numbers in number theoretic algorithms*: Proc. Internat. Cong. Math., Zürich, Switzerland, 1994, vol. 1 (S. D. Chatterji, ed.), Birkhaäuser-Verlag, Basel, 1995, 411–422.

21. C. Pomerance, J. W. Smith and R. Tuler, *A pipeline architecture for factoring large integers with the quadratic sieve algorithm*, SIAM J. Comput. 17 (1988), 387–403.

22. O. Schirokauer, D. Weber and T. Denny, *Discrete logarithms: The effectiveness of the index calculus method*: Algorithmic Number Theory, (Talence, 1996) 337–361, Lecture Notes in Comput. Sci., 1122, Springer, Berlin, 1996.

23. H. C. Williams and J. O. Shallit, *Factoring integers before computers*: Mathematics of Computation 1943–1993, Fifty Years of Computational Mathematics (W. Gautschi, ed.), Proc. Symp. Appl. Math. 48, American Mathematical Society, Providence, 1994, pp. 481–531.

Originally appeared as:
Pomerance, Carl. "A Tale of Two Sieves." *Notices of the AMS.* (December 1996): pp. 1473–1485.

Second Helping

For more recent progress on Fermat numbers, factoring and in fact computational number theory in general, see:

1. J. P. Buhler and P. Stevenhagen (eds.), Surveys in algorithmic number theory, to appear.

2. R. Crandall and C. Pomerance, Prime numbers: a computational perspective, second ed., Springer, New York, 2005.

Part III: Irrationality and Continued Fractions

One of the first theorems concerning irrational numbers is that $\sqrt{2}$ is irrational. Few proofs are as direct and succinct as the one given by Tom Apostol in "Irrationality of the square root of two– A geometric proof" (*Amer. Math. Monthly*, vol. 107, no. 9 (November 2000), pp. 841–842), done by a picture and not nearly 1000 words. For the more algebraically inclined, we present Harley Flanders' "Math Bite: irrationality of $\sqrt{m}$ " (*Math. Magazine*, vol. 72, no. 3 (June 1999), p. 235).

It is apparently easier to prove the irrationality of $\sqrt{2}$ than that of such transcendental numbers as e and π. Yet Ivan Niven, in "A simple proof that π is irrational" (*Bulletin of the AMS*, vol. 53 (1947), p. 509) gives a one-page proof of the irrationality of π using nothing more than elementary calculus. Almost 40 years later, in "π, e, and other irrational numbers" (*Amer. Math. Monthly*, vol. 93, no. 9 (November 1986), pp. 722–723), Alan Parks generalizes Niven's argument to prove that x is irrational whenever $0 < x \leq \pi$ and $\cos x$ and $\sin x$ are rational and whenever $x > 0$ and $\ln x$ is rational.

Since the number e is irrational, it has a nonterminating continued fraction expansion. In fact, its continued fraction expansion has the beautiful repetitive structure $e = [2, 1, 2, 1, 1, 4, 1, 1, 6, 1, 1, 8, ...]$. This was first proved by Euler, but in "A short proof of the simple continued fraction of e" (*Amer. Math. Monthly*, vol. 113, no. 1 (January 2006), pp. 57–61), Henry Cohn provides a very short proof requiring little more than the product rule from calculus.

We continue our discussion of continued fractions with Ed Burger's "Diophantine Olympics and World Champions: Polynomials and Primes Down Under" (Amer. Math. Monthly, vol. 107, no. 9 (November 2000), pp. 822–829). Given an irrational number, its continued fraction generates a sequence of best rational approximations. For example, the golden ratio generates the sequence $\frac{1}{1}, \frac{2}{1}, \frac{3}{2}, \frac{5}{3}, \frac{8}{5}, \frac{13}{8}, ...$, whose denominators are all Fibonacci numbers. Burger proves that there exist irrational numbers whose sequences of best rational approximations have denominators that are only perfect squares. The result is also true if we replace "perfect squares" with perfect cubes, prime numbers, or many other special sets of numbers. The paper is written with considerable humor and wit, and was inspired by the 2000 Olympics. For this paper, Burger received the Chauvenet Prize, MAA's "gold medal" for mathematical exposition.

The international appeal of Johan Wästlund's "An elementary proof of the Wallis product formula for pi," (*Amer. Math. Monthly*, vol. 114, no. 10 (December 2007), pp. 914–917) about the most famous constant in all of mathematics, is worth the price of admission. A Swede writes about a proof, originally in Norwegian, of an identity due to an Englishman, using a principle from the Greeks and tools from Hindus and Persians, referring to a paper in Russian, and appearing in an American journal.

Were you ever in a grove of trees and could not see out of the grove because the trees were too thick, and did you wonder how thin those trees would have to be so that you could see out if you were looking in the right direction? Our next Biscuit, "The Orchard Problem," is from Ross Honsberger's collection (*Mathematical Gems* (pp. 43–53, Dolciani Mathematical Expositions, No. 1, MAA, 1973). Honsberger begins with an intriguing restatement of the trees-and-forest problem and continues with a gentle and highly readable introduction to the geometry of numbers, which deals with interactions between convex sets and lattice points. He begins by proving two results by Hans Blichfeldt, continues with Hermann Minkowski's fundamental theorem, and ends with a solution to the Orchard Problem. As a bonus, he includes a list of seven problems.

Irrationality of the Square Root of Two—a Geometric Proof

Tom M. Apostol

This note presents a remarkably simple proof of the irrationality of $\sqrt{2}$ that is a variation of the classical Greek geometric proof.

By the Pythagorean theorem, an isosceles right triangle of edge-length 1 has hypotenuse of length $\sqrt{2}$. If $\sqrt{2}$ is rational, some positive integer multiple of this triangle must have three sides with integer lengths, and hence there must be a *smallest* isosceles right triangle with this property. But inside any isosceles right triangle whose three sides have integer lengths we can always construct a smaller one with the same property, as shown below. Therefore $\sqrt{2}$ cannot be rational.

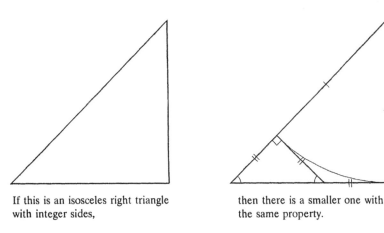

If this is an isosceles right triangle with integer sides,

then there is a smaller one with the same property.

Figure 1

Construction. A circular arc with center at the uppermost vertex and radius equal to the vertical leg of the triangle intersects the hypotenuse at a point, from which a perpendicular to the hypotenuse is drawn to the horizontal leg. Each line segment in the diagram has integer length, and the three segments with double tick marks have equal lengths. (Two of them are tangents to the circle from the same point.) Therefore the smaller isosceles right triangle with hypotenuse on the horizontal base also has integer sides.

The reader can verify that similar arguments establish the irrationality of $\sqrt{n^2 + 1}$ and $\sqrt{n^2 - 1}$ for any integer $n > 1$. For $\sqrt{n^2 + 1}$ use a right triangle with legs of lengths 1 and n. For $\sqrt{n^2 - 1}$ use a right triangle with hypotenuse n and one leg of length 1.

Originally appeared as:

Apostol, Tom M. "Irrationality of the Square Root of Two—A Geometric Proof." *American Mathematical Monthly*. vol. 107, no. 9 (November 2000): pp. 841–842.

Math Bite: Irrationality of $\sqrt{m}$

Harley Flanders

A recent *Gazette* article [3] reminded me of Theodor Estermann's proof of the irrationality of $\sqrt{2}$. As his proof is not well known, I take the liberty of publicizing it with a slight "generalization." The really interesting thing about this proof is that it doesn't use divisibility, just mathematical induction in its "$\mathbb{Z}$ is well-ordered" form.

THEOREM. *Suppose m is not a perfect square. Then $\sqrt{m}$ is irrational.*

Proof. Let n be the integer with $n < \sqrt{m} < n+1$. It suffices to prove that $\alpha = \sqrt{m} - n$ is irrational. Suppose not. As $0 < \alpha < 1$, we have $\alpha = p/q$ where $0 < p < q$. We may assume that q is *as small as possible* (Estermann's key idea). Then we have

$$\frac{q}{p} = \frac{1}{\sqrt{m}-n} = \frac{\sqrt{m}+n}{m-n^2} = \frac{\alpha+2n}{m-n^2}.$$

We solve for α:

$$\alpha = \frac{(m-n^2)q}{p} - 2n = \frac{(m-n^2)q - 2np}{p} = \frac{r}{p}.$$

Thus α is a fraction with even smaller denominator, a contradiction.

REFERENCES

1. T. Estermann, The irrationality of $\sqrt{2}$, *Math. Gazette* 59 (1975), 110.
2. K. Roth and R. C. Vaughn, Obituary of Theodor Estermann, *Bull. London Math. Soc.* 26 (1994), 593–606.
3. M.-K. Siu, Estermann and Pythagoras, *Math. Gazette* 82 (1998), 92–93.

Originally appeared as:
Flanders, Harley. "Math Bite: Irrationality of $\sqrt{m}$." *Mathematics Magazine*. vol. 72, no. 3 (June 1999): p. 235

A Simple Proof that π is Irrational

Ivan Niven

Let $\pi = a/b$, the quotient of positive integers. We define the polynomials

$$f(x) = \frac{x^n(a - bx)^n}{n!},$$

$$F(x) = f(x) - f^{(2)}(x) + f^{(4)}(x) - \cdots + (-1)^n f^{(2n)}(x),$$

the positive integer n being specified later. Since $n!f(x)$ has integral coefficients and terms in x of degree not less than n, $f(x)$ and its derivatives $f^{(i)}(x)$ have integral values for $x=0$; also for $x=\pi=a/b$, since $f(x) = f(a/b - x)$. By elementary calculus we have

$$\frac{d}{dx}\{F'(x) \sin x - F(x) \cos x\} = F''(x) \sin x + F(x) \sin x = f(x) \sin x$$

and

(1) $$\int_0^\pi f(x) \sin x \, dx = [F'(x) \sin x - F(x) \cos x]_0^\pi = F(\pi) + F(0).$$

Now $F(\pi) + F(0)$ is an *integer*, since $f^{(i)}(\pi)$ and $f^{(i)}(0)$ are integers. But for $0 < x < \pi$,

$$0 < f(x) \sin x < \frac{\pi^n a^n}{n!},$$

so that the integral in (1) is *positive, but arbitrarily small* for n sufficiently large. Thus (1) is false, and so is our assumption that π is rational.

Originally appeared as:

Niven, Ivan. "A Simple Proof that π is Irrational." *Bulletin of the AMS*. vol. 53, (1947): p. 509.

π, e and Other Irrational Numbers

Alan E. Parks

In [1] Niven gave a clever, short proof that π is irrational. We would like to show how his proof can be generalized to prove substantially more:

(a) If $0 < |r| \leq \pi$ and if $\cos(r)$ and $\sin(r)$ are rational, then r is irrational.
(b) If r is positive and rational, $r \neq 1$, then $\ln(r)$ is irrational.

For example, (a) shows that π is irrational. If $a^2 + b^2 = c^2$ for rational numbers a, b, c, with $bc \neq 0$, then (a) shows that $\arccos(a/c)$ is irrational. Taking the contrapositive in (b) with $r = e$, we see that e is irrational.

Of course, by a famous theorem of Lindemann, all the numbers in (a) and (b) are not just irrational, but transcendental. The novelty of our argument is not so much the conclusion, but that our proof is elementary and can be effectively presented to students of calculus, for we require nothing beyond integration by parts and knowledge of the limit $\lim_{k \to \infty} M^k/k!$. Then again, at the beginning level, the fact that certain real numbers which occur naturally are irrational is interesting enough to present for its own sake.

We will apply the following theorem to prove (a) and (b). Its proof is a generalization of Niven's argument in [1].

THEOREM. *Let c be a positive real number and let $f(x)$ be a continuous function on $[0, c]$, positive on $(0, c)$. Suppose there are (antiderivatives) $f_1(x), f_2(x), \ldots$ with $f_1'(x) = f(x)$ and with $f_k'(x) = f_{k-1}(x)$ for all $k \geq 2$, and such that $f_k(0), f_k(c)$ are integers for all $k \geq 1$. Then c is irrational.*

Proof. Let P be the set of all polynomials $g(x)$ with real coefficients such that $g(0), g(c), g'(0), g'(c), \ldots, g^{(k)}(0), g^{(k)}(c), \ldots$ are all integers.

CLAIM 1. *If $g(x)$ is in P, then $\int_0^c f(x)g(x)\,dx$ is an integer.*

Proof. Successive integrations by parts give

$$\int_0^c f(x)g(x)\,dx = \left[f_1 \cdot g - f_2 \cdot g' + f_3 \cdot g'' - \cdots + (-1)^d f_{d+1} \cdot g^{(d)} \right]_0^c,$$

where d is the degree of $g(x)$. This proves the claim.

We will also need the following easy fact.

(1) If $g(x)$ and $h(x)$ are in P, then so is $g(x)h(x)$.

Now assume that c is rational, and write $c = m/n$, where m, n are positive integers. Then one verifies:

(2) $m - 2nx$ is in P.

Let $g_k(x) = x^k(m - nx)^k/k!$ for $k = 0, 1, 2, \ldots$.

CLAIM 2. *$g_k(x)$ is in P for all k.*

Proof. Induction on k: $g_0(x) = 1$ is an element of P. For $k \geq 1$,

$$g_k'(x) = g_{k-1}(x)(m - 2nx).$$

By induction, g_{k-1} is in P, by (2) $m - 2nx$ is in P, and thus by (1) g'_k is in P. Since also $g_k(0)$ and $g_k(c)$ are 0, we have that g_k is in P.

Observe that $g_k(x) > 0$ on $(0, c)$, a property shared by $f(x)$, so that $\int_0^c f(x) g_k(x)\, dx > 0$. By Claim 1, the integral is also an integer; therefore

(3) $\int_0^c f(x) g_k(x)\, dx \geq 1$ for all k.

Let M be the maximum for $x(m - nx)$ on $[0, c]$, and L that for $f(x)$, then

$$\int_0^c f(x) g_k(x)\, dx \leq \int_0^c L \cdot \frac{M^k}{k!}\, dx = c \cdot L \cdot \frac{M^k}{k!}.$$

But $\lim_{k \to \infty} M^k / k! = 0$, contradicting (3). We are forced to conclude that c is irrational.

To prove the statement (a) mentioned at the beginning, observe that if $\cos(r)$ and $\sin(r)$ are rational, so are $\cos(|r|)$ and $\sin(|r|)$, and thus we can find a positive integer n such that $n \cdot \sin(|r|)$ and $n \cdot \cos(|r|)$ are integers. Apply the theorem, with $c = |r|$ and $f(x) = n \cdot \sin(x)$, to conclude that $|r|$ is irrational, hence that r is irrational.

To prove (b), observe that $r > 1$ without loss of generality, so that $\ln(r) > 0$. Write $r = m/n$ for some positive integers m, n, and apply the theorem with $c = \ln(r)$ and $f(x) = n \cdot e^x$.

Reference

1. I. Niven, A simple proof of the irrationality of π, Bull. Amer. Math. Soc., 53 (1947) 509.

Originally appeared as:

Parks, Alan E. "π, e and Other Irrational Numbers." *American Mathematical Monthly*. vol. 93, no. 9 (November 1986): pp. 722–723.

A Short Proof of the Simple Continued Fraction of e

Henry Cohn

1. INTRODUCTION. In [3], Euler analyzed the Ricatti equation to prove that the number e has the continued fraction expansion

$$e = [2, 1, 2, 1, 1, 4, 1, 1, 6, 1, 1, 8, \ldots] = 2 + \cfrac{1}{1 + \cfrac{1}{2 + \cfrac{1}{1 + \cfrac{1}{1 + \cfrac{1}{4 + \cdots}}}}}.$$

The pattern becomes more elegant if one replaces the initial 2 with 1, 0, 1, which yields the equivalent continued fraction

$$[1, 0, 1, 1, 2, 1, 1, 4, 1, 1, 6, 1, 1, 8, 1, 1, 10, 1, \ldots] \tag{1}$$

because

$$1 + \cfrac{1}{0 + \cfrac{1}{1 + \cdots}} = 2 + \cdots.$$

One of the most interesting proofs is due to Hermite; it arose as a byproduct of his proof of the transcendence of e in [5]. (See [6] for an exposition by Olds.) The purpose of this note is to present an especially short and direct variant of Hermite's proof and to explain some of the motivation behind it.

Consider any continued fraction $[a_0, a_1, a_2, \ldots]$. Its ith convergent is defined to be the continued fraction $[a_0, a_1, \ldots, a_i]$. One of the most fundamental facts about continued fractions is that the ith convergent equals p_i/q_i, where p_i and q_i can be calculated recursively using

$$p_n = a_n p_{n-1} + p_{n-2}, \qquad q_n = a_n q_{n-1} + q_{n-2},$$

starting from the initial conditions $p_0 = a_0$, $q_0 = 1$, $p_1 = a_0 a_1 + 1$, and $q_1 = a_1$. For a proof see [4, p. 130] or any introduction to continued fractions.

2. PROOF OF THE EXPANSION. Let $[a_0, a_1, a_2, \ldots]$ be the continued fraction (1). In other words, $a_{3i+1} = 2i$ and $a_{3i} = a_{3i+2} = 1$, so p_i and q_i are as follows:

i	0	1	2	3	4	5	6	7	8
p_i	1	1	2	3	8	11	19	87	106
q_i	1	0	1	1	3	4	7	32	39

(Note that $q_1 = 0$ so p_1/q_1 is undefined, but that will not be a problem.) Then p_i and q_i satisfy the recurrence relations

$$\begin{aligned} p_{3n} &= p_{3n-1} + p_{3n-2}, & q_{3n} &= q_{3n-1} + q_{3n-2}, \\ p_{3n+1} &= 2np_{3n} + p_{3n-1}, & q_{3n+1} &= 2nq_{3n} + q_{3n-1}, \\ p_{3n+2} &= p_{3n+1} + p_{3n}, & q_{3n+2} &= q_{3n+1} + q_{3n}. \end{aligned}$$

To verify that the continued fraction (1) equals e, we must prove that

$$\lim_{i \to \infty} \frac{p_i}{q_i} = e.$$

Define the integrals

$$A_n = \int_0^1 \frac{x^n(x-1)^n}{n!} e^x \, dx,$$

$$B_n = \int_0^1 \frac{x^{n+1}(x-1)^n}{n!} e^x \, dx,$$

$$C_n = \int_0^1 \frac{x^n(x-1)^{n+1}}{n!} e^x \, dx.$$

Proposition 1. *For $n \geq 0$, $A_n = q_{3n}e - p_{3n}$, $B_n = p_{3n+1} - q_{3n+1}e$, and $C_n = p_{3n+2} - q_{3n+2}e$.*

Proof. In light of the recurrence relations cited earlier, we need only verify the initial conditions $A_0 = e - 1$, $B_0 = 1$, and $C_0 = 2 - e$ (which are easy to check) and prove the three recurrence relations

$$A_n = -B_{n-1} - C_{n-1}, \tag{2}$$

$$B_n = -2nA_n + C_{n-1}, \tag{3}$$

$$C_n = B_n - A_n. \tag{4}$$

Of course, (4) is trivial. To prove (2) (i.e., $A_n + B_{n-1} + C_{n-1} = 0$) integrate both sides of

$$\frac{x^n(x-1)^n}{n!}e^x + \frac{x^n(x-1)^{n-1}}{(n-1)!}e^x + \frac{x^{n-1}(x-1)^n}{(n-1)!}e^x = \frac{d}{dx}\left(\frac{x^n(x-1)^n}{n!}e^x\right),$$

which follows immediately from the product rule for derivatives. To prove (3) (i.e., $B_n + 2nA_n - C_{n-1} = 0$) integrate both sides of

$$\frac{x^{n+1}(x-1)^n}{n!}e^x + 2n\frac{x^n(x-1)^n}{n!}e^x - \frac{x^{n-1}(x-1)^n}{(n-1)!}e^x = \frac{d}{dx}\left(\frac{x^n(x-1)^{n+1}}{n!}e^x\right),$$

which follows from the product rule and some additional manipulation. This completes the proof. ∎

The recurrences (2) and (3) can also be proved by integration by parts.

Theorem 1. $e = [1, 0, 1, 1, 2, 1, 1, 4, 1, 1, 6, 1, 1, 8, 1, 1, 10, 1, \dots]$.

Proof. Clearly A_n, B_n, and C_n tend to 0 as $n \to \infty$. It follows from Proposition 1 that

$$\lim_{i \to \infty} q_i e - p_i = 0.$$

Because $q_i \geq 1$ when $i \geq 2$, we see that

$$e = \lim_{i \to \infty} \frac{p_i}{q_i} = [1, 0, 1, 1, 2, 1, 1, 4, 1, \dots],$$

as desired. ∎

3. MOTIVATION. The most surprising aspect of this proof is the integral formulas, which have no apparent motivation. The difficulty is that the machinery that led to them has been removed from the final proof. Hermite wrote down the integrals while studying Padé approximants, a context in which it is easier to see how one might think of them.

Padé approximants are certain rational function approximations to a power series. They are named after Hermite's student Padé, in whose 1892 thesis [7] they were studied systematically. However, for the special case of the exponential function they are implicit in Hermite's 1873 paper [5] on the transcendence of e.

We will focus on the power series

$$e^z = \sum_{k \geq 0} \frac{z^k}{k!},$$

since it is the relevant one for our purposes. A *Padé approximant* to e^z of type (m, n) is a rational function $p(z)/q(z)$ with $p(z)$ and $q(z)$ polynomials such that $\deg p(z) \leq m$, $\deg q(z) \leq n$, and

$$\frac{p(z)}{q(z)} = e^z + O\left(z^{m+n+1}\right)$$

as $z \to 0$. In other words, the first $m + n + 1$ coefficients in the Taylor series of $p(z)/q(z)$ agree with those for e^z. One cannot expect more agreement than that: $p(z)$ has $m + 1$ coefficients and $q(z)$ has $n + 1$, so there are $m + n + 2$ degrees of freedom in toto, one of which is lost because $p(z)$ and $q(z)$ can be scaled by the same factor without changing their ratio. Thus, one expects to be able to match $m + n + 1$ coefficients. (Of course this argument is not rigorous.)

It is easy to see that there can be only one Padé approximant of type (m, n): if $r(z)/s(z)$ is another, then

$$\frac{p(z)s(z) - q(z)r(z)}{q(z)s(z)} = \frac{p(z)}{q(z)} - \frac{r(z)}{s(z)} = O\left(z^{m+n+1}\right).$$

Because $p(z)s(z) - q(z)r(z)$ vanishes to order $m + n + 1$ at $z = 0$ but has degree at most $m + n$ (assuming $\deg r(z) \leq m$ and $\deg s(z) \leq n$), it must vanish identically. Thus, $p(z)/q(z) = r(z)/s(z)$.

We have no need to deal with the existence of Padé approximants here, because it will follow from a later argument. One can compute the Padé approximants of e^z by solving simultaneous linear equations to determine the coefficients of the numerator and denominator, after normalizing so the constant term of the denominator is 1.

The usefulness of Padé approximants lies in the fact that they provide a powerful way to approximate a power series. Those of type $(m, 0)$ are simply the partial sums of the series, and the others are equally natural approximations. If one is interested in approximating e, it makes sense to plug $z = 1$ into the Padé approximants for e^z and see what happens.

Let $r_{m,n}(z)$ denote the Padé approximant of type (m, n) for e^z. Computing continued fractions reveals that

$$r_{1,1}(1) = [2, 1],$$
$$r_{1,2}(1) = [2, 1, 2],$$
$$r_{2,1}(1) = [2, 1, 2, 1],$$
$$r_{2,2}(1) = [2, 1, 2, 1, 1],$$
$$r_{2,3}(1) = [2, 1, 2, 1, 1, 4],$$
$$r_{3,2}(1) = [2, 1, 2, 1, 1, 4, 1],$$
$$r_{3,3}(1) = [2, 1, 2, 1, 1, 4, 1, 1],$$

etc. In other words, when we set $z = 1$, the Padé approximants of types (n, n), $(n, n + 1)$, and $(n + 1, n)$ appear to give the convergents to the continued fraction of e. There is no reason to think Hermite approached the problem quite this way, but his paper [5] does include some numerical calculations of approximations to e, and it is plausible that his strategy was informed by patterns he observed in the numbers.

It is not clear how to prove this numerical pattern directly from the definitions. However, Hermite found an ingenious way to derive the Padé approximants from integrals, which can be used to prove it. In fact, it will follow easily from Proposition 1.

It is helpful to reformulate the definition as follows. For a Padé approximant of type (m, n), we are looking for polynomials $p(z)$ and $q(z)$ of degrees at most m and n, respectively, such that $q(z)e^z - p(z) = O\left(z^{m+n+1}\right)$ as $z \to 0$. In other words, the function

$$z \mapsto \frac{q(z)e^z - p(z)}{z^{m+n+1}}$$

must be holomorphic. (Here that is equivalent to being bounded for z near 0. No complex analysis is needed in this article.)

One way to recognize a function as being holomorphic is to write it as a suitable integral. For example, because

$$\frac{(z-1)e^z + 1}{z^2} = \int_0^1 x e^{zx}\, dx,$$

it is clear that $z \mapsto ((z-1)e^z + 1)/z^2$ is holomorphic. Of course, that is unnecessary for such a simple function, but Hermite realized that this technique was quite powerful.

A Short Proof of the Simple Continued Fraction of e

It is not clear how he thought of it, but everyone who knows calculus has integrated an exponential times a polynomial, and one can imagine he simply remembered that the answer has exactly the form we seek.

Lemma 1. *Let $r(x)$ be a polynomial of degree k. Then there are polynomials $q(z)$ and $p(z)$ of degree at most k such that*

$$\int_0^1 r(x)e^{zx}\,dx = \frac{q(z)e^z - p(z)}{z^{k+1}}.$$

Specifically,

$$q(z) = r(1)z^k - r'(1)z^{k-1} + r''(1)z^{k-2} - \cdots$$

and

$$p(z) = r(0)z^k - r'(0)z^{k-1} + r''(0)z^{k-2} - \cdots.$$

Proof. Integration by parts implies that

$$\int_0^1 r(x)e^{zx}\,dx = \frac{r(1)e^z - r(0)}{z} - \frac{1}{z}\int_0^1 r'(x)e^{zx}\,dx,$$

from which the desired result follows by induction. ∎

To get a Padé approximant $p(z)/q(z)$ of type (m, n), we want polynomials $p(z)$ and $q(z)$ of degrees m and n, respectively, such that

$$z \mapsto \frac{q(z)e^z - p(z)}{z^{m+n+1}}$$

is holomorphic. That suggests we should take $k = m + n$ in the lemma. However, if $r(x)$ is not chosen carefully, then the degrees of $q(z)$ and $p(z)$ will be too high. To choose $r(x)$, we examine the explicit formulas

$$q(z) = r(1)z^{m+n} - r'(1)z^{m+n-1} + r''(1)z^{m+n-2} - \cdots$$

and

$$p(z) = r(0)z^{m+n} - r'(0)z^{m+n-1} + r''(0)z^{m+n-2} - \cdots.$$

The condition that $\deg q(z) \le n$ simply means $r(x)$ has a root of order m at $x = 1$, and similarly $\deg p(z) \le m$ means $r(x)$ has a root of order n at $x = 0$. Since $\deg r(x) = m + n$, our only choice (up to a constant factor) is to take

$$r(x) = x^n(x-1)^m,$$

and that polynomial works. Thus,

$$\int_0^1 x^n(x-1)^m e^{zx}\,dx = \frac{q(z)e^z - p(z)}{z^{m+n+1}},$$

where $p(z)/q(z)$ is the Padé approximant of type (m, n) to e^z.

Setting $z = 1$ recovers the integrals used in the proof of the continued fraction expansion of e, except for the factor of $1/n!$, which simply makes the answer prettier. Fundamentally, the reason why the factorial appears is that

$$\frac{d^n}{dx^n}\left(\frac{x^n(x-1)^n}{n!}\right)$$

has integral coefficients. Note also that up to a change of variables, this expression is the Rodrigues formula for the Legendre polynomial of degree n [1, p. 99].

Natural generalizations of these integrals play a fundamental role in Hermite's proof of the transcendence of e. See [2, p. 4] for an especially short version of the proof or chapter 20 of [8] for a more leisurely account (although the integrals used there are slightly different from those in this paper).

REFERENCES

1. G. Andrews, R. Askey, and R. Roy, *Special Functions*, Cambridge University Press, Cambridge, 1999.
2. A. Baker, *Transcendental Number Theory*, Cambridge University Press, Cambridge, 1975.
3. L. Euler, De fractionibus continuis dissertatio, *Comm. Acad. Sci. Petropol.* **9** (1744) 98–137; also in *Opera Omnia*, ser. I, vol. 14, Teubner, Leipzig, 1925, pp. 187–215; English translation by M. Wyman and B. Wyman, An essay on continued fractions, *Math. Systems Theory* **18** (1985) 295–328.
4. G. H. Hardy and E. M. Wright, *An Introduction to the Theory of Numbers*, 5th ed., Oxford University Press, Oxford, 1979.
5. C. Hermite, Sur la fonction exponentielle, *C. R. Acad. Sci.* **77** (1873) 18–24, 74–79, 226–233, and 285–293; also in *Œuvres*, vol. 3, Gauthier-Villars, Paris, 1912, pp. 150–181.
6. C. D. Olds, The simple continued fraction expansion of e, this MONTHLY **77** (1970) 968–974.
7. H. Padé, Sur la représentation approchée d'une fonction par des fractions rationelles, *Ann. Sci. École Norm. Sup.*, 3ième Série **9** (1892) 3–93 (supplément); also available at http://www.numdam.org.
8. M. Spivak, *Calculus*, 2nd ed., Publish or Perish, Inc., Houston, 1980.

Microsoft Research, Redmond, WA 98052-6399
cohn@microsoft.com

Originally appeared as:
Cohn, Henry. "A Short Proof of the Simple Continued Fraction of e." *American Mathematical Monthly*. vol. 113, no. 1 (January 2006): pp. 57–61.

Diphantine Olympics and World Champions: Polynomials and Primes Down Under

Edward B. Burger

1. LET THE GAMES BEGIN: THE OPENING CEREMONIES. For those who think globally, "down under" may provoke thoughts of Australia—the home of the 2000 Olympic Games. For those who think rationally, "down under" may provoke thoughts of denominators of fractions. In this paper, we hope to provoke both.

In basic diophantine approximation, the name of the game is to tackle the following: How close do integer multiples of an irrational number get to integral values? Specifically, if α is an irrational number and the function $\|\cdot\|$ on $\mathbb{R}$ gives the *distance to the nearest integer* (that is, $\|x\| = \min\{|x - m| : m \in \mathbb{Z}\}$), then the game really is a competition among all integers n to minimize the value $\|\alpha n\|$.

Suppose someone serves us an irrational number α. We write $q_1, q_2, q_3, \ldots$ for the (winning) sequence of integers that is able to make it over the following three hurdles:

(i) $0 < q_1 < q_2 < q_3 < \cdots < q_i < \cdots$;
(ii) $\|\alpha q_1\| > \|\alpha q_2\| > \|\alpha q_3\| > \cdots > \|\alpha q_i\| > \cdots$;
(iii) if q is any integer such that $1 \le q < q_n$, $q \ne q_{n-1}$, then $\|\alpha q\| > \|\alpha q_{n-1}\|$.

We now award such a sequence of integers the title: *The world champion approximation sequence for α* or simply say that the q_n's form *the team of world champions for α*. Let us momentarily suppress the natural desire to ask the obvious two questions:

- Do world champions exist for each irrational α?
- If a world champion sequence does exist, how would we find it?

Instead, let's uncover the connections between the world champions and the Olympic Games "down under".

If we let p_n denote the nearest integer to αq_n, then

$$\|\alpha q_n\| = |\alpha q_n - p_n| = q_n \left| \alpha - \frac{p_n}{q_n} \right|,$$

and thus we see the q_n "down under" in the fraction p_n/q_n. Suppose that $1 \le q < q_n$ and $\|\alpha q\| = |\alpha q - p|$. Then properties (ii) and (iii) ensure that

$$\left| \alpha - \frac{p}{q} \right| > \left| \alpha - \frac{p_n}{q_n} \right|.$$

Hence we see that p_n/q_n is the *best* rational approximation to α having a q "down under" not exceeding our champion q_n.

It is clear that once we know q_n, the value of p_n is completely determined: it must be the nearest integer to αq_n. Thus in order to find world champion (best) *rational* approximations to α, we need focus our attention only on finding the q_n's—that is, the world champions for α.

Now what do these world champion sequences look like? Table 1 lists some popular numbers and the first few (in fact *a perfect* 10) terms in their associated world champion sequences (μ denotes *Mahler's number*: $\mu = 0.12345678910111213141516171819 20\ldots$).

TABLE 1

	$\dfrac{1+\sqrt{5}}{2}$	e	$\sqrt[3]{2}$	π	μ
q_1	1	1	3	7	73
q_2	2	3	4	106	81
q_3	3	4	23	113	12075796
q_4	5	7	27	33102	12075877
q_5	8	32	50	33215	24151673
q_6	13	39	227	66317	36227550
q_7	21	71	277	99532	169061873
q_8	34	465	504	265381	205289423
q_9	55	536	4309	364913	374351296
q_{10}	89	1001	4813	1360120	579640719

Even the casual spectator might not be able to refrain from making several interesting observations. One such observation is that the world champions for $(1 + \sqrt{5})/2$ appear to be the complete list of Fibonacci numbers—a perennial favorite sequence among number theory fans. A slightly less visible observation is that the sequences of champions seem to satisfy a recurrence relation of the form: $q_n = a_n q_{n-1} + q_{n-2}$ for some positive integer a_n. In fact, as we'll mention again in the next section, this observation holds for all α's.

Our observations show that any sequence of world champions must grow very fast (on the order of exponential growth), and the slowest growing sequence of champions is the Fibonacci sequence where the coefficients a_n are all equal to 1.

All these thoughts inspire the question we pose and consider here. Suppose we are given a sequence of increasing integers. Must they be the complete team of world champions in the eyes of some irrational number? That is, given a sequence, does there always exist a number α that has the given sequence as its sequence of world champions? We know from the previous paragraph that the answer is "no" since slow growing sequences can never make the cut. But what if the sequence were to really work at it, get in shape, and trim down? That is, is it possible that there is always a *subsequence* of any given sequence that contains all the world champions for an irrational number? In particular, do there exist α's for which *all* their world champions are perfect squares? How about perfect cubes? How about primes? After some warm-up's in the next section, we take on these questions and perform some basic routines in the hopes of discovering the thrill of victory. *Let the games begin.*

2. WARMING UP: SOME DIOPHANTINE MENTALROBICS. Our training program begins with a classic feat by Dirichlet from 1842 that still holds the record for being best possible.

Theorem 1. *Let α be a real number and let $Q \geq 1$ be an integer. Then there exists an integer q such that $1 \leq q \leq Q$ and*

$$\|\alpha q\| \leq \frac{1}{Q+1}.$$

The pigeonhole principle allows one to give a beautifully executed one line proof of Dirichlet's theorem. In fact, here is the line:

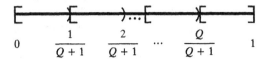

Figure 1

which has been partitioned into $Q + 1$ subintervals each of length $1/(Q + 1)$. We now toss in $Q + 2$ points: $\alpha_n = \alpha n - [\alpha n]$ for $n = 0, 1, 2, \ldots, Q$ and $\alpha_{Q+1} = \alpha 0 - [-1]$, where $[x]$ denotes the *integer part of x*. Thus there must exist two points that land in the same subinterval, and hence there are indices $m_1 < m_2$ such that

$$|\alpha_{m_2} - \alpha_{m_1}| = |\alpha q - p| = \|\alpha q\| \leq \frac{1}{Q + 1},$$

where $1 \leq q \leq Q$, and thus we happily find ourselves at the finish line of the proof. ∎

Of course if α is an irrational number, then, try as it might, $\|\alpha q\|$ can never equal 0. Therefore by letting Q sprint off to infinity, we immediately have the following.

Corollary 2. *If α is an irrational real number, then there exist infinitely many distinct integers q satisfying*

$$\|\alpha q\| < \frac{1}{q}. \tag{2.1}$$

We now want to generate an infinite roster of q's that are fit enough to satisfy the (2.1) challenge. We first write $\alpha = a_0 + \alpha_0$, where $a_0 = [\alpha]$ and α_0 denotes the *fractional part of α*. Thus $1/\alpha_0 > 1$ and so after a double flip we see

$$\alpha = a_0 + \frac{1}{1/\alpha_0} = a_0 + \frac{1}{a_1 + \alpha_1},$$

where $a_1 = [1/\alpha_0]$ and α_1 is the fractional part of $1/\alpha_0$. Since α is irrational, we can repeat this game forever and discover that

$$\alpha = a_0 + \cfrac{1}{a_1 + \cfrac{1}{a_2 + \cfrac{1}{\ddots}}},$$

where all the a_n's are integers and $a_n > 0$ for all $n > 0$. Such an expansion is the *continued fraction expansion for α*, which we write simply as $\alpha = [a_0, a_1, a_2, \ldots]$ so as to allow players to print their expansions on the back of their team jerseys. We often hear fans yell out "*partial quotients!*" whenever the a_n's make their appearance. If we decide to call a time-out during the continued fraction game, then our halted process would produce the rational number $[a_0, a_1, \ldots, a_n]$, which we denote by p_n/q_n (where p_n and q_n are relatively prime). Those who are true number theory mathletes refer to p_n/q_n as the *nth convergent of α*. Lagrange, in 1770, thrilled the fans when he showed that for $n > 0$, the q_n's appearing "down under" in the convergents meet all the requirements to be named the world champion sequence for α. In fact, the denominators q_n form the *entire* team of world champions for α; for the play-by-play details, see [3] or [5].

By letting some 2×2 matrices enter into the arena, we can get a better feel for how the q_n's interact with each other. In particular, using induction and some simple linear algebra gymnastics, we can verify the fact that for all $n \geq 0$,

$$\begin{pmatrix} a_0 & 1 \\ 1 & 0 \end{pmatrix} \begin{pmatrix} a_1 & 1 \\ 1 & 0 \end{pmatrix} \cdots \begin{pmatrix} a_n & 1 \\ 1 & 0 \end{pmatrix} = \begin{pmatrix} p_n & p_{n-1} \\ q_n & q_{n-1} \end{pmatrix}.$$

If we lob determinants back and forth, we see that $p_n q_{n-1} - p_{n-1} q_n = \pm 1$ and hence we conclude that q_{n-1} and q_n must always be relatively prime. Although consecutive players do not like to share common factors, the matrix product does reveal that any three consecutive members of the team *can* play together in the sense that

$$q_n = a_n q_{n-1} + q_{n-2}. \tag{2.2}$$

These two observations are crucial as sequences prepare for the Olympic Games.

3. TRAINING TO BE THE BEST: HOW TO SHED UNWANTED TERMS. Let's now tackle our question: Within every increasing sequence of integers, is there a subsequence that has what it takes to be the complete team of world champions for some irrational number? Sadly, the answer is "no": There is no subsequence of $2, 4, 6, 8, 10, 12, \ldots$ that can be a world champion sequence since, as we've seen at the end of the previous section, consecutive members from a team of world champions must be relatively prime. Thus there can never be a world champion sequence in any sequence of integers for which, from some point onward, all the terms share a common factor. Plainly we should not bother to consider such pathetic sequences that cannot make even the first cut. Unfortunately, there also are 'non-trivial' sequences that never can make it to the Olympic Games. Consider, for example, the sorry sequence beginning with

$$2, 3, 4, 13, 168, 177, 1584, 6396, 83317, 1000128, \ldots. \tag{3.1}$$

The terms in this sequence were carefully recruited so as to satisfy an ever growing list of anonymous congruences (sponsors who wish not to be mentioned here). Notice that in the first ten terms, there does not exist a triple $r < s < t$ such that $t = as + r$ for any positive integer a. Thus in view of the line-up from (2.2), we see that the first ten terms in (3.1) do not contain even three players from a world champion team. By our secret recruiting process, this pattern continues for the entire roster of terms in the list—ah, the agony of defeat.

We now bring on some sequences that are better known by the fans. First, let's have sequences generated by polynomial functions take the field.

Theorem 3. *Let $f(x)$ be a nonconstant polynomial with integer coefficients whose leading coefficient is positive. Then there exists a sequence of integers $n_1, n_2, n_3, \ldots$ such that $f(n_1), f(n_2), f(n_3), \ldots$ is a complete world champion sequence for some irrational number if and only if there exist integers n_1 and n_2 such that*

$$0 < f(n_1) < f(n_2) \quad \text{and} \quad f(n_2) \equiv 1 \bmod f(n_1).$$

Proof: If for all $i \geq 1$, $f(n_i) = q_i$ represents a team of world champions of some irrational number, then, as $q_0 = 1$, we must have $0 < q_1 < q_2$ and $q_2 = a_2 q_1 + 1$. Thus we see that $0 < f(n_1) < f(n_2)$ and $f(n_2) \equiv 1 \bmod f(n_1)$.

Conversely, suppose that $0 < f(n_1) < f(n_2)$ and $f(n_2) \equiv 1 \bmod f(n_1)$. We set $a_1 = f(n_1)$ and let a_2 be the positive integer such that $f(n_2) = a_2 f(n_1) + 1$. Let's now assume that $n_1, n_2, \ldots, n_I$ have been defined, are warmed up, and are ready to play, where $I \geq 2$. From the benches, we now select a positive integer c_{I+1} that is so large that if n_{I+1} is defined by $n_{I+1} = c_{I+1} f(n_I) + n_{I-1}$, then $f(n_I) < f(n_{I+1})$. Such integers c_{I+1} exist since $f(x)$ tends to infinity as x pumps up. Notice that

$$f(n_{I+1}) = f(c_{I+1} f(n_I) + n_{I-1}) \equiv f(n_{I-1}) \bmod f(n_I).$$

Therefore there exists a positive integer, say a_{I+1}, such that

$$f(n_{I+1}) = a_{I+1} f(n_I) + f(n_{I-1}).$$

Thus if we let $q_i = f(n_i)$ for all $i \geq 1$, then the q_i's are the team of world champions for the irrational number $[0, a_1, a_2, a_3, \ldots]$ and we've won the game. ∎

As an immediate consequence of Theorem 3, we see that there are irrational numbers whose world champion sequences contain only perfect squares; there are other irrationals whose world champion sequences contain only cubes, or for that matter, any power. Consider, for example, the number

$\alpha = 0.7599451328616293768517377161220850549970853922385659989630374553\ldots$
$= [0, 1, 3, 6, 29, 739, 538810, 290287122557, 84266613096281243920895, \ldots].$

The first few world champions for α are given in Table 2.

TABLE 2

n	q_n
0.	1
1	1
2	4
3	25
4	729
5	538756
6	290287121089
7	84266613096281242843329
8	7100862082718357559748563880517485796441580544

It is immediately apparent to any calculator that all the entries in the right column are indeed perfect squares—thus those "..."s can be replaced by real, red-blooded digits to produce an α that has only perfect square world champions!

Another potentially amusing example can be found if we take on the polynomial $f(x) = Ax + 1$, for some integer $A > 0$. Selecting $n_1 = 1$ and $n_2 = A + 1$, we immediately conclude that there exist real numbers α such that each element q_n of its world champion sequence satisfies $q_n \equiv 1 \bmod A$. Thus there are complete teams of world champions agile enough to dodge having a particular factor. In this spirit of dodging factors, we now describe a winning strategy for a similar game involving the ever popular and timeless team of prime numbers.

Theorem 4. *There exist irrational numbers α that have only prime numbers as their world champions.*

Proof: As always, we set $q_0 = 1$. Next, we set $q_1 = 2$ and $q_2 = 3$, so $a_1 = 2$ and $a_2 = 1$. Suppose now that the primes $q_1 < q_2 < \cdots < q_I$ have all been defined for $I \geq 2$. We now consider the arithmetic progression $\{Aq_I + q_{I-1} : A = 1, 2, \ldots\}$, and apply another important result of Dirichlet, which states that if r and s are relatively prime positive integers, then the arithmetic progression $r + s, 2r + s, 3r + s, \ldots$ contains infinitely many primes (see [4] for an instant re-play of this major AP upset). Thus we know that there exists a positive integer A, call it a_{I+1}, such that $a_{I+1}q_I + q_{I-1}$ is a prime, let's name it q_{I+1}. Therefore if we let $\alpha = [0, a_1, a_2, a_3, \ldots]$, then its world champion sequence consists solely of prime numbers—this places us right in the middle of the winner's circle. ∎

As an illustration, we now introduce the number

$\alpha = 0.38547782732324065153134100625493772881752720832373553581742207176582\ldots$
$= [0, 2, 1, 1, 2, 6, 2, 10, 18, 20, 16, \ldots]$

and its first ten world champions:

TABLE 3

n	q_n
0	1
1	2
2	3
3	5
4	13
5	83
6	179
7	1873
8	33893
9	679733
10	10909621

One doesn't really require Olympic-like factorization skills to verify that the champions in Table 3 are all prime numbers.

Finally we remark that in the proofs of Theorems 3 and 4 we are able, without the use of steroids, to make the world champions grow as fast as we wish. This observation in turn implies that the partial quotients, a_n, in the continued fraction expansion for the associated irrational α can grow at record breaking speeds (in fact we saw this particular event live during our perfect square example). It turns out that if there is a subsequence of a_n's that grow amazingly fast, then the associated number α must, in fact, be a transcendental number. This remarkable fact follows from a record setting 1844 result due to Liouville, who showed that algebraic numbers cannot be approximated too well by rational numbers whose sizes "down under" are modest; see [3] or [5] for flashbacks to this history-making score. Thus we close our games with the commentary that we may find *transcendental* numbers that accomplish all that is demanded of them in both Theorems 3 and 4. In fact, the α debuting just after Theorem 3 is transcendental.

4. THE 2004 OLYMPICS: CAN OUR FAVORITE CHAMPS ARISE FROM THE BAD? With any pursuit—athletic or intellectual—we should always look ahead toward challenges for future thrill-seekers. Thus in our closing ceremonies we look ahead to the next Olympic Games and hope to inspire others to go for the gold.

The α's that are able to give birth to interesting world champions as described in Theorems 3 and 4 can have arbitrarily large a_n values in their continued fraction expansions. Is it possible to find an α that satisfies either theorem and for which all its partial quotients are *bounded*? Numbers α having bounded partial quotients are known as *badly approximable numbers*. This title has been awarded since it can be shown that a number α has bounded partial quotients if and only if (2.2) cannot be improved in the sense that there exists a constant $c = c(\alpha) > 0$ such that $\frac{c}{q} < \|\alpha q\|$ for *all* integers q; see [3] or [5] for the proof of why these numbers deserve the name *bad*.

A completely new game plan would be required if one wanted to attempt to show that there are badly approximable numbers whose world champions satisfy Theorems 3 or 4. In the proof of Theorem 3, the partial quotients are born and bred to race off at record speeds to infinity. Is there another training technique that prevents the partial quotients from running away?

To adopt the strategy of the proof of Theorem 4, one would, at the very least, need to know that there exists a constant C such that each arithmetic progression of the form $\{Ar + s : A = 1, 2, \ldots\}$, where $\gcd(r, s) = 1$, contains a prime $p \leq Cr$. Such an assertion appears to be ridiculously optimistic as the world record-holder in this direction is Heath-Brown [2] who in 1992 nearly caused a riot among fans when he produced the *startling* result that every arithmetic progression $\{Ar + s : A = 1, 2, \ldots\}$ with $\gcd(r, s) = 1$ contains a prime $p \leq Cr^{5.5}$. The previous world champion exponent was a whopping 13.5, found by Chen and Liu [1] back in 1989. The exponent on r is known as *Linnik's constant*.

These closing remarks may invite both the number theory athlete and fan to train and take on the challenge of showing that there do not exist badly approximable numbers whose world champions are all perfect squares or are all primes—two potentially difficult goals, but certainly in the true tradition of the Olympic spirit.

Figure 2. The author overlooking Sydney—the site of the 2000 Olympic Games, where he was inspired to consider turning sequences into world champions.

ACKNOWLEDGMENTS. The remarks and observations made here were inspired while the author was a Visiting Fellow at Macquarie University, "down under" in Sydney, Australia. He thanks the Mathematics Department for its warm hospitality.

REFERENCES

1. J. R. Chen and J. M Liu, On the least prime in an arithmetical progression. IV, *Sci. China Ser. A* **32** (1989) 792–807.
2. D. R. Heath-Brown, Zero-free regions for Dirichlet L-functions, and the least prime in an arithmetic progression, *Proc. London Math. Soc.* (3) **64** (1992) 265–338.
3. E. B. Burger, *Exploring the Number Jungle: An Interactive Journey into Diophantine Analysis*, AMS Student Mathematical Library Series, Providence, 2000.
4. I. Niven, H. Zuckerman, and H. Montgomery, *An Introduction to the Theory of Numbers* (5th Ed.), Wiley, New York, 1991.
5. W. M. Schmidt, *Diophantine Approximation*, Springer Lecture Notes in Mathematics **785**, Springer, New York, 1980.

Originally appeared as:

Burger, Edward B. "Diophantine Olympics and World Champions: Polynomials and Primes Down Under." *American Mathematical Monthly*. vol. 107, no. 9 (November 2000): pp. 822–842.

An Elementary Proof of the Wallis Product Formula for Pi

Johan Wästlund

1. THE WALLIS PRODUCT FORMULA. In 1655, John Wallis wrote down the celebrated formula

$$\frac{2}{1} \cdot \frac{2}{3} \cdot \frac{4}{3} \cdot \frac{4}{5} \cdots = \frac{\pi}{2}. \tag{1}$$

Most textbook proofs of (1) rely on evaluation of some definite integral like

$$\int_0^{\pi/2} (\sin x)^n \, dx$$

by repeated partial integration. The topic is usually reserved for more advanced calculus courses. The purpose of this note is to show that (1) can be derived using only the mathematics taught in elementary school, that is, basic algebra, the Pythagorean theorem, and the formula $\pi \cdot r^2$ for the area of a circle of radius r.

Viggo Brun gives an account of Wallis's method in [1] (in Norwegian). Yaglom and Yaglom [2] give a beautiful proof of (1) which avoids integration but uses some quite sophisticated trigonometric identities.

2. A NUMBER SEQUENCE. We define a sequence of numbers by $s_1 = 1$, and for $n \geq 2$,

$$s_n = \frac{3}{2} \cdot \frac{5}{4} \cdots \frac{2n-1}{2n-2}.$$

The partial products of (1) with an odd number of factors can be written as

$$o_n = \frac{2^2 \cdot 4^2 \cdots (2n-2)^2 \cdot (2n)}{1 \cdot 3^2 \cdots (2n-1)^2} = \frac{2n}{s_n^2}, \tag{2}$$

while those with an even number of factors are of the form

$$e_n = \frac{2^2 \cdot 4^2 \cdots (2n-2)^2}{1 \cdot 3^2 \cdots (2n-3)^2 \cdot (2n-1)} = \frac{2n-1}{s_n^2}. \tag{3}$$

Here $e_1 = 1$ should be interpreted as an empty product. Clearly $e_n < e_{n+1}$ and $o_n > o_{n+1}$, and by comparing (2) and (3) we see that $e_n < o_n$. Therefore we must have

$$e_1 < e_2 < e_3 < \cdots < o_3 < o_2 < o_1.$$

Thus if $1 \leq i \leq n$,

$$\frac{2i}{s_i^2} = o_i \geq o_n$$

and

$$\frac{2i-1}{s_i^2} = e_i \le e_n,$$

from which it follows that

$$\frac{2i-1}{e_n} \le s_i^2 \le \frac{2i}{o_n}. \tag{4}$$

It will be convenient to define $s_0 = 0$. Notice that with this definition, (4) holds also for $i = 0$. We denote the difference $s_{n+1} - s_n$ by a_n. Observe that $a_0 = 1$, and for $n \ge 1$,

$$a_n = s_{n+1} - s_n = s_n\left(\frac{2n+1}{2n} - 1\right) = \frac{s_n}{2n} = \frac{1}{2} \cdot \frac{3}{4} \cdots \frac{2n-1}{2n}.$$

We first derive the identity

$$a_i a_j = \frac{j+1}{i+j+1} a_i a_{j+1} + \frac{i+1}{i+j+1} a_{i+1} a_j. \tag{5}$$

Proof. After the substitutions

$$a_{i+1} = \frac{2i+1}{2(i+1)} a_i$$

and

$$a_{j+1} = \frac{2j+1}{2(j+1)} a_j,$$

the right hand side of (5) becomes

$$a_i a_j \left(\frac{2j+1}{2(j+1)} \cdot \frac{j+1}{i+j+1} + \frac{2i+1}{2(i+1)} \cdot \frac{i+1}{i+j+1}\right) = a_i a_j. \quad \blacksquare$$

If we start from a_0^2 and repeatedly apply (5), we obtain the identities

$$1 = a_0^2 = a_0 a_1 + a_1 a_0 = a_0 a_2 + a_1^2 + a_2 a_0 = \cdots$$
$$= a_0 a_n + a_1 a_{n-1} + \cdots + a_n a_0. \tag{6}$$

Proof. By applying (5) to every term, the sum $a_0 a_{n-1} + \cdots + a_{n-1} a_0$ becomes

$$\left(a_0 a_n + \frac{1}{n} a_1 a_{n-1}\right) + \left(\frac{n-1}{n} a_1 a_{n-1} + \frac{2}{n} a_2 a_{n-2}\right) + \cdots + \left(\frac{1}{n} a_{n-1} a_1 + a_n a_0\right). \tag{7}$$

After collecting terms, this simplifies to $a_0 a_n + \cdots + a_n a_0$. $\quad \blacksquare$

3. A GEOMETRIC CONSTRUCTION. We divide the positive quadrant of the xy-plane into rectangles by drawing the straight lines $x = s_n$ and $y = s_n$ for all n. Let $R_{i,j}$ be the rectangle with lower left corner (s_i, s_j) and upper right corner (s_{i+1}, s_{j+1}). The area of $R_{i,j}$ is $a_i a_j$. Therefore the identity (6) states that the total area of the rectangles $R_{i,j}$ for which $i + j = n$ is 1. We let P_n be the polygonal region consisting of all

An Elementary Proof of the Wallis Product Formula for Pi

rectangles $R_{i,j}$ for which $i + j < n$. Hence the area of P_n is n (see Figure 1).

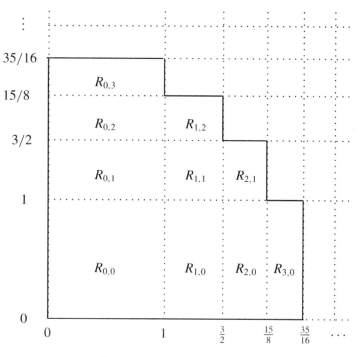

Figure 1. The region P_4 of area 4.

The outer corners of P_n are the points (s_i, s_j) for which $i + j = n + 1$ and $1 \leq i, j \leq n$. By the Pythagorean theorem, the distance of such a point to the origin is

$$\sqrt{s_i^2 + s_j^2}.$$

By (4), this is bounded from above by

$$\sqrt{\frac{2(i+j)}{o_n}} = \sqrt{\frac{2(n+1)}{o_n}}.$$

Similarly, the inner corners of P_n are the points (s_i, s_j) for which $i + j = n$ and $0 \leq i, j \leq n$. The distance of such a point to the origin is bounded from below by

$$\sqrt{\frac{2(i+j-1)}{e_n}} = \sqrt{\frac{2(n-1)}{e_n}}.$$

Therefore P_n contains a quarter circle of radius $\sqrt{2(n-1)/e_n}$, and is contained in a quarter circle of radius $\sqrt{2(n+1)/o_n}$. Since the area of a quarter circle of radius r is equal to $\pi r^2/4$ while the area of P_n is n, this leads to the bounds

$$\frac{(n-1)\pi}{2e_n} < n < \frac{(n+1)\pi}{2o_n},$$

from which it follows that

$$\frac{(n-1)\pi}{2n} < e_n < o_n < \frac{(n+1)\pi}{2n}.$$

It is now clear that as $n \to \infty$, e_n and o_n both approach $\pi/2$.

REFERENCES

1. V. Brun, Wallis's og Brounckers formler for π (in Norwegian), *Norsk matematisk tidskrift* **33** (1951) 73-81.
2. A. M. Yaglom and I. M. Yaglom, An elementary derivation of the formulas of Wallis, Leibnitz and Eule for the number π (in Russian), *Uspechi matematiceskich nauk.* (N. S.) **57** (1953) 181–187.

Department of Mathematics, Chalmers University of Technology, S-412 96 Göteborg, Sweden
wastlund@gmail.com

Originally appeared as:
Wästlund, Johan. "An Elementary Proof of the Wallis Product Formula for Pi." *American Mathematical Monthly*. vol. 114, no. 10 (December 2007): pp. 914–917.

The Orchard Problem

Ross Honsberger

In this essay we consider an intriguing problem and the beautiful mathematics involved in its solution. We note that the points in a coordinate plane which have both coordinates integers are called lattice points.

The Orchard Problem: A tree is planted at each lattice point in a circular orchard which has center at the origin and radius 50. (All trees are taken to be exact vertical cylinders of the same radius.) If the radius of the trees exceeds $1/50$ of a unit, show that from the origin one is unable to see out of the orchard no matter in what direction he looks; show, however that if the trees are shrunk to a radius less than $1/\sqrt{2501}$, one can see out if he looks in the right direction.

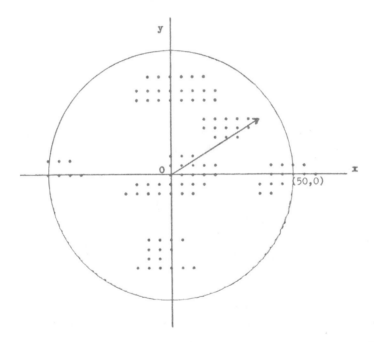

The second part of the problem is quite easy. The solution of the first part turns on an interesting theorem of Herman Minkowski (1864–1909), a close friend of Hilbert. In order to prove this theorem, we establish first a result called Blichfeldt's Lemma, after an American Mathematician.

Blichfeldt's Lemma. *Suppose we are given a plane region R which has an area in excess of n units, n a positive integer. Then, no matter where R occurs in the plane, it is always possible to translate it (i.e., to slide it without turning it) to a position where it covers at least n + 1 lattice points.*

133

For example, if R has an area of $8\frac{1}{4}$, then it can be translated to cover at least 9 lattice points.

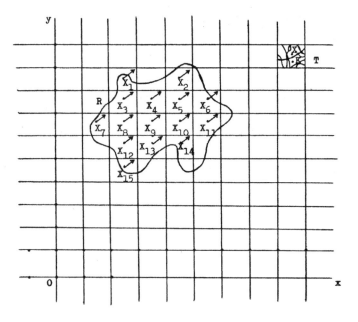

Proof. The lines whose equations are $x = a$ and $y = b$, a and b integers, we shall call "lattice lines" since their intersections are the lattice points. Let us cut the plane along each lattice line, thus chopping it up into squares of unit size. This also cuts our region R into pieces. Let us suppose that the region R is painted red and that the rest of the plane is unpainted. Some of the unit squares, then, may be all red, some partly red, and others completely unpainted.

Let us take all the squares which have any red at all on them and pile them up, without turning them, one on top of the other, in some distant square T. Consider now any point K of the base T. On top of T is piled layer after layer, each one covering the point K with some point of the layer. Sometimes K may occur under a red point in the layer, sometimes under an unpainted point. Among the vertical column of points that pile up on top of K we are interested in how many are red.

It is our claim that some point of the base T must get covered with a red point by at least $n + 1$ different layers. (Recall that the integer n enters our considerations through the fact that the area of R exceeds n units.) To prove this, consider the contrary. Suppose that no point of T lies under more than n red points in its column. Some points may occur at the bottom of a column containing exactly n red points, some fewer, but none have more than n red points.

Now let us calculate the amount of red paint in the pile. The area of the base T is one square unit, and even if every point of T had the maximum number, n, of red points in the column above it, this would provide only enough red paint to give each square n coats. (Each point of T could be given n coats with the n red points above it.) Thus, at most there is only enough paint in the pile to cover n units of area. But all of R is present in the pile, and it has an area greater than n units. Thus there must be enough paint in the pile to cover more than n units, a contradiction. Consequently, some point X of T is covered at least $n + 1$ times with points of R.

Now drive a needle straight down through all the layers above the point X. This marks a point in each layer, and, because of the above, it must occur in the red part of at least $n + 1$ layers. Let us denote these red points $X_1, X_2, \ldots, X_m$. Here m is at least $n + 1$. Finally, return all the squares to their original places in the plane, thus reconstructing R.

Now each point X_i occurs in its square in the same relative position. Consequently, any translation of R which carries one X_i to cover a lattice point will also carry every other X_i to cover a lattice point. But each X_i is red, and is therefore a point of R. And there are at least $n + 1$ of them. Such a translation, then, carries R to cover at least $n + 1$ lattice points.

Blichfeldt's Lemma is very important in our proof of Minkowski's Theorem. However, it does not figure directly. Rather, we use an easy corollary which arises in the case $n = 1$.

Corollary. *If R is a plane region with area exceeding 1, then some pair of distinct points A and B of R have a run and a rise which are both integers.* (*The run between the points (x_1, y_1) and (x_2, y_2) is $x_2 - x_1$ and the rise is $y_2 - y_1$.*)

Proof. Notice that there is no claim that the points A and B, themselves, are lattice points. The corollary holds whether or not R covers any lattice points at all.

We know by Blichfeldt's Lemma that R can be translated so that (at least) $n + 1 = 2$ of its points, say A and B are carried onto lattice points A_1 and B_1. Since the coordinates of lattice points are integers, the run and rise between any two of them are integers. However, a translation, changing only the position and not the direction of a line, does not change runs or rises. The run and rise between A and B before they are moved by the translation are exactly the same as the integral run and the integral rise between the lattice points A_1 and B_1 onto which they are carried.

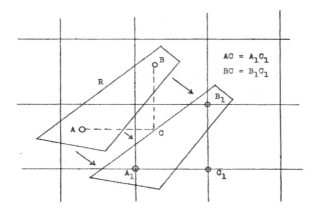

Now we are ready to prove an intuitively obvious, but not logically evident, theorem of Minkowski.

Minkowski's Theorem. *A plane, convex region with area exceeding 4, which is symmetric about the origin, covers a lattice point besides the origin.*

(A convex set contains the entire segment AB for every pair of points A and B that it contains.)

Proof. If a region is symmetric about the origin, then for each point P that it contains, it contains also the point P' which is obtained from P by reflection in the origin, that is, by sending it through the origin the same distance on the other side. If P has coordinates (x, y), then P' has coordinates $(-x, -y)$. We observe, then, that if the region covers one additional lattice point besides the origin, it must cover also a second lattice point which is symmetrical with the origin.

Let R denote the given region, and let us shrink R towards the origin O until it is just half as large in every direction; that is, draw every point of R towards the origin until its

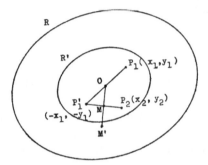

new position, while remaining in line with the old, is just half as far from the origin. This transformation is called a *dilatation* with center O and ratio $1/2$. Suppose R is taken into the region R' by this dilatation. Now dilatations carry lines into parallel lines (in the diagram, let A', B' denote the images of A, B; then $A'B'$ is parallel to AB, and the points C of the segment AB go into the points C' of $A'B'$). As a result, a dilatation does not change the size of the angles in a figure, and therefore it does not alter the shape of a figure. Consequently, R' has the same shape as R, only it is smaller. That is to say, R' also is a plane, convex region which is symmetric about the origin. It is symmetric about the origin O because O is the center of symmetry of R and also the center of the dilatation.

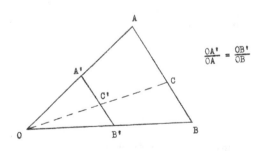

But what is the area of R'? Its width in every direction is just half what it used to be. If R were a rectangle, then R' would be a rectangle with half the length and half the width. In all cases, however, reducing the linear dimensions of a plane figure in the ratio $1:2$ reduces the area in the ratio $(1:2)^2 = 1:4$. Since R began with an area exceeding 4, we conclude that R' still has an area exceeding 1. Thus we may apply the Corollary to Blichfeldt's Lemma.

Accordingly, R' contains two points $P_1(x_1, y_1)$ and $P_2(x_2, y_2)$ whose run and rise, $x_2 - x_1$ and $y_2 - y_1$, are both integers. Because the origin O is the center of R', the point $P_1'(-x_1, -y_1)$ which is symmetric to P_1 in the origin must also lie in R'. Thus P_1 and P_2 are two points which belong to R'.

Now we use the fact that R' is convex. By definition, this assures us that every point of the segment $P_1' P_2$ belongs to R'. In particular, its midpoint $M((x_2 - x_1)/2, (y_2 - y_1)/2)$ occurs in R'.

Now let us reverse our dilatation by subjecting R' to the dilatation with center O and ratio $2:1$. This stretches R' back to its original state, namely R. Under this dilatation, every point of R' is moved to a point twice as far from the origin. The point M, then, is carried into the point $M'(x_2 - x_1, y_2 - y_1)$. Thus M' is a point of R. But, by Blichfeldt's Corollary, M' is a lattice point. And it is not the origin, itself, for that would mean that P_1 and P_2 coincide, in contradiction to Blichfeldt's Corollary, which asserts that they are different points.

Now let us attack the first part of the orchard problem. We wish to show that if the radius r of the trees exceeds $\frac{1}{50}$ of a unit, then there is no way to see out of the orchard from the origin. Let AOB denote an arbitrary diameter of the orchard. Suppose the radius of the trees is $r = \frac{1}{50} + q$. We observe that if r is greater than $1/2$, the trees will be growing into each other. Thus, while q is positive, it is not a very large number. Now let p denote any number which is bigger than $1/50$ but less than r, say $\frac{1}{50} + \frac{1}{2}q$. At A and B on the boundary of the orchard construct tangents, and cut off along these tangents in both directions points C, D, E, F at a distance p from A and B.

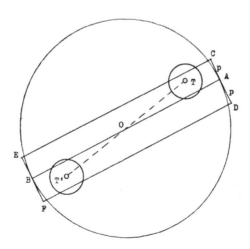

Thus a rectangle $EFDC$ is determined which has the origin O as center. (The diagram greatly exaggerates the size of this rectangle.) The length of the rectangle is $FD = AB = 100$, and the width is $2p$, giving an area of $200p$. Because p exceeds $1/50$, this area exceeds 4. And *EFDC* is a plane, convex region which is symmetric about the origin. By Minkowski's theorem, then, our rectangle contains some lattice point T other than the origin. The tree which is planted at T has radius r, which exceeds $p = CA$. Thus the tree at T extends far enough to cross the line *OA*, blocking one's view in this direction. By symmetry, the rectangle contains also the symmetric lattice point T', at which is planted a tree that blocks the view along *OB*. Thus it would appear that we may conclude that one cannot see out from the origin. However, there is a difficulty in our argument which must be overcome (do you see what it is?).

A very small part of the rectangle sticks out of the orchard at each corner. If the lattice point T happens to occur in one of these parts of the rectangle, then there is no tree planted there to block the view. We need to show that T does not occur in a part of *EFDC* which lies outside the orchard. We proceed indirectly. For any point in the rectangle the maximum distance from the origin is the half-diagonal $OC = \sqrt{50^2 + p^2}$. Since $p < 1$, we have that $OT \leq OC < \sqrt{2501}$. However, for T outside the orchard we have $OT > 50$. Hence
$$2500 < OT^2 < 2501.$$
If T is the lattice point (x, y), then $OT^2 = x^2 + y^2$, where x and y are integers, implying that OT^2 is an integer. But there is no integer between 2500 and 2501. Thus T cannot occur outside the orchard.

We complete the solution by showing that if the radius is diminished to anything less than $1/\sqrt{2501}$, one can see out of the orchard along the line joining the origin to the point $N(50, 1)$. The length of the segment *ON* is $\sqrt{2501}$. Since the lattice points occur in rows and columns, it is easy to see that the lattice point in the orchard which is nearest to the line *ON* is the point $M(1, 0)$, or the equally close $K(49, 1)$. Let L denote the point $(50, 0)$.

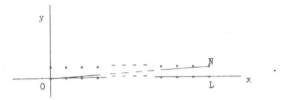

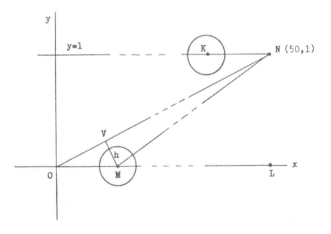

Now the area $\triangle OMN$ can be obtained in two ways. First, it is given by $\frac{1}{2} \cdot OM \cdot LN$, which yields $\frac{1}{2} \cdot 1 \cdot 1 = \frac{1}{2}$. Using ON as base and the altitude $h = MV$, we also get $\frac{1}{2} \cdot ON \cdot h$, which yields $\frac{1}{2} h \sqrt{2501}$. Consequently, we have

$$\frac{1}{2} h \sqrt{2501} = \frac{1}{2}, \quad \text{giving } h = \frac{1}{\sqrt{2501}}.$$

Thus h exceeds the radius of the trees. As a result, the tree at M is not big enough to intersect the line ON; similarly, neither is the tree at K. Since the closest trees do not cross ON, then no tree blocks the view in this direction.

Exercises

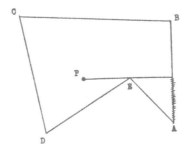

1. The figure shows a point P inside a polygon from which part of the side AB and all of AE are not visible. Construct a polygon G with a point P inside it such that no side of G is completely visible from P. Construct a polygon H with a point P outside it such that no side of H is completely visible from P.

2. Given a circle, center O, and a point P outside it, construct a straight line PQR (whenever possible), Q and R on the circle, with Q the midpoint of PR. (Hint: Consider the dilatation with center P and ratio 1/2.)

3. In a square lattice, show that (a) no matter what three lattice points are joined, an equilateral triangle never results, (b) no matter what five lattice points are joined, a regular pentagon never results.

4. Show that a convex region of area 1 can be covered by some parallelogram of area not greater than 2.

5. A circle C_0 of radius $R_0 = 1$ km. is tangent to a line L at Z. A circle C_1 of radius $R_1 = 1$ mm. is drawn tangent to C_0 and L, on the right-hand side of C_0. A family of circles C_i is constructed outwardly to the right side so that each C_i is tangent to C_0, L, and to the previous circle C_{i-1}. Eventually the members become so big that it is impossible to enlarge the family further. How many circles can be drawn before this happens? (See Figure 9.)

6. Look up pages 43 and 44 in Hilbert's *Geometry and the Imagination* for an application of Minkowski's theorem to the problem of approximating real numbers by sequences of rational numbers.

7. L. G. Schnirelman (Russian: 1905–1935) proved that some four points on a closed convex curve are the vertices of a square. Use this result to show that every convex curve of perimeter less than four can be positioned on a coordinate plane so as to cover no lattice points at all.

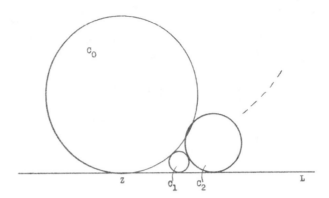

References

1. Yaglom and Yaglom, *"Challenging Mathematical Problems with Elementary Solutions,"* vol. II, Holden-Day, San Francisco, 1967.

Originally appeared as:
Honsberger, Ross. "The Orchard Problem." *Mathematical Gems*. Chapter 4, Mathematical Association of America, 1973. pp. 43–53.

Part IV: Sums of Squares and Polygonal Numbers

This chapter contains five Biscuits about sums of squares and of polygonal numbers. The Pythagoreans were the first to be intrigued by the triangular numbers $n(n + 1)/2$, the squares n^2, and polygonal numbers in general. Since that time, sums of such numbers have fascinated such giants as Fermat, Euler, Lagrange, Legendre, Gauss, and Cauchy. We have included Don Zagier's astonishing one-sentence proof of Fermat's Two Squares Theorem, Proofs Without Words of theorems about sums of squares from Martin Gardner, Dan Kalman and Roger Nelsen, Melvyn Nathanson's short elementary proof of Cauchy's Polygonal Number Theorem, and A. Hall's remarkable genealogical chart containing all primitive Pythagorean triples.

A common first reaction to "A one-sentence proof that every prime $p \equiv 1 \pmod 4$ is a sum of two squares" (*Amer. Math. Monthly*, vol. 97, no. 2 (February 1990), p. 144) is very likely "Wow!" It is a marvel of elegance coupled with economy. The proof is a refinement of an idea of Liouville. For a prime $p \equiv 1 \pmod 4$, Zagier defines a mapping f on a set S of triples of natural numbers mod p, states that f is an involution with exactly one fixed point, deduces that another mapping is an involution, and concludes that p is a sum of two squares! He suggests that the reader verify all the claims in the paper, including the combinatorial theorem that if f is an involution on a finite set S, then S and the set of fixed points of f have the same parity. A straightforward and enlightening task is to go through those verifications. A good exercise or project for a number theory class might be to understand and explain the geometry of the set of triples in the paper. This introduction is already longer than Zagier's paper!

Terence Jackson gives us a tasty follow-up to Zagier's work, not in our collection but well worth reading, with a similar proof that every prime $p \equiv 3 \pmod 8$ is of the form $x^2 + 2y^2$ (*Amer. Math. Monthly*, vol. 107, no. 5 (May 2000), p. 448). Jackson's proof is also based on the parity of the number of fixed points of a cleverly-constructed involution.

Sums of squares are naturals for Proofs Without Words. In our next two Biscuits, Martin Gardner and Dan Kalman (Sums of squares II, in *Proofs Without Words: Exercises in Visual Thinking*, p. 78; also in *Scientific American*, vol. 229, no. 4, (1973), p.115 (Gardner) and *College Math. Journal* vol. 22 (1991), p.124 (Kalman)) and Roger Nelsen (Sums of squares VIII, in *Proofs Without Words II: More Exercises in Visual Thinking*, MAA, Classroom Resource Materials, p. 88) have shown us examples. Read and enjoy!

You all know about Archimedes' famous "Eureka!" moment. (Yes, you do: remember the bathtub?) Gauss also had such a moment at the tender age of nineteen, when he wrote in his diary these words: EYPHKA! num = $\Delta + \Delta + \Delta$. What excited him was that he had proved Fermat's conjecture that every positive integer can be written as a sum of at most three triangular numbers. This is part of a more general conjecture, proved by Cauchy in

1813, that every number is a sum of three triangular numbers, four squares, five pentagonal numbers, and so on. In our next Biscuit, Melvyn B. Nathanson describes a completely elementary proof of this general theorem in "A short proof of Cauchy's polygonal number theorem" (*Proceedings of the AMS*, vol. 99, no. 1 (January 1987), pp. 22–24). Again, working through this paper will pay handsome dividends.

A (primitive) Pythagorean triple is a triple of (relatively prime) positive integers (a, b, c) such that $a^2 + b^2 = c^2$. Methods of generating such triples were known in Euclid's time, and in his 1970 paper from the *Mathematical Gazette*, A. Hall shows that all such triples are related. He does this by presenting a truly remarkable "family tree" containing all primitive Pythagorean triples together with rules for producing, at any point in the tree, the "next generation" of triples. Along the way, Hall points out a variety of interesting details about the tree—and the ancestor of all triples is, of course, (3, 4, 5).

A One-Sentence Proof that Every Prime $p \equiv 1 \pmod{4}$ is a Sum of Two Squares

D. Zagier

The involution on the finite set $S = \{(x,y,z) \in \mathbb{N}^3 : x^2 + 4yz = p\}$ defined by

$$(x,y,z) \mapsto \begin{cases} (x+2z,\, z,\, y-x-z) & \text{if } x < y-z \\ (2y-x,\, y,\, x-y+z) & \text{if } y-z < x < 2y \\ (x-2y,\, x-y+z,\, y) & \text{if } x > 2y \end{cases}$$

has exactly one fixed point, so $|S|$ is odd and the involution defined by $(x,y,z) \mapsto (x,z,y)$ also has a fixed point. □

This proof is a simplification of one due to Heath-Brown [1] (inspired, in turn, by a proof given by Liouville). The verifications of the implicitly made assertions—that S is finite and that the map is well-defined and involutory (i.e., equal to its own inverse) and has exactly one fixed point—are immediate and have been left to the reader. Only the last requires that p be a prime of the form $4k+1$, the fixed point then being $(1,1,k)$.

Note that the proof is not constructive: it does not give a method to actually find the representation of p as a sum of two squares. A similar phenomenon occurs with results in topology and analysis that are proved using fixed-point theorems. Indeed, the basic principle we used: "The cardinalities of a finite set and of its fixed-point set under any involution have the same parity," is a combinatorial analogue and special case of the corresponding topological result: "The Euler characteristics of a topological space and of its fixed-point set under any continuous involution have the same parity."

For a discussion of constructive proofs of the two-squares theorem, see the Editor's Corner elsewhere in this issue.

REFERENCE

1. D. R. Heath-Brown, Fermat's two-squares theorem, *Invariant* (1984) 3–5.

Originally appeared as:
Zagier, D. "A One-Sentence Proof that every Prime $p \equiv 1$ (mod 4) is a Sum of Two Squares." *American Mathematical Monthly.* vol. 97, no. 2 (February 1990): p. 144.

Sums of Squares II

*Martin Gardner &
Dan Kalman*

$$3(1^2 + 2^2 + \cdots + n^2) = (2n + 1)(1 + 2 + \cdots + n)$$

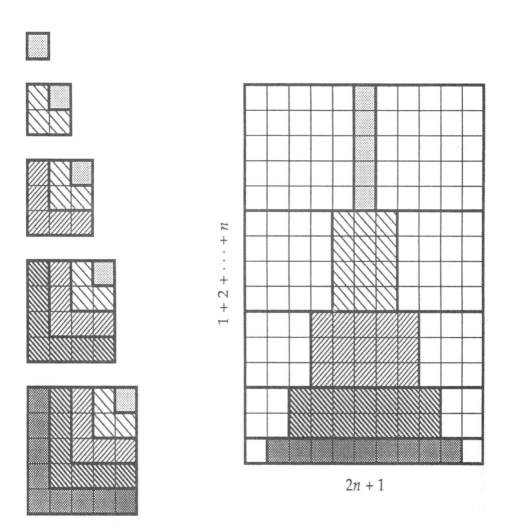

Originally appeared as:
Gardner, Martin and Dan Kalman. "Sum of Squares II." *Proof Without Words: Exercises in Visual Thinking*. Ed. Roger B. Nelsen, Mathematical Association of America, 1993. p. 78.

Sums of Squares VIII

Roger B. Nelsen

$$k^2 = 1 + 3 + \cdots + (2k-1) \Rightarrow \sum_{k=1}^{n} k^2 = \frac{n(n+1)(2n+1)}{6}$$

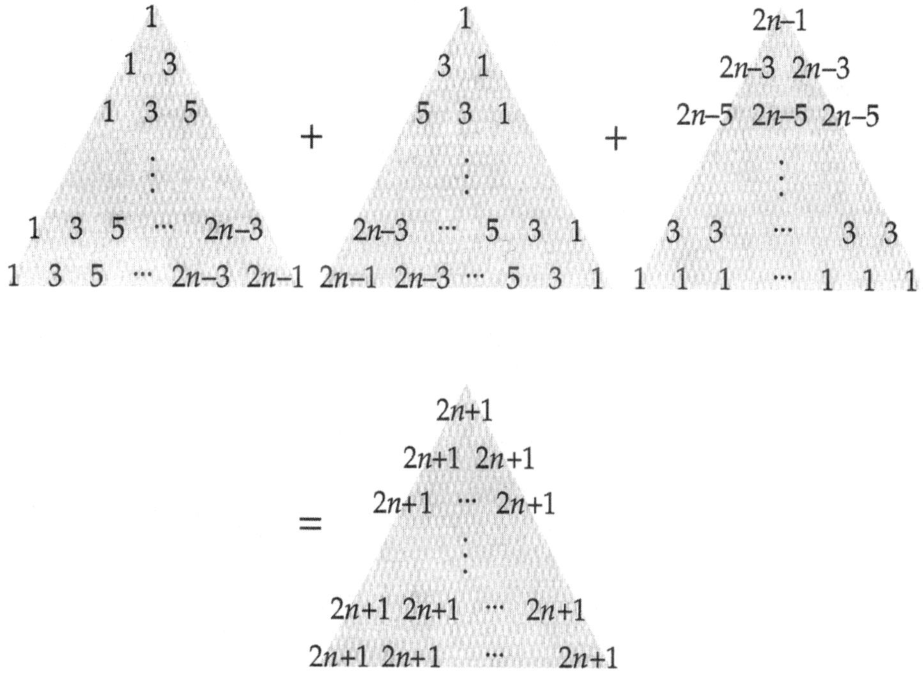
1
1 3
1 3 5
1 3 5 ⋯ 2n−3
1 3 5 ⋯ 2n−3 2n−1
+
1
3 1
5 3 1
2n−3 ⋯ 5 3 1
2n−1 2n−3 ⋯ 5 3 1
+
2n−1
2n−3 2n−3
2n−5 2n−5 2n−5
3 3 ⋯ 3 3
1 1 1 ⋯ 1 1 1
=
2n+1
2n+1 2n+1
2n+1 ⋯ 2n+1
2n+1 2n+1 ⋯ 2n+1
2n+1 2n+1 ⋯ 2n+1

$$3\left(1^2 + 2^2 + \cdots + n^2\right) = (2n+1)(1 + 2 + \cdots + n)$$

$$\therefore 1^2 + 2^2 + \cdots + n^2 = \frac{2n+1}{3} \cdot \frac{n(n+1)}{2}$$

Originally appeared as:

Nelsen, Roger B. "Sum of Squares VIII." *Proof Without Words II: Exercises in Visual Thinking*. Ed. Roger B. Nelsen, Mathematical Association of America, 2000. p. 88.

A Short Proof of Cauchy's Polygonal Number Theorem

Melvyn B. Nathanson

Abstract. This paper presents a simple proof that every nonnegative integer is the sum of $m + 2$ polygonal numbers of order $m + 2$.

Let $m \geq 1$. The polygonal numbers of order $m + 2$ are the integers

$$p_m(k) = \frac{m}{2}\left(k^2 - k\right) + k$$

for $k = 0, 1, 2, \ldots$. Fermat [3] asserted that every nonnegative integer is the sum of $m + 2$ polygonal numbers of order $m + 2$. For $m = 2$, Lagrange [5] proved that every nonnegative integer is the sum of four squares $p_2(k) = k^2$. For $m = 1$, Gauss [4] proved that every nonnegative integer is the sum of three triangular numbers $p_1(k) = (k^2 + k)/2$, or equivalently, that every positive integer $n \equiv 3 \pmod{8}$ is the sum of three odd squares. Cauchy [1] proved Fermat's statement for all $m \geq 3$, and Legendre [6] refined and extended this result. For $m \geq 3$ and $n \leq 120m$, Pepin [8] published tables of explicit representations of n as a sum of $m + 2$ polygonal numbers of order $m + 2$, at most four of which are different from 0 or 1. Dickson [2] prepared similar tables. Pall [7] obtained important related results on sums of values of a quadratic polynomial.

Uspensky and Heaslet [9, p. 380] and Weil [10, p. 102] have written that there is no short and easy proof of Cauchy's polygonal number theorem. The object of this note is to present a short and easy proof.

Because of Pepin's and Dickson's tables, it suffices to consider only $n \geq 120m$. For completeness, I also include a proof of Cauchy's lemma.

Cauchy's Lemma. *Let a and b be odd positive integers such that $b^2 < 4a$ and $3a < b^2 + 2b + 4$. Then there exist nonnegative integers s, t, u, v such that*

(1) $$a = s^2 + t^2 + u^2 + v^2,$$

(2) $$b = s + t + u + v.$$

Proof. Since a and b are odd, it follows that $4a - b^2 \equiv 3 \pmod{8}$, and so, by Gauss's triangular number theorem, there exist odd integers $x \geq y \geq z > 0$ such that

(3) $$4a - b^2 = x^2 + y^2 + z^2.$$

Choose the sign of $\pm z$ so that $b + x + y \pm z \equiv 0 \pmod 4$. Define integers s, t, u, v by

$$s = \frac{b+x+y\pm z}{4}, \quad t = \frac{b+x}{2} - s = \frac{b+x-y\mp z}{4}$$

$$u = \frac{b+y}{2} - s = \frac{b-x+y\mp z}{4}, \quad v = \frac{b\pm z}{2} - s = \frac{b-x-y\pm z}{4}.$$

149

Then equations (1) and (2) are satisfied, and $s \geq t \geq u \geq v$. To show these integers are nonnegative, it suffices to prove that $v \geq 0$, or $v > -1$. This is true if $b - x - y - z > -4$, or, equivalently, if $x + y + z < b + 4$. The maximum value of $x + y + z$ subject to the constraint (3) is $\sqrt{12a - 3b^2}$, and the inequality $3a < b^2 + 2b + 4$ implies that $x + y + z \leq \sqrt{12a - 3b^2} < b + 4$. This proves the lemma.

Theorem 1. *Let $m \geq 3$ and $n \geq 120m$. Then n is the sum of $m + 1$ polygonal numbers of order $m + 2$, at most four of which are different from 0 or 1.*

Proof. Let b_1 and b_2 be consecutive odd integers. The set of numbers of the form $b + r$, where $b \in \{b_1, b_2\}$ and $r \in \{0, 1, \ldots, m-3\}$, contains a complete set of residue classes modulo m, and so $n \equiv b + r \pmod{m}$ for some $b \in \{b_1, b_2\}$ and $r \in \{0, 1, \ldots, m-3\}$. Define

(4) $$a = 2\left(\frac{n-b-r}{m}\right) + b = \left(1 - \frac{2}{m}\right)b + 2\left(\frac{n-r}{m}\right).$$

Then a is an odd integer, and

(5) $$n = \frac{m}{2}(a-b) + b + r.$$

If $0 < b < \frac{2}{3} + \sqrt{8(n/m) - 8}$, then the quadratic formula implies that

$$b^2 - 4a = b^2 - 4\left(1 - \frac{2}{m}\right)b - 8\left(\frac{n-r}{m}\right) < 0$$

and so $b^2 < 4a$. Similarly, if $b > \frac{1}{2} + \sqrt{6(n/m) - 3}$, then $3a < b^2 + 2b + 4$. Since the length of the interval

(6) $$I = \left(\frac{1}{2} + \sqrt{6\left(\frac{n}{m}\right) - 3}, \frac{2}{3} + \sqrt{8\left(\frac{n}{m}\right) - 8}\right)$$

is greater than 4, it follows that I contains two consecutive odd positive integers b_1 and b_2. Thus, there exist odd positive integers a and b that satisfy (5) and the inequalities $b^2 < 4a$ and $3a < b^2 + 2b + 4$. Cauchy's Lemma implies that there exist s, t, u, v satisfying (1) and (2), and so

$$n = \frac{m}{2}(a-b) + b + r = \frac{m}{r}(s^2 - s) + s + \cdots + \frac{m}{2}(v^2 - v) + v + r$$
$$= p_m(s) + p_m(t) + p_m(u) + p_m(v) + r.$$

This completes the proof.

Note that this result is slightly stronger than Cauchy's theorem. Legendre [6] proved that every sufficiently large integer is the sum of five polygonal numbers of order $m + 2$, one of which is either 0 or 1. This can also be easily proved.

Theorem 2. *Let $m \geq 3$. If m is odd, then every sufficiently large integer is the sum of four polygonal numbers of order $m + 2$. If m is even, then every sufficiently large integer is the sum of five polygonal numbers of order $m + 2$, one of which is either 0 or 1.*

Proof. There is an absolute constant c such that if $n > cm^3$, then the length of the interval I defined in (6) is greater than $2m$, and so I contains at least m consecutive odd integers.

If m is odd, these form a complete set of residues modulo m, and so $n \equiv b \pmod{m}$ for some odd number $b \in I$. Let $r = 0$. Define a by formula (4).

If m is even and $n > cm^3$, then $n \equiv b + r \pmod{m}$ for some odd integer $b \in I$ and $r \in \{0, 1\}$. Define a by (4).

In both cases, the theorem follows immediately from Cauchy's Lemma.

References

1. A. Cauchy, "Démonstration du théorème général de Fermat sur les nombres polygones," *Mém. Sci. Math. Phys. Inst. France* (1) 14 (1813–15), 177–220 = Oeuvres (2), vol. 6, 320–353.

2. L. E. Dickson, "All positive integers are sums of values of a quadratic function of x," *Bull. Amer. Math. Soc.* 33 (1927), 713–720.

3. P. Fermat, quoted in T. L. Heath, *Diophantus of Alexandria*, Dover New York, 1964, p. 188.

4. C. F. Gauss, *Disquisitiones Arithmeticae*, Yale Univ. Press, New Haven, Conn., and London, 1966.

5. J. L. Lagrange, "Démonstration d'un théorème d'arithmetique," *Nouveaux Memoires de L'Acad. Royale des Sci. et Belles-L. de Berlin*, 1770, pp. 123–133 = Oeuvres, vol. 3, pp. 189–201.

6. A. M. Legendre, *Théorie des Nombres*, 3rd ed., vol. 2, 1830, pp. 331–356.

7. G. Pall, "Large positive integers are sums of four or five values of a quadratic function," *Amer. J. Math.* 54 (1932), 66–78.

8. T. Pepin, "Démonstration du théorème de Fermat sur les nombres polygones," *Atti Accad. Pont. Nuovi Lincei* 46 (1892–93), 119–131.

9. J. V Uspensky and M. A. Heaslet, *Elementary Number Theory*, McGraw-Hill, New York and London, 1939.

10. A. Weil, Number Theory, *An Approach Through History from Hammurabi to Legendre*, Birkhauser, Boston, Mass., 1983.

Originally appeared as:
Nathanson, Melvyn B. "A Short Proof of Cauchy's Polygonal Number Theorem." *Proceedings of the AMS*. vol. 99, no. 1 (January 1987), pp. 22–24.

Genealogy of Pythagorean Triads

A. Hall

The infinite set of primitive Pythagorean triads may be shown to form a symmetrically organised family, in which each triad has three offspring, the progenitor being (3, 4, 5). The first three generations are shown below.

The offspring of any triad (x, y, z) are, in the order shown,

$$\begin{pmatrix} 1 & -2 & 2 \\ 2 & -1 & 2 \\ 2 & -2 & 3 \end{pmatrix} \begin{pmatrix} x \\ y \\ z \end{pmatrix}, \quad \begin{pmatrix} 1 & 2 & 2 \\ 2 & 1 & 2 \\ 2 & 2 & 3 \end{pmatrix} \begin{pmatrix} x \\ y \\ z \end{pmatrix} \quad \text{and} \quad \begin{pmatrix} -1 & 2 & 2 \\ -2 & 1 & 2 \\ -2 & 2 & 3 \end{pmatrix} \begin{pmatrix} x \\ y \\ z \end{pmatrix}.$$

If $x = m^2 - n^2$, $y = 2mn$, $z = m^2 + n^2$, the triad thus derived from (m, n) gives rise to triads derived from $(2m - n, m)$, $(2m + n, m)$ and $(m + 2n, n)$.

If m and n are co-prime and of unlike parity, then these pairs are also co-prime and of unlike parity. This ensures that every triad in the family is a primitive triad.

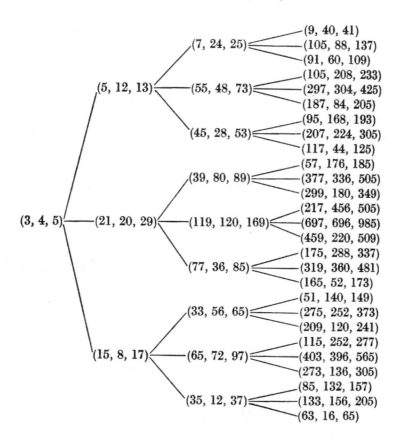

It remains to be shown that if M and N are co-prime and of unlike parity, with $M > N$, then the triad derived from (M, N) has its place in the family. There are three possibilities:

(a) $N < M < 2N$ Let $m = N$, $n = 2N - M$;
then $M = 2m - n$, $N = m$.

(b) $2N < M < 3N$ Let $m = N$, $n = M - 2N$;
then $M = 2m + n$, $N = m$.

(c) $3N < M$ Let $m = M - 2N$, $n = N$;
then $M = m + 2n$, $N = n$.

In every case, $m > n$, m and n are co-prime and of unlike parity, and $m + n < M + N$. In every case, M and N fit the formulae for offspring of (m, n). This process may be repeated indefinitely, with the sum of m and n diminished each time, leading back inevitably to $(2, 1)$.

The family tree is therefore a complete representation of the infinite set of primitive Pythagorean triads. It is interesting to trace certain important lines of descent.

The central line contains all triads in which x and y are consecutive integers. These were discussed in Classroom Note 147.

The extreme top line contains all triads in which y and z are consecutive integers, while the extreme lower line contains all in which x and z are consecutive odd numbers.

Further sequences may be traced by alternating branches up, across and down (U, A and D):

$UAUA$ ... and $AUAU$ ... contain triads for which m and n are in the Fibonacci sequence 1, 1, 2, 3, 5, 8, 13, ... In these triads, x and z are alternately members of the same sequence. The smaller acute angle tends alternately towards $\tan^{-1} \frac{1}{2}$ and $\sin^{-1} \frac{2}{3}$.

$DUDU$ ... gives the other sequence of triads mentioned in Classroom Note 147, representing triangles in which the hypotenuse and twice the shortest side are consecutive integers, thus tending to form half an equilateral triangle. $UDUD$ ... tends to produce the same type of triangle.

$DADA$ ... and $ADAD$ ... tend to produce triangles similar to those produced by $UAUA$ ... and $AUAU$ ..., and this symmetrical tendency in lines of descent seems to apply generally.

The difference between x and y in any triad is equal to either the difference or the sum of x and y in the "parent triad." For this reason, the difference between the two shorter sides of any primitive Pythagorean triangle must be either 1 (the difference of 3 and 4) or the sum of the shorter sides in another primitive Pythagorean triangle. The possible differences are therefore: 1, 7, 17, 23, 31, 41, 47, 49, ... It was the investigation of these, possible differences, suggested by Mr. P. I. Wyndham, which led to the results in this note.

Originally appeared as:
Hall, A. "Genealogy of Pythagorean Triads." *Mathematical Gazette*. vol. 54, no. 390 (December 1970), pp. 377–379.

Part V: Fibonacci Numbers

One of the joys of studying number theory is investigating the properties of special numbers that possess almost magical qualities. Perhaps no numbers have been subjected to greater scrutiny, and deservedly so, than the Fibonacci numbers: 1, 1, 2, 3, 5, 8, 13, 21,.... Numerous books have been written about these beautiful numbers and there even exists a research journal, *The Fibonacci Quarterly*, that is devoted to exploring their special properties.

In the first article in this chapter, "A dozen questions about Fibonacci numbers" (*Math Horizons*, vol. 12, no. 3 (February 2005) pp. 5–9), James Tanton presents more than a dozen questions (somewhere between 13 and 21) whose answers are Fibonacci numbers. Many of the questions are combinatorial, but some involve probability, "Fibodecimals," divisibility patterns, and a simple derivation of Binet's formula that expresses Fibonacci numbers in terms of an irrational-looking formula involving $\sqrt{5}$ and the golden ratio.

In "The Fibonacci numbers—exposed" (*Math. Magazine*, vol. 76, no. 3 (June 2003), pp. 167–181), Dan Kalman and Robert Mena convincingly argue that the Fibonacci numbers are not all that special. Because they belong to a class of sequences, namely those satisfying a second order linear recurrence with constant coefficients, nearly all of the beautiful formulas and identities satisfied by the Fibonacci and Lucas numbers are special cases of equally beautiful formulas satisfied by all sequences in the class. In addition to the Fibonacci and Lucas numbers, this class contains all constant sequences, all arithmetic progressions, the Fermat sequence, the Mersenne sequence, the repunit numbers, and more. Using difference operators acting on the real vector space of real sequences, matrix algebra, and Binet-like formulas, they establish generalizations of Fibonacci formulas like

$$F_n = F_m F_{n-m+1} + F_{m-1} F_{n-m}, \quad \sum_{k=1}^{n} F_k = F_{n+2} - 1,$$

$$\sum_{k=1}^{n} F_k^2 = F_n F_{n+1}, \quad F_{n-1} F_{n+1} - F_n^2 = (-1)^n,$$

and $\gcd(F_n, F_m) = F_{\gcd(n, m)}$, to name a few.

In a companion article, "The Fibonacci numbers—exposed more discretely" (*Math. Magazine*, vol. 76, no. 3 (June 2003), pp. 182–192), Arthur Benjamin and Jennifer Quinn show how all of the generalized identities can also be obtained by elementary counting arguments. Specifically, Fibonacci numbers count the ways to tile strips of a given length with squares and dominoes. To obtain the generalized versions, one needs only to add a splash of color to these combinatorial arguments.

A Dozen Questions about Fibonacci Numbers

James Tanton

In his famous 1202 text *Liber Abaci* the great Leonardo of Pisa, a.k.a. Fibonacci, poses, and solves the following rabbit breeding problem: *Find the number of rabbit pairs present in any month, if, starting from a single pair, each pairs of rabbits born one month produces another pair of rabbits for each month after the next.* The count of rabbit pairs from month-to-month gives the famous Fibonacci sequence 1, 1, 2, 3, 5, 8, 13, 21, 34, 55, 89, ... If F_n denotes the nth Fibonacci number, then we have $F_n = F_{n-1} + F_{n-2}$ with $F_1 = F_2 = 1$. (Surprisingly it was 17th century Dutch mathematician Albert Girard who first wrote down this recurrence relation, not Fibonacci.)

Any problem whose "nth case" solution is the sum of the two previous case solutions gives rise to the Fibonacci (or at least a Fibonacci-like) sequence. For instance, the number of rabbit pairs in any month equals the number of rabbits present the previous month, plus the offspring of all the rabbit pairs that were present two months before. The number of ways to climb a set of stairs one or two steps at a time is a Fibonacci number as is seen by considering separately the possibilities of starting with a single step or with a double-step.

The Fibonacci numbers possess astounding mathematical properties and new results about them are still being discovered today.

Like many a scholar, I have become intrigued by the mathematical delights of the Fibonacci numbers. Here, for your amusement, is a collection of some of my favorite Fibonacci tidbits, some classic, some new. I hope you enjoy thinking about them as much as I did.

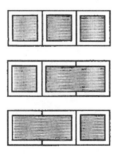

Figure 1. The three ways to tile a 1×3 strip using squares and dominoes.

Question 1: A Tiling Classic

Using only 1×1 square tiles and 1×2 dominoes one can tile a 1×3 strip of squares three different ways, as in Figure 1. In how many different ways can one tile a 1×15 strip of squares using only square tiles and dominoes? How many of these tilings have a single domino covering the 6th and 7th cells of the strip? How many don't?

Question 2: Honeycomb Walk

A bee, starting in cell one of the simple honeycomb design shown in Figure 2, wishes to stroll to cell number eleven via a path of connected cells. Each step of the journey must head to the right (that is, the bee can only move from one cell to a neighboring cell of a higher number). In how many different ways could the bee travel from cell one to cell eleven? In general, how many different paths are there from cell one to cell N?

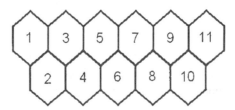

Figure 2. How many ways can a bee walk from cell one to cell n?

Question 3: Ordered Partitions

a) A *partition* of a number N is any collection of positive integers that sum to N. If the order of the terms in a sum is considered important, then we say that we have an *ordered partition* of N. For example, there are four ordered partitions of the number 3, namely, $1 + 1 + 1, 1 + 2, 2 + 1$, and 3 itself. How many ordered partitions are there for the number 5? Find a formula for the number of ordered partitions of a positive integer N.

b) Sally is particularly fond of the number one and likes to use either a blue crayon or a black crayon to write this number. (All other numbers she writes in black pen.) Sally noticed that with two different types of one there are now thirteen different ordered partitions of the number 3:

$$3, 21, 2\mathbf{1}, 12, \mathbf{1}2, 111, 11\mathbf{1}, 1\mathbf{1}1, \mathbf{1}11, 1\mathbf{11}, \mathbf{1}1\mathbf{1}, \mathbf{11}1, \mathbf{111}$$

Find the number of ordered partitions of the number 5 with two different types of one. How many such partitions are there of a number N?

Question 4: "ABABA"

The language of "ABABA" uses only two letters: A and B. No word in this language contains two consecutive Bs, but any string of letters avoiding two neighboring Bs is indeed a word. For example, "AABAAAAB" and "BAA" are both words, but "ABBA" is not. How many 10-lettered words are there in this language? How many N-lettered words does ABABA possess?

Question 5: "ABEEBA"

Another language, called "ABEEBA," uses three letters of the alphabet: A, B, and E. No word in this language contains an "A" immediately followed by an "E," but all other combinations of letters avoiding this situation do indeed form words in this language. For instance, "ABBBEAA" is a word, but "BBAEA" is not. Count how many 1-, 2-, and 3-lettered words this language possesses. What do you notice? How many N-lettered words are there in this language?

Question 6: Restricted Permutations

There are three ways to rearrange the letters ABC so that each letter, if it moves, shifts at most one place to the left or at most one place to the right. These permutations are: ABC, ACB, and BAC. (Notice, in "CAB" for instance, the letter C has moved two places to the left.) Let's call a rearrangement of letters satisfying this rule a "restricted permutation." How many restricted permutations are there of a string of N letters?

Question 7: Avoiding Heads

If I toss a coin ten times, what is the probably that I won't see two heads in a row?

Question 8: "Fibodecimals"

Consider the decimal 0.11235955056... obtained by adding the Fibonacci numbers expressed as the nonzero parts of smaller and smaller decimals:

$$
\begin{aligned}
& 0.1 \\
+ & 0.01 \\
+ & 0.002 \\
+ & 0.0003 \\
+ & 0.00005 \\
+ & 0.000008 \\
+ & 0.0000013 \\
+ & 0.00000021 \\
+ & 0.000000034 \\
+ & 0.0000000055 \\
+ & 0.00000000089 \\
+ & 0.000000000144 \\
+ & \ldots
\end{aligned}
$$

Notice that in calculating this infinite sum, one must "carry digits" an infinite number of times. Prove, that once all the digits have been carried, the resulting decimal is a repeating decimal.

Question 9: Divisibility Patterns?

Three is a factor of six, and $F_3 = 2$ is a factor of $F_6 = 8$. Seven is a factor of twenty-one, and $F_7 = 13$ is a factor of $F_{21} = 10{,}946$. Fourteen is a factor of fifty-six and, lo and behold, $F_{14} = 377$ is a factor of $F_{56} = 225{,}851{,}433{,}717$. Is this a coincidence, or is it always the case that if a divides b then F_a divides F_b?

Question 10: Foul ABABA

Any word in the language of ABABA (Question 4) that begins and ends with an A is actually a swear word in the language. For instance, "A," "AA," and "ABA" are all swear words, as is the name of the language itself. How many N-lettered swear words are there? How many of these N-letter swear words possess precisely k Bs? Use foul ABABA to prove:

$$\binom{N-1}{0} + \binom{N-2}{1} + \binom{N-3}{2} + \binom{N-4}{3} + \cdots = F_N$$

(Interpret the binomial coefficient $\binom{a}{b} = \frac{a!}{b!(a-b)!}$ as equal to zero if $a < b$.)

Question 11: The Joy of Subsets

a) How many subsets of $\{1, 2, 3, \ldots, N\}$ contain no two consecutive numbers?

b) How many subsets $\{1, 2, 3, \ldots, N\}$ consist only of pairs of consecutive numbers? (By this, we mean that the elements of the subsets can be subdivided into disjoint pairs, with each pair containing consecutive numbers. For example, $\{1, 2, 4, 5, 8, 9\}$ is such a subset of $\{1, 2, 3, \ldots, 10\}$, but $\{4, 5, 6, 8, 9\}$ is not.)

c) If S is a subset of $\{1, 2, 3, \ldots, N\}$, let "$S+1$" denote the corresponding set with "1" added to each of the elements of S. For instance, if, $S = \{1, 4, 5, 8\}$, then $S+1 = \{2, 5, 6, 9\}$. How many subsets S of $\{1, 2, 3, \ldots, N\}$ have the property that $S \cup (S+1) = \{1, 2, 3, \ldots, N, N+1\}$? (For example, there are five such subsets of $\{1, 2, 3, 4, 5\}$, namely, $\{1, 3, 4, 5\}$, $\{1, 3, 5\}$, $\{1, 2, 4, 5\}$, $\{1, 2, 3, 5\}$, and $\{1, 2, 3, 4, 5\}$ itself.)

Question 12: A Fibonacci Power Sequence

a) Show that there are two distinct numbers x and y with the property that the corresponding sequences of powers $1, x, x^2, x^3, x^4, \ldots$ and $1, y, y^2, y^3, y^4, \ldots$ each behave like the Fibonacci sequence (namely, that each term in the sequence, except the first and the second, is the sum of the two preceding terms).

b) Use this to find a formula for the nth Fibonacci number!

Answers, Comments, and Further Questions

1. Consider the general situation first. Set $T(N)$ to be the number of ways to tile a $1 \times N$ strip of squares using only square tiles and dominoes. Clearly, $T(1) = 1$ and $T(2) = 2$. We have also seen that $T(3) = 3$. Consider now the challenge of tiling a $1 \times N$ strip for $N > 3$. There are two ways to start. Using a square tile first leaves the challenge of tiling $N - 1$ spaces with more square tiles and dominoes–there are $T(N - 1)$ ways to do this–whereas starting with a domino first leaves $N - 2$ spaces to tile, which can be done in $T(N - 2)$ different ways. Thus a $1 \times N$ strip can be tiled in $T(N) = T(N - 1) + T(N - 2)$ different ways.

As this sequence of tiling numbers starts as the sequence of Fibonacci numbers (with index shifted by one), and satisfies the same recursive relation as the Fibonacci numbers, then it must be the case that the sequence continues to match the Fibonacci numbers. We have then that $T(n) = F_{n+1}$. In particular, $T(15) = F_{16} = 987$.

Placing a domino in the 6th and 7th squares of a 1×15 strip divides the strip into two pieces, one five cells long, and the second eight cells long. These can be tiled, respectively, $F_6 = 8$ and $F_9 = 34$ different ways, yielding $8 \times 34 = 272$ tilings in all for the 1×15 strip with a domino in that position. Any tiling that lacks a domino in this position has a "break" between the 6th and 7th positions and so can be regarded as a tiling of a 1×6 strip adjoined with a tiling of a 1×9 strip. There are $F_7 \times F_{10} = 13 \times 55 = 715$ such tilings.

Comment: This trick of counting the number of tilings of a string of square cells that do, and don't, have a domino in a certain position yields an interesting formula for the Fibonacci numbers. Precisely, consider a strip $a + b$ cells long. One can either view this strip as two separate strips, respectively, a and b cells long, each to be tiled separately (and there are $F_{a+1}F_{b+1}$ different ways to tile these strips), or if a domino is to be placed in the ath and $(a + 1)$st cells, as two strips $a - 1$ and $b - 1$ cells long to be tiled separately (and there are $F_a F_b$ ways to accomplish this). As there are F_{a+b+1} possible tilings in all, it must be the case then that $F_{a+b+1} = F_{a+1}F_{b+1} + F_a F_b$.

2. To arrive at cell 11, the bee can either first reach cell 10, and then step to 11, or first reach cell 9, and then step straight to cell 11. (In fact any path to cell 11 must be one of these two types.) If $H(N)$ denotes the number of honeycomb paths to cell N, then we have that $H(11) = H(10) + H(9)$. In general, the same reasoning shows that $H(N) = H(N-1) + H(N-2)$ with $H(1) = 1$ (there is one way to arrive at cell 1, namely, do nothing!) and $H(2) = 1$. Thus the sequence of numbers $H(N)$ matches the Fibonacci numbers precisely. In particular, $H(11) = F_{11} = 89$.

Comment: Each honeycomb walk can be recorded as a sequence of square tiles and dominoes! A square tile represents a diagonal step switching rows and a domino a horizontal step within a row.

3. a) There are 2^{N-1} ordered partitions of the number N. To see this, write the number N as the sum of N 1s using $N-1$ plus signs between them. To form an ordered partition of the number N one can simply delete some, all, or none of these plus signs and "amalgamate" any free-floating 1s. (For instance, deleting the first, fourth and fifth plus signs from $1 + 1 + 1 + 1 + 1 + 1$ yields " $11 + 1 + 111$" representing the ordered partition "$2 + 1 + 3$" of the number 6.) As each sign represents a choice—to delete or not to delete—and there are $N-1$ choices, there are 2^{N-1} possible outcomes in all.

b) As shown in Figure 3, interpret an ordered partition of N as instructions for a honeycomb walk from the first cell, cell 1, to the cell N places over along the top row, cell number $2N + 1$. As there are F_{2N+1} paths to cell $2N + 1$, there are F_{2N+1} ordered partitions of N with two different types of "1." In particular, there are $F_{11} = 89$ such ordered partitions of the number five (as opposed to just 16 with a single type of "1").

Taking it Further 1. Can anything be said about the number of ordered partitions of a number N if the number "1" can be colored in any one of k different colors, $k \geq 3$?

Taking it Further 2. If paths that end in a cell along the top row of the honeycomb represent ordered partitions of numbers, what do paths that end on cells in the bottom row represent?

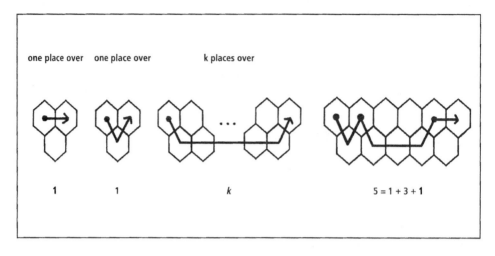

Figure 3. Every ordered partition of n with two colors of ones can be represented as a honeycomb walk.

4. Let $W(N)$ be the number of N-letter strings of the letters A and B containing no two consecutive Bs. To form such a string one may either begin with the letter "A" and complete the remaining $N-1$ spaces with any string of As and Bs that avoids two consecutive Bs. There are $W(N-1)$ ways to do this. Alternatively, one may begin with the letter B, which must be followed by the letter A. This leaves $N-2$ spaces, which may be filled with any string of letters containing no two consecutive Bs. There are $W(N-2)$ such strings. This shows that $W(N) = W(N-1) + W(N-2)$ with $W(1) = 2 = F_3$ and $W(2) = 3 = F_4$. We have then that $W(N) = F_{N+2}$. In particular, $W(10) = F_{12} = 144$.

5. Let $A(N)$ be the number of N-lettered words in ABEEBA. We have $A(1) = 3$ and, as one can check, $A(2) = 8$. ("AE" is the only two-letter string that is not permitted.) If you had the patience, you may have found that $A(3) = 21$, suggesting that $A(N)$ represents every second Fibonacci number:

$$A(N) = F_{2N+2}.$$

By considering the last letter, we have $A(N) = 3A(N-1) - A(N-2)$ since from any legal string of length $N-1$, we can attach the letter A, B, or E, but we have to throw away the $A(N-2)$ cases where the string ended with an A and we attached an E. Notice that the sequence of numbers F_{2N+2} satisfy the same recursive relation:

$$\begin{aligned}F_{2N+2} &= F_{2N} + F_{2N+1} = F_{2N} + F_{2N} + F_{2N-1}\\ &= F_{2N} + F_{2N} + (F_{2N} - F_{2N-2}) = 3F_{2N} - F_{2N-2}.\end{aligned}$$

It must indeed be the case that $A(N)$ and F_{2N+2} match.

6. In any restricted permutation, if a letter X moves to the position of its neighbor Y, then Y must move to the position of X. Thus letters that move do so in pairs by switching places. In a string of N letters, place a domino over any pair that are to switch locales and a square tile over any letter that is to remain fixed. Thus the number of restricted permutations on N letters is the same as the number of tilings of a $1 \times N$ strip with square tiles and dominoes. There are thus F_{N+1} restricted permutations on N letters.

Taking it Further. Let a and b be positive integers. Can anything be said about the number of permutations in which letters are allowed to move at most a places to the left or at most b places to the right? Suppose there is a "wrap around" effect where letters may move from the beginning of the word to the end of the word and vice versa?

7. If T represents the toss of tails, and H the toss of heads, then there are $2^{10} = 1024$ strings of Ts and Hs ten letters long, representing all the possible outcomes of ten tosses of the coin. According to Question 4, $F_{12} = 144$ of these strings do not contain two consecutive Hs. Thus the probability of not seeing two heads in a row in ten tosses of a coin is $144/1024 \approx 14.1\%$.

Taking it Further. If one is willing to toss a coin indefinitely, two consecutive heads will eventually appear. (Why?) They might appear on the first and second toss, or, if not then, on the second and third toss, or, if not then, on the third and fourth toss, and so on. Use this idea to establish:

$$\frac{F_1}{4} + \frac{F_2}{8} + \frac{F_3}{16} + \cdots = 1$$

8. Let's evaluate the decimal. Set $d = 0.11235955056\ldots = F_1/10 + F_2/10^2 + F_3/10^3 + \ldots$ and consider $10d + d$. We have:

$$11d = F_1 + \frac{F_2 + F_1}{10} + \frac{F_3 + F_2}{10} + \frac{F_4 + F_3}{10} + \cdots$$
$$= 1 + \frac{F_3}{10} + \frac{F_4}{10^2} + \frac{F_5}{10^3} + \cdots$$
$$= 1 + 100d - 10F_1 - F_2$$
$$= 100d - 10$$

Thus $89d = 10$ and so $d = 10/89$. Since d is a fraction its decimal representation must repeat.

Taking it Further. See Curtis Bennett's article on Fibonacci Decimals on page 28 of this magazine.

9. The observation persists: for positive integers n and k it is always the case that F_{nk} is a multiple of F_n. This can be proved by induction on k. It is certainly true for $k = 1$. Notice that, by the comment after question 1, we have $F_{n(k+1)} = F_{nk+(n-1)+1} = F_{nk+1}F_n + F_{nk}F_{n-1}$. If F_{nk} is a multiple of F_n, then we see that $F_{n(k+1)}$ is a sum of two multiples F_n, and so too is a multiple of F_n. This establishes the inductive step.

Taking it Further. If d is the greatest common divisor of two integers a and b, show that F_d is the greatest common divisor of F_a and F_b.

10. By deleting the beginning and ending As we see that the number of N-lettered swear words in ABABA equals the count of all $(N-2)$-lettered words. By Question 4, there are $W(N-2) = F_n$ such swear words.

To form a swear word of N letters with precisely k Bs, one must insert this many Bs in the $N-k-1$ spaces between the $N-k$ As. There are $\binom{N-k-1}{k}$ ways to do this.

The sum in the question simply gives the total number of words by counting those with a particular number of Bs.

11. a) Any N-lettered word in the language of ABABA corresponds to a subset of $\{1, 2, \ldots, N\}$ containing no two consecutive numbers: the placement of the Bs in the word dictate the subset. By Question 4 there are thus F_{n+2} such subsets.

b) Any tiling of a $1 \times N$ strip of squares using square tiles and dominoes corresponds to a subset of $\{1, 2, \ldots, N\}$ consisting solely of pairs of consecutive terms: each placement of a domino corresponds to the selection of a pair. By Question 1 there are thus F_{n+1} such subsets (including the empty subset).

c) Any such subset must contain the numbers 1 and N and cannot omit two consecutive integers. Thus each desired subset corresponds to an N-letter swear word in ABABA: the placement of the As correspond to elements of the subset. By Question 10 there are thus F_n subsets with the desired property.

Taking it Further. For a positive integer k and subset S of $\{1, 2, 3, \ldots, N\}$ let "$S+k$" denote the corresponding set with k added to each of its elements. Show that the number of subsets S of $\{1, 2, 3, \ldots, N\}$ with the property $S \cup (S+k) = \{1, 2, 3, \ldots, N+k\}$ is a product of k Fibonacci numbers.

12. a) We need to find a number x such that $x^{n+2} = x^{n+1} + x^n$ for each natural number n. This is equivalent to solving the quadratic $x^2 = x + 1$. There are two distinct solutions: the golden ratio $\varphi = \frac{1+\sqrt{5}}{2}$ and the quantity $\tau = \frac{1-\sqrt{5}}{2}$. Take x and y to be these two values.

b) Subtract the two sequences to obtain a new Fibonacci-like sequence whose nth term is $(x-y)F_n$ (Why?). Dividing this sequence by $x - y = \sqrt{5}$ gives us the Fibonacci sequence. Thus

$$F_n = \frac{x^n - y^n}{x-y} = \frac{1}{\sqrt{5}}\left(\left(\frac{1+\sqrt{5}}{2}\right)^n - \left(\frac{1-\sqrt{5}}{2}\right)^n\right)$$

This is Binet's formula, discovered by French mathematician Jacques-Phillippe-Marie Binet (1786–1855).

Taking it Further 1. Use Binet's formula to prove

$$\lim_{n \to \infty} \frac{F_{n+1}}{F_n} = \varphi.$$

Taking it Further 2. Notice that $2 = \varphi + \varphi^{-2}$ which, in "base φ," would be written: 10.01. Also, "3," in base φ, is 100.01, "4" is 101.01, and "11" is 10101.0101. Prove that every positive integer has a base φ representation using only the digits "0" and "1," with no two consecutive 1's. Are such representations unique?

Originally appeared as:
Tanton, James. "A Dozen Questions about Fibonacci Numbers." *Math Horizons*. vol. 12, no. 3 (February 2005), pp. 5–9.

A Second Helping

I am particularly fond of the use of "language" in this article: ABABA, foul ABABA, and ABEEBA prove to be particularly fruitful models for determining Fibonacci identities. (Can you indeed answer the second challenge stated at the end of the solution to question 3 to show that paths that end in the second row of the honeycomb walk match precisely words in ABEEBA?) Their link to colored partitions is intriguing and yearns for further analysis.

Question 11 can be generalized considerably. (See "Fibonacci numbers, generating sets and the hexagonal property." *The Fibonacci Quarterly*, 38 (2000), 299–309, for one possible direction.)

Proving that numbers have unique representations in unusual bases is subtle, if not outright difficult! For example, every number does have a unique base one-and-a-half representation using the "digits" 0, 1 and $\frac{1}{2}$.

$$\text{(For example, } 10 = 1 \times \left(\frac{3}{2}\right)^4 + \frac{1}{2} \times \left(\frac{3}{2}\right)^3 + 1 \times \left(\frac{3}{2}\right)^2 + 0 \times \left(\frac{3}{2}\right) + 1\text{)},$$

and proving so takes some work. Working in base φ (as described in the solutions to question 12) is just mean.

The Fibonacci Numbers—Exposed

Dan Kalman and
Robert Mena

Among numerical sequences, the Fibonacci numbers F_n have achieved a kind of celebrity status. Indeed, Koshy gushingly refers to them as one of the "two shining stars in the vast array of integer sequences" [**16**, p. xi]. The second of Koshy's "shining stars" is the *Lucas* numbers, a close relative of the Fibonacci numbers, about which we will say more below. The Fibonacci numbers are famous for possessing wonderful and amazing properties. Some are well known. For example, the sums and differences of Fibonacci numbers are Fibonacci numbers, and the ratios of Fibonacci numbers converge to the golden mean. Others are less familiar. Did you know that any four consecutive Fibonacci numbers can be combined to form a Pythagorean triple? Or how about this: The greatest common divisor of two Fibonacci numbers is another Fibonacci number. More precisely, the gcd of F_n and F_m is F_k, where k is the gcd of n and m.

With such fabulous properties, it is no wonder that the Fibonacci numbers stand out as a kind of super sequence. But what if it is not such a special sequence after all? What if it is only a rather pedestrian sample from an entire race of super sequences? In this case, the home world is the planet of two-term recurrences. As we shall show, its inhabitants are all just about as amazing as the Fibonacci sequence.

The purpose of this paper is to demonstrate that many of the properties of the Fibonacci numbers can be stated and proved for a much more general class of sequences, namely, second-order recurrences. We shall begin by reviewing a selection of the properties that made Fibonacci numbers famous. Then there will be a survey of second-order recurrences, as well as general tools for studying these recurrences. A number of the properties of the Fibonacci numbers will be seen to arise simply and naturally as the tools are presented. Finally, we will see that Fibonacci connections to Pythagorean triples and the gcd function also generalize in a natural way.

Famous Fibonacci properties

The Fibonacci numbers F_n are the terms of the sequence $0, 1, 1, 2, 3, 5, \ldots$ wherein each term is the sum of the two preceding terms, and we get things started with 0 and 1 as F_0 and F_1. You cannot go very far in the lore of Fibonacci numbers without encountering the companion sequence of Lucas numbers L_n, which follows the same recursive pattern as the Fibonacci numbers, but begins with $L_0 = 2$ and $L_1 = 1$. The first several Lucas numbers are therefore 2, 1, 3, 4, 7.

Regarding the origins of the subject, Koshy has this to say:

> The sequence was given its name in May of 1876 by the outstanding French mathematician François Edouard Anatole Lucas, who had originally called it "the series of Lamé," after the French mathematician Gabriel Lamé [**16**, p. 5].

Although Lucas contributed a great deal to the study of the Fibonacci numbers, he was by no means alone, as a perusal of Dickson [4, Chapter XVII] reveals. In fact, just about all the results presented here were first published in the nineteenth century. In particular, in his foundational paper [17], Lucas, himself, investigated the generalizations that interest us. These are sequences A_n defined by a recursive rule $A_{n+2} = aA_{n+1} + bA_n$ where a and b are fixed constants. We refer to such a sequence as a *two-term recurrence*.

The popular lore of the Fibonacci numbers seems not to include these generalizations, however. As a case in point, Koshy [16] has devoted nearly 700 pages to the properties of Fibonacci and Lucas numbers, with scarcely a mention of general two-term recurrences. Similar, but less encyclopedic sources are Hoggatt [9], Honsberger [11, Chapter 8], and Vajda [21]. There has been a bit more attention paid to so-called *generalized Fibonacci numbers*, A_n, which satisfy the same recursive formula $A_{n+2} = A_{n+1} + A_n$, but starting with arbitrary initial values A_0 and A_1, particularly by Horadam (see for example Horadam [12], Walton and Horadam [22], as well as Koshy [16, Chapter 7]). Horadam also investigated the same sort of sequences we consider, but he focused on different aspects from those presented here [14, 15]. In [14] he includes our Examples 3 and 7, with an attribution to Lucas's 1891 *Théorie des Nombres*. With Shannon, Horadam also studied Pythagorean triples, and their paper [20] goes far beyond the connection to Fibonacci numbers considered here. Among more recent references, Bressoud [3, chapter 12] discusses the application of generalized Fibonacci sequences to primality testing, while Hilton and Pedersen [8] present some of the same results that we do. However, none of these references share our general point of emphasis, that in many cases, properties commonly perceived as unique to the Fibonacci numbers, are actually shared by large classes of sequences.

It would be impossible to make this point here in regard to all known Fibonacci properties, as Koshy's tome attests. We content ourselves with a small sample, listed below. We have included page references from Koshy [16].

Sum of squares $\sum_1^n F_i^2 = F_n F_{n+1}$. (Page 77.)

Lucas-Fibonacci connection $L_{n+1} = F_{n+2} + F_n$. (Page 80.)

Binet formulas The Fibonacci and Lucas numbers are given by

$$F_n = \frac{\alpha^n - \beta^n}{\alpha - \beta} \quad \text{and} \quad L_n = \alpha^n + \beta^n,$$

where

$$\alpha = \frac{1 + \sqrt{5}}{2} \quad \text{and} \quad \beta = \frac{1 - \sqrt{5}}{2}.$$

(Page 79.)

Asymptotic behavior $F_{n+1}/F_n \to \alpha$ as $n \to \infty$. (Page 122.)

Running sum $\sum_1^n F_i = F_{n+2} - 1$. (Page 69.)

Matrix form We present a slightly permuted form of what generally appears in the literature. Our version is

$$\begin{bmatrix} 0 & 1 \\ 1 & 1 \end{bmatrix}^n = \begin{bmatrix} F_{n-1} & F_n \\ F_n & F_{n+1} \end{bmatrix}.$$

(Page 363.)

Cassini's formula $F_{n-1}F_{n+1} - F_n^2 = (-1)^n$. (Page 74)

Convolution property $F_n = F_m F_{n-m+1} + F_{m-1} F_{n-m}$. (Page 88, formula 6.)

Pythagorean triples If w, x, y, z are four consecutive Fibonacci numbers, then $(wz, 2xy, yz - wx)$ is a Pythagorean triple. That is, $(wz)^2 + (2xy)^2 = (yz - wx)^2$. (Page 91, formula 88.)

Greatest common divisor $\gcd(F_m, F_n) = F_{\gcd(m,n)}$. (Page 198.)

This is, as mentioned, just a sample of amazing properties of the Fibonacci and Lucas numbers. But they all generalize in a natural way to classes of two-term recurrences. In fact, several of the proofs arise quite simply as part of a general development of the recurrences. We proceed to that topic next.

Generalized Fibonacci and Lucas numbers

Let a and b be any real numbers. Define a sequence A_n as follows. Choose initial values A_0 and A_1. All succeeding terms are determined by

$$A_{n+2} = aA_{n+1} + bA_n. \tag{1}$$

For fixed a and b, we denote by $\mathcal{R}(a, b)$ the set of all such sequences. To avoid a trivial case, we will assume that $b \neq 0$.

In $\mathcal{R}(a, b)$, we define two distinguished elements. The first, F, has initial terms 0 and 1. In $\mathcal{R}(1, 1)$, F is thus the Fibonacci sequence. In the more general case, we refer to F as the (a, b)-Fibonacci sequence. Where no confusion will result, we will suppress the dependence on a and b. Thus, in every $\mathcal{R}(a, b)$, there is an element F that begins with 0 and 1, and this is the Fibonacci sequence for $\mathcal{R}(a, b)$.

Although F is the primordial sequence in $\mathcal{R}(a, b)$, there is another sequence L that is of considerable interest. It starts with $L_0 = 2$ and $L_1 = a$. As will soon be clear, L plays the same role in $\mathcal{R}(a, b)$ as the Lucas numbers play in $\mathcal{R}(1, 1)$. Accordingly, we refer to L as the (a, b)-Lucas sequence. For the most part, there will be only one a and b under consideration, and it will be clear from context which $\mathcal{R}(a, b)$ is the home for any particular mention of F or L. In the rare cases where some ambiguity might occur, we will use $F^{(a,b)}$ and $L^{(a,b)}$ to indicate the F and L sequences in $\mathcal{R}(a, b)$.

In the literature, what we are calling F and L have frequently been referred to as *Lucas sequences* (see Bressoud [3, chapter 12] and Weisstein [23, p. 1113]) and denoted by U and V, the notation adopted by Lucas in 1878 [17]. We prefer to use F and L to emphasize the idea that there are Fibonacci and Lucas sequences in each $\mathcal{R}(a, b)$, and that these sequences share many properties with the traditional F and L. In contrast, it has sometimes been the custom to attach the name *Lucas* to the L sequence for a particular $\mathcal{R}(a, b)$. For example, in $\mathcal{R}(2, 1)$, the elements of F have been referred to as *Pell numbers* and the elements of L as *Pell-Lucas* numbers [23, p. 1334].

Examples Of course, the most familiar example is $\mathcal{R}(1, 1)$, in which F and L are the famous Fibonacci and Lucas number sequences. But there are several other choices of a and b that lead to familiar examples.

Example 1: $\mathcal{R}(11, -10)$. The Fibonacci sequence in this family is $F = 0, 1, 11, 111, 1111, \ldots$ the sequence of repunits, and $L = 2, 11, 101, 1001, 10001, \ldots$. The initial 2, which at first seems out of place, can be viewed as the result of putting two 1s in the same position.

Example 2: $\mathcal{R}(2, -1)$. Here F is the sequence of whole numbers $0, 1, 2, 3, 4, \ldots$, and L is the constant sequence $2, 2, 2, \ldots$. More generally, $\mathcal{R}(2, -1)$ consists of all the arithmetic progressions.

Example 3: $\mathcal{R}(3, -2)$. $F = 0, 1, 3, 7, 15, 31, \ldots$ is the Mersenne sequence, and $L = 2, 3, 5, 9, 17, 33, \ldots$ is the Fermat sequence. These are just powers of 2 plus or minus 1.

Example 4: $\mathcal{R}(1, -1)$. $F = 0, 1, 1, 0, -1, -1, 0, 1, 1, \ldots$ and $L = 2, 1, -1, -2, -1, 1, 2, 1, -1, \ldots$. Both sequences repeat with period 6, as do all the elements of $\mathcal{R}(1, -1)$.

Example 5: $\mathcal{R}(3, -1)$. $F = 0, 1, 3, 8, 21, \ldots$ and $L = 2, 3, 7, 18, \ldots$. Do you recognize these? They are the *even-numbered* Fibonacci and Lucas numbers.

Example 6: $\mathcal{R}(4, 1)$. $F = 0, 1, 4, 17, 72, \ldots$ and $L = 2, 4, 18, 76, \ldots$. Here, L gives every third Lucas number, while F gives 1/2 of every third Fibonacci number.

Example 7: $\mathcal{R}(2, 1)$. $F = 0, 1, 2, 5, 12, 29, 70, \ldots$ and $L = 2, 2, 6, 14, 34, 82, \ldots$. These are the Pell sequences, mentioned earlier. In particular, for any n, $(x, y) = (F_{2n} + F_{2n-1}, F_{2n})$ gives a solution to Pell's Equation $x^2 - 2y^2 = 1$. This extends to the more general Pell equation, $x^2 - dy^2 = 1$, when $d = k^2 + 1$. Then, using the F sequence in $\mathcal{R}(2k, 1)$, we obtain solutions of the form $(x, y) = (kF_{2n} + F_{2n-1}, F_{2n})$. Actually, equations of this type first appeared in the Archimedean cattle problem, and were considered by the Indian mathematicians Brahmagupta and Bhaskara [2, p. 221]. Reportedly, Pell never worked on the equations that today bear his name. Instead, according to Weisstein [23], "while Fermat deserves the credit for being the first [European] to extensively study the equation, the erroneous attribution to Pell was perpetrated by none other than Euler."

Coincidentally, the even terms F_{2n} in $\mathcal{R}(a, 1)$ also appear in another generalized Fibonacci result, related to an identity discussed elsewhere in this issue of the MAGAZINE [6]. The original identity for normal Fibonacci numbers is

$$\arctan\left(\frac{1}{F_{2n}}\right) = \arctan\left(\frac{1}{F_{2n+1}}\right) + \arctan\left(\frac{1}{F_{2n+2}}\right).$$

For $F \in \mathcal{R}(a, 1)$ the corresponding result is

$$\arctan\left(\frac{1}{F_{2n}}\right) = \arctan\left(\frac{a}{F_{2n+1}}\right) + \arctan\left(\frac{1}{F_{2n+2}}\right).$$

The wonderful world of two-term recurrences

The Fibonacci and Lucas sequences are elements of $\mathcal{R}(1, 1)$, and many of their properties follow immediately from the recursive rule that each term is the sum of the two preceding terms. Similarly, it is often easy to establish corresponding properties for elements of $\mathcal{R}(a, b)$ directly from the fundamental identity (1). For example, in $\mathcal{R}(1, 1)$, the *Sum of Squares* identity is

$$F_1^2 + F_2^2 + \cdots + F_n^2 = F_n F_{n+1}.$$

The generalization of this to $\mathcal{R}(a, b)$ is

$$b^n F_0^2 + b^{n-1} F_1^2 + \cdots + b F_{n-1}^2 + F_n^2 = \frac{F_n F_{n+1}}{a}. \tag{2}$$

This can be proved quite easily using (1) and induction.

Many of the other famous properties can likewise be established by induction. But to provide more insight about these properties, we will develop some analytic methods, organized loosely into three general contexts. First, we can think of $\mathcal{R}(a, b)$ as a subset of $\mathbb{R}^\infty$, the real vector space of real sequences, and use the machinery of difference operators. Second, by deriving Binet formulas for elements of $\mathcal{R}(a, b)$, we obtain explicit representations as linear combinations of geometric progressions. Finally, there is a natural matrix formulation which is tremendously useful. We explore each of these contexts in turn.

Difference operators We will typically represent elements of $\mathbb{R}^\infty$ with uppercase roman letters, in the form

$$A = A_0, A_1, A_2, \ldots.$$

There are three fundamental linear operators on $\mathbb{R}^\infty$ to consider. The first is the *left-shift*, Λ. For any real sequence $A = A_0, A_1, A_2, \ldots$, the shifted sequence is $\Lambda A = A_1, A_2, A_3, \ldots$.

This shift operator is a kind of discrete differential operator. Recurrences like (1) are also called difference equations. Expressed in terms of Λ, (1) becomes

$$(\Lambda^2 - a\Lambda - b)A = 0.$$

This is analogous to expressing a differential equation in terms of the differential operator, and there is a theory of difference equations that perfectly mirrors the theory of differential equations. Here, we have in mind linear constant coefficient differential and difference equations.

As one fruit of this parallel theory, we see at once that $\Lambda^2 - a\Lambda - b$ is a linear operator on $\mathbb{R}^\infty$, and that $\mathcal{R}(a, b)$ is its null space. This shows that $\mathcal{R}(a, b)$ is a subspace of $\mathbb{R}^\infty$. We will discuss another aspect of the parallel theories of difference and differential equation in the succeeding section on Binet formulas.

Note that any polynomial in Λ is a linear operator on $\mathbb{R}^\infty$, and that all of these operators commute. For example, the *forward difference* operator Δ, defined by $(\Delta A)_k = A_{k+1} - A_k$, is given by $\Delta = \Lambda - 1$. Similarly, consider the k-term sum, Σ_k, defined by $(\Sigma_k A)_n = A_n + A_{n+1} + \cdots + A_{n+k-1}$. To illustrate, $\Sigma_2(A)$ is the sequence $A_0 + A_1, A_1 + A_2, A_2 + A_3, \ldots$. These sum operators can also be viewed as polynomials in Λ: $\Sigma_k = 1 + \Lambda + \Lambda^2 + \cdots + \Lambda^{k-1}$.

Because these operators commute with Λ, they are operators on $\mathcal{R}(a, b)$, as well. In general, if Ψ is an operator that commutes with Λ, we observe that Ψ also commutes with $\Lambda^2 - a\Lambda - b$. Thus, if $A \in \mathcal{R}(a, b)$, then $(\Lambda^2 - a\Lambda - b)\Psi A = \Psi(\Lambda^2 - a\Lambda - b)A = \Psi 0 = 0$. This shows that $\Psi A \in \mathcal{R}(a, b)$. In particular, $\mathcal{R}(a, b)$ is closed under differences and k-term sums.

This brings us to the second fundamental operator, the *cumulative sum* Σ. It is defined as follows: $\Sigma(A) = A_0, A_0 + A_1, A_0 + A_1 + A_2, \ldots$. This is not expressible in terms of Λ, nor does it commute with Λ, in general. However, there is a simple relation connecting the two operators:

$$\Delta \Sigma = \Lambda. \tag{3}$$

This is a sort of discrete version of the fundamental theorem of calculus. In the opposite order, we have

$$(\Sigma \Delta A)_n = A_{n+1} - A_0,$$

a discrete version of the other form of the fundamental theorem. It is noteworthy that Leibniz worked with these sum and difference operators as a young student, and later identified this work as his inspiration for calculus (Edwards [**5**, p. 234]).

The final fundamental operator is the *k-skip*, Ω_k, which selects every kth element of a sequence. That is, $\Omega_k(A) = A_0, A_k, A_{2k}, A_{3k}, \ldots$. By combining these operators with powers of Λ, we can sample the terms of a sequence according to any arithmetic progression. For example,

$$\Omega_5 \Lambda^3 A = A_3, A_8, A_{13}, \ldots.$$

Using the context of operators and the linear space $\mathcal{R}(a, b)$, we can derive useful results. First, it is apparent that once A_0 and A_1 are chosen, all remaining terms are completely determined by (1). This shows that $\mathcal{R}(a, b)$ is a two-dimensional space. Indeed, there is a natural basis $\{E, F\}$ where E has starting values 1 and 0, and F, with starting values 0 and 1, is the (a, b)-Fibonacci sequence. Thus

$$E = 1, 0, b, ab, a^2 b + b^2, \ldots$$
$$F = 0, 1, a, a^2 + b, a^3 + 2ab, \ldots.$$

Clearly, $A = A_0 E + A_1 F$ for all $A \in \mathcal{R}(a, b)$. Note further that $\Lambda E = bF$, so that we can easily express any A just using F:

$$A_n = bA_0 F_{n-1} + A_1 F_n \tag{4}$$

As an element of $\mathcal{R}(a, b)$, L can thus be expressed in terms of F. From (4), we have

$$L_n = 2b F_{n-1} + a F_n.$$

But the fundamental recursion (1) then leads to

$$L_n = bF_{n-1} + F_{n+1}. \tag{5}$$

This is the analog of the Lucas-Fibonacci connection stated above.

Recall that the difference and the k-term sum operators all preserve $\mathcal{R}(a, b)$. Thus, ΔF and $\Sigma_k F$ are elements of $\mathcal{R}(a, b)$ and can be expressed in terms of F using (4). The case for Σ is a more interesting application of operator methods. The question is this: If $A \in \mathcal{R}(a, b)$, what can we say about ΣA?

As a preliminary step, notice that a sequence is constant if and only if it is annihilated by the difference operator Δ. Now, suppose that $A \in \mathcal{R}(a, b)$. That means $(\Lambda^2 - a\Lambda - b)A = 0$, and so too

$$\Lambda(\Lambda^2 - a\Lambda - b)A = 0.$$

Now commute Λ with the other operator, and use (3) to obtain

$$(\Lambda^2 - a\Lambda - b)\Delta \Sigma A = 0.$$

Finally, since Δ and Λ commute, pull Δ all the way to the front to obtain

$$\Delta(\Lambda^2 - a\Lambda - b)\Sigma A = 0.$$

This shows that while $(\Lambda^2 - a\Lambda - b)\Sigma A$ may not be 0 (indicating $\Sigma A \notin \mathcal{R}(a, b)$), at worst it is constant. Now it turns out that there are two cases. If $a + b \neq 1$, it can be shown that ΣA differs from an element of $\mathcal{R}(a, b)$ by a constant. That tells us at once that there is an identity of the form

$$(\Sigma A)_n = c_0 F_n + c_1 F_{n-1} + c_2,$$

which corresponds to the running sum property for Fibonacci numbers. We will defer the determination of the constants c_i to the section on Binet formulas.

Here is the verification that ΣA differs from an element of $\mathcal{R}(a, b)$ by a constant when $a + b \neq 1$. We know that $(\Lambda^2 - a\Lambda - b)\Sigma A$ is a constant c_1. Suppose that we can find another constant, c, such that $(\Lambda^2 - a\Lambda - b)c = c_1$. Then we would have $(\Lambda^2 - a\Lambda - b)(\Sigma A - c) = 0$, hence $\Sigma A - c \in \mathcal{R}(a, b)$. It is an exercise to show c can be found exactly when $a + b \neq 1$.

When $a + b = 1$ we have the second case. A little experimentation with (1) will show you that in this case $\mathcal{R}(a, b)$ includes all the constant sequences. The best way to analyze this situation is to develop some general methods for solving difference equations. We do that next.

Binet formulas We mentioned earlier that there is a perfect analogy between linear constant coefficient difference and differential equations. In the differential equation case, a special role is played by the exponential functions, $e^{\lambda t}$, which are eigenvectors for the differential operator: $De^{\lambda t} = \lambda \cdot e^{\lambda t}$. For difference equations, the analogous role is played by the geometric progressions, $A_n = \lambda^n$. These are eigenvectors for the left-shift operator: $\Lambda \lambda^n = \lambda \cdot \lambda^n$. Both differential and difference equations can be formulated in terms of polynomials in the fundamental operator, Λ or D, respectively. These are in fact characteristic polynomials—the roots λ are eigenvalues and correspond to eigenvector solutions to the differential or difference equation. Moreover, except in the case of repeated roots, this leads to a basis for the space of all solutions.

We can see how this all works in detail in the case of $\mathcal{R}(a, b)$, which is viewed as the null space of $p(\Lambda) = \Lambda^2 - a\Lambda - b$. When is the geometric progression $A_n = \lambda^n$ in this null space? We demand that $A_{n+2} - aA_{n+1} - bA_n = 0$, so the condition is

$$\lambda^{n+2} - a\lambda^{n+1} - b\lambda^n = 0.$$

Excluding the case $\lambda = 0$, which produces only the trivial solution, this leads to $p(\lambda) = 0$ as a necessary and sufficient condition for $\lambda^n \in \mathcal{R}(a, b)$. Note also that the roots of p are related to the coefficients in the usual way. Thus, if the roots are λ and μ, then

$$\lambda + \mu = a \tag{6}$$

$$\lambda\mu = -b. \tag{7}$$

Now if λ and μ are distinct, then λ^n and μ^n are independent solutions to the difference equation. And since we already know that the null space is two dimensional, that makes $\{\lambda^n, \mu^n\}$ a basis. In this case, $\mathcal{R}(a, b)$ is characterized as the set of linear combinations of these two geometric progressions. In particular, for $A \in \mathcal{R}(a, b)$, we can express A in the form

$$A_n = c_\lambda \lambda^n + c_\mu \mu^n. \tag{8}$$

The constants c_λ and c_μ are determined by the initial conditions

$$A_0 = c_\lambda + c_\mu$$

$$A_1 = c_\lambda \lambda + c_\mu \mu.$$

We are assuming λ and μ are distinct, so this system has the solution

$$c_\lambda = \frac{A_1 - \mu A_0}{\lambda - \mu}$$

$$c_\mu = \frac{\lambda A_0 - A_1}{\lambda - \mu}.$$

Now let us apply these to the special cases of F and L. For F, the initial values are 0 and 1, so $c_\lambda = 1/(\lambda - \mu)$ and $c_\mu = -1/(\lambda - \mu)$. For L the initial terms are 2 and $a = \lambda + \mu$. This gives $c_\lambda = c_\mu = 1$. Thus,

$$F_n = \frac{\lambda^n - \mu^n}{\lambda - \mu} \tag{9}$$

$$L_n = \lambda^n + \mu^n. \tag{10}$$

These are the *Binet Formulas* for $\mathcal{R}(a, b)$.

When $\lambda = \mu$, the fundamental solutions of the difference equation are $A_n = \lambda^n$ and $B_n = n\lambda^n$. Most of the results for $\mathcal{R}(a, b)$ have natural extensions to this case. For example, in the case of repeated root λ, the Binet formulas become

$$F_n = n\lambda^{n-1}$$

$$L_n = 2\lambda^n.$$

Extensions of this sort are generally quite tractable, and we will not typically go into the details. Accordingly, we will assume from now on that p has distinct roots, or equivalently, that $a^2 + 4b \neq 0$.

Another special case of interest occurs when one root is 1. In this case, the geometric progression 1^n is constant, and $\mathcal{R}(a, b)$ contains all the constant sequences. As we saw earlier, the condition for this is $a + b = 1$. Now the Binet representation gives a new way of thinking about this result. It is an exercise to verify that $a + b = 1$ if and only if 1 is a root of p.

If both roots equal 1, the fundamental solutions are $A_n = 1$ and $B_n = n$. This shows that $\mathcal{R}(2, -1)$ consists of all the arithmetic progressions, confirming our earlier observation for Example 2.

Let us revisit the other examples presented earlier, and consider the Binet formulas for each.

Example 0: $\mathcal{R}(1, 1)$. For the normal Fibonacci and Lucas numbers, $p(t) = t^2 - t - 1$, and the roots are α and β as defined earlier. The general version of the Binet formulas reduce to the familiar form upon substitution of α and β for λ and μ.

Example 1: $\mathcal{R}(11, -10)$. Here, with $p(t) = t^2 - 11t + 10$, the roots are 10 and 1. In this case the Binet formulas simply tell us what is already apparent: $F_n = (10^n - 1)/9$ and $L_n = 10^n + 1$.

Example 3: $\mathcal{R}(3, -2)$. In this example, $p(t) = t^2 - 3t + 2$, with roots 2 and 1. The Binet formulas confirm the pattern we saw earlier: $F_n = 2^n - 1$ and $L_n = 2^n + 1$.

Example 4: $\mathcal{R}(1, -1)$. Now $p(t) = t^2 - t + 1$. Note that $p(t)(t + 1) = t^3 + 1$, so that roots of p are cube roots of -1 and hence, sixth roots of 1. This explains the periodic nature of F and L. Indeed, since $\lambda^6 = \mu^6 = 1$ in this case, every element of $\mathcal{R}(1, -1)$ has period 6 as well.

The Fibonacci Numbers—Exposed

Example 5: $\mathcal{R}(3, -1)$. The roots in this example are α^2 and β^2. The Binet formulas involve only even powers of α and β, hence the appearance of the even Fibonacci and Lucas numbers.

Example 6: $\mathcal{R}(4, 1)$. This example is similar to the previous one, except that the roots are α^3 and β^3.

Example 7: $\mathcal{R}(2, 1)$. For this example $p(t) = t^2 - 2t - 1$, so the roots are $1 \pm \sqrt{2}$. The Binet formulas give

$$F_n = \frac{(1 + \sqrt{2})^n - (1 - \sqrt{2})^n}{2\sqrt{2}} \quad \text{and} \quad L_n = (1 + \sqrt{2})^n + (1 - \sqrt{2})^n.$$

Characterizing $\mathcal{R}(a, b)$ in terms of geometric progressions has immediate applications. For example, consider the ratio of successive terms of a sequence in $\mathcal{R}(a, b)$. Using (8), we have

$$\frac{A_{n+1}}{A_n} = \frac{c_\lambda \lambda^{n+1} + c_\mu \mu^{n+1}}{c_\lambda \lambda^n + c_\mu \mu^n}.$$

Now assume that $|\lambda| > |\mu|$, and divide out λ^n:

$$\frac{A_{n+1}}{A_n} = \frac{c_\lambda \lambda + c_\mu \mu (\mu^n/\lambda^n)}{c_\lambda + c_\mu (\mu^n/\lambda^n)}.$$

Since $(\mu/\lambda)^n$ will go to 0 as n goes to infinity, we conclude

$$\frac{A_{n+1}}{A_n} \to \lambda \quad \text{as} \quad n \to \infty.$$

In words, the ratio of successive terms of a sequence in $\mathcal{R}(a, b)$ always tends toward the dominant eigenvalue as n goes to infinity. That is the general version of the asymptotic behavior we observed for Fibonacci numbers.

As a second example, if $A_n = c_\lambda \lambda^n + c_\mu \mu^n$, then $\Omega_k A_n = c_\lambda \lambda^{kn} + c_\mu \mu^{kn}$. This is a linear combination of two geometric progressions as well, with eigenvalues λ^k and μ^k. Consequently, $\Omega_k A \in \mathcal{R}(a', b')$ for some a' and b'. Now using the relationship between roots and coefficients again, we deduce that $a' = \lambda^k + \mu^k$, and by (10) that gives $a' = L_k^{(a,b)}$. Similarly, we find $b' = -(\lambda\mu)^k = -(-b)^k$. Thus,

$$\Omega_k : \mathcal{R}(a, b) \to \mathcal{R}(L_k^{(a,b)}, -(-b)^k). \tag{11}$$

We can extend this slightly. If $A \in \mathcal{R}(a, b)$, then so is $\Lambda^d A$ for any positive integer d. Thus, $\Omega_k \Lambda^d A \in \mathcal{R}(a', b')$. In other words, when $A \in \mathcal{R}(a, b)$, the sequence $B_n = A_{kn+d}$ is in $\mathcal{R}(a', b')$. This corresponds to sampling A at the terms of an arithmetic progression.

In the particular case of F and L, we can use the preceding results to determine the effect of Ω_k explicitly. For notational simplicity, we will again denote by a' and b' the values $L_k^{(a,b)}$ and $-(-b)^k$, respectively. We know that $\Omega_k F^{(a,b)} \in \mathcal{R}(a', b')$, and begins with the terms 0 and $F_k^{(a,b)}$. This is necessarily a multiple of $F^{(a',b')}$, and in particular, gives

$$\Omega_k F^{(a,b)} = F_k^{(a,b)} \cdot F^{(a',b')}. \tag{12}$$

Similarly, $\Omega_k L^{(a,b)}$ begins with 2 and $L_k^{(a,b)}$. But remember that the latter of these is exactly $a' = L_1^{(a',b')}$. Thus,

$$\Omega_k L^{(a,b)} = L^{(a',b')}. \tag{13}$$

Of course, this last equation is easily deduced directly from the Binet formula for $L^{(a,b)}$, as well. The observations in Examples 5 and 6 are easily verified using (12) and (13).

For one more example, let us return to the analysis of ΣA for $A \in \mathcal{R}(a, b)$. Again using the expression $A_n = c_\lambda \lambda^n + c_\mu \mu^n$ we find the terms of ΣA as

$$\Sigma A_n = c_\lambda \frac{\lambda^{n+1} - 1}{\lambda - 1} + c_\mu \frac{\mu^{n+1} - 1}{\mu - 1}.$$

Evidently, this is invalid if either λ or μ equals 1. So, as before, we exclude that possibility by assuming $a + b \neq 1$.

Under this assumption, we found earlier that ΣA must differ from an element of $\mathcal{R}(a, b)$ by a constant. Now we can easily determine the value of that constant. Rearranging the preceding equation produces

$$\Sigma A_n = \frac{c_\lambda \lambda}{\lambda - 1} \lambda^n + \frac{c_\mu \mu}{\mu - 1} \mu^n - \left(\frac{c_\lambda}{\lambda - 1} + \frac{c_\mu}{\mu - 1} \right).$$

This clearly reveals ΣA as the sum of an element of $\mathcal{R}(a, b)$ with the constant $C = -(c_\lambda/(\lambda - 1) + c_\mu/(\mu - 1))$.

In general, the use of this formula requires expressing A in terms of λ and μ. But in the special case of F, we can express the formula in terms of a and b. Recall that when $A = F$, $c_\lambda = 1/(\lambda - \mu)$ and $c_\mu = -1/(\lambda - \mu)$. Substituting these in the earlier formula for C, leads to

$$C = -\frac{1}{\lambda - \mu} \left(\frac{1}{\lambda - 1} - \frac{1}{\mu - 1} \right)$$

$$= -\frac{1}{\lambda - \mu} \frac{\mu - \lambda}{(\lambda - 1)(\mu - 1)} = \frac{1}{(\lambda \mu - \lambda - \mu + 1)}.$$

Once again using (6) and (7), this yields

$$C = \frac{1}{1 - a - b}. \tag{14}$$

As an example, let us consider ΣF for $\mathcal{R}(2, 3)$. In the table below, the first several terms of F and ΣF are listed.

n	0	1	2	3	4	5
F_n	0	1	2	7	20	61
ΣF_n	0	1	3	10	30	91

In this example, we have $C = 1/(1 - 2 - 3) = -1/4$. Accordingly, adding 1/4 to each term of ΣF should produce an element of $\mathcal{R}(2, 3)$. Carrying out this step produces

$$\Sigma F + \frac{1}{4} = \frac{1}{4}(1, 5, 13, 41, 121, 365, \ldots).$$

As expected, this is an element of $\mathcal{R}(2, 3)$.

Applying a similar analysis in the general case (with the assumption $a + b \neq 1$) leads to the identity

$$\Sigma F_n = \frac{1}{a+b-1}(F_{n+1} + bF_n - 1).$$

This reduces to the running sum property for Fibonacci numbers when $a = b = 1$. A similar analysis applies in the case $a + b = 1$. We leave the details to the reader.

In the derivation of the Binet formulas above, a key role was played by the eigenvectors and eigenvalues of the shift operator. It is therefore not surprising that there is a natural matrix formulation of these ideas. That topic is the third general context for tool development.

Matrix formulation Using the natural basis $\{E, F\}$ for $\mathcal{R}(a, b)$, we can represent Λ by a matrix M. We already have seen that $\Lambda E = bF$, so the first column of M has entries 0 and b. Applying the shift to F produces $(1, a, \ldots) = E + aF$. This identifies the second column entries of M as 1 and a, so

$$M = \begin{bmatrix} 0 & 1 \\ b & a \end{bmatrix}. \tag{15}$$

Now if $A \in \mathcal{R}(a, b)$, then relative to the natural basis it is represented by $[A] = [A_0 \ A_1]^T$. Similarly, the basis representation of $\Lambda^n A$ is $[A_n \ A_{n+1}]^T$. On the other hand, we can find the same result by applying M n times to $[A]$. Thus, we obtain

$$\begin{bmatrix} 0 & 1 \\ b & a \end{bmatrix}^n \begin{bmatrix} A_0 \\ A_1 \end{bmatrix} = \begin{bmatrix} A_n \\ A_{n+1} \end{bmatrix}. \tag{16}$$

Premultiplying by $[1 \ 0]$ then gives

$$\begin{bmatrix} 1 & 0 \end{bmatrix} \begin{bmatrix} 0 & 1 \\ b & a \end{bmatrix}^n \begin{bmatrix} A_0 \\ A_1 \end{bmatrix} = A_n. \tag{17}$$

This gives a matrix representation for A_n.

Note that in general, the ith column of a matrix M can be expressed as the product $M\mathbf{e}_i$, where $\mathbf{e}_i$ is the ith standard basis element. But here, the standard basis elements are representations $[E]$ and $[F]$. In particular, $M^n[E] = [E_n \ E_{n+1}]^T$ and $M^n[F] = [F_n \ F_{n+1}]^T$. This gives us the columns of M^n, and therefore

$$M^n = \begin{bmatrix} E_n & F_n \\ E_{n+1} & F_{n+1} \end{bmatrix}.$$

Then, using $\Lambda E = bF$, we have

$$\begin{bmatrix} 0 & 1 \\ b & a \end{bmatrix}^n = \begin{bmatrix} bF_{n-1} & F_n \\ bF_n & F_{n+1} \end{bmatrix}. \tag{18}$$

This is the general version of the matrix form for Fibonacci numbers.

The matrix form leads immediately to two other properties. First, taking the determinant of both sides of (18), we obtain

$$bF_{n-1}F_{n+1} - bF_n^2 = (-b)^n.$$

Simplifying,

$$F_{n-1}F_{n+1} - F_n^2 = (-1)^n b^{n-1},$$

the general version of Cassini's formula.

Second, start with $M^n = M^m M^{n-m}$, expressed explicitly in the form

$$\begin{bmatrix} bF_{n-1} & F_n \\ bF_n & F_{n+1} \end{bmatrix} = \begin{bmatrix} bF_{m-1} & F_m \\ bF_m & F_{m+1} \end{bmatrix} \begin{bmatrix} bF_{n-m-1} & F_{n-m} \\ bF_{n-m} & F_{n-m+1} \end{bmatrix}.$$

By inspection, we read off the 1, 2 entry of both sides, obtaining

$$F_n = F_m F_{n-m+1} + bF_{m-1}F_{n-m}, \qquad (19)$$

generalizing the *Convolution Property* for regular Fibonacci numbers. As a special case, replace n with $2n + 1$ and m with $n + 1$, producing

$$F_{2n+1} = F_{n+1}^2 + bF_n^2. \qquad (20)$$

This equation will be applied in the discussion of Pythagorean triples.

This concludes our development of general tools. Along the way, we have found natural extensions of all but two of our famous Fibonacci properties. These extensions are all simple and direct consequences of the basic ideas in three general contexts: difference operators, Binet formulas, and matrix methods. Establishing analogs for the remaining two properties is just a bit more involved, and we focus on them in the next section.

The last two properties

Pythagorean triples In a way, the connection with Pythagorean triples is trivial. The well-known parameterization $(x^2 - y^2, 2xy, x^2 + y^2)$ expresses primitive Pythagorean triples in terms of quadratic polynomials in two variables. The construction using Fibonacci numbers is similar. To make this clearer, note that if w, x, y, z are four consecutive Fibonacci numbers, then we may replace w with $y - x$ and z with $y + x$. With these substitutions, the Fibonacci parameterization given earlier for Pythagorean triples becomes

$$(wz, 2xy, yz - wx) = \big((y - x)(y + x), 2yx, y(y + x) - x(y - x)\big).$$

Since we can reduce the parameterization to a quadratic combination of two parameters in this way, the ability to express Pythagorean triples loses something of its mystery. In fact, if w, x, y, z are four consecutive terms of any sequence in $\mathcal{R}(a, b)$, we may regard x and y as essentially arbitrary, and so use them to define a Pythagorean triple $(x^2 - y^2, 2xy, x^2 + y^2)$. Thus, we can construct a Pythagorean triple using *just two* consecutive terms of a Fibonacci-like sequence.

Is that cheating? It depends on what combinations of the sequence elements are considered legitimate. The Fibonacci numbers have been used to parameterize Pythagorean triples in a variety of forms. The version given above, $(wz, 2xy, yz - wx)$, appears in Koshy [16] with a 1968 attribution to Umansky and Tallman. Here and below we use consecutive letters of the alphabet rather than the original subscript formulation, as a notational convenience. Much earlier, Raine [19] gave it this way: $(wz, 2xy, t)$, where, if w is F_n then t is F_{2n+3}. Boulger [1] extended Raine's results and observed that the triple can also be expressed $(wz, 2xy, x^2 + y^2)$. Horadam [13] reported it in the form $(xw, 2yz, 2yz + x^2)$. These combinations use a variety of different quadratic monomials, including both yz and x^2. So, if those are permitted, why not simply use the classical $(x^2 - y^2, 2xy, x^2 + y^2)$ and be done with it? The more complicated parameterizations we have cited then seem to be merely exercises in complexification.

In light of these remarks, it should be no surprise that the Fibonacci parameterization of Pythagorean triples can be generalized to $\mathcal{R}(a,b)$. For example, Shannon and Horadam [20] give the following version: $((a/b^2)xw, 2Pz(Pz - x), x^2 + 2Pz(Pz - x))$ where $P = (a^2 - -b)/2b^2$.

Using a modified version of the diophantine equation, we can get closer to the simplicity of Raine's formulation. For $\mathcal{R}(a,b)$ we replace the Pythagorean identity with

$$X^2 + bY^2 = Z^2 \tag{21}$$

and observe that the parameterization

$$(X, Y, Z) = (v^2 - bu^2, 2uv, v^2 + bu^2)$$

always produces solutions to (21). Now, if $w, x, y,$ and z are four consecutive terms of $A \in \mathcal{R}(a,b)$, then we can express the first and last as

$$w = \frac{1}{b}(y - ax)$$

$$z = bx + ay.$$

Define constants $c = b/a$ and $d = c - a$. Then a calculation verifies that

$$(X, Y, Z) = (cwz - dxy, 2xy, xz + bwy) \tag{22}$$

is a solution to (21). In fact, with $u = x$ and $v = y$, it is exactly the parameterization given above.

In the special case that $x = F_n^{(a,b)}$, we can also express (22) in the form

$$(X, Y, Z) = (cwz - dxy, 2xy, t)$$

where $t = F_{2n+1}$. This version, which generalizes the Raine result, follows from (20). Note, also, that when $a = b = 1$, (22) becomes $(wz, 2xy, xz + wy)$, which is another variant on the Fibonacci parameterization of Pythagorean triples.

Greatest common divisor The Fibonacci properties considered so far make sense for real sequences in $\mathcal{R}(a,b)$. Now, however, we will consider divisibility properties that apply to integer sequences. Accordingly, we henceforth assume that a and b are integers, and restrict our attention to sequences $A \in \mathcal{R}(a,b)$ for which the initial terms A_0 and A_1 are integers, as well. Evidently, this implies A is an integer sequence. In order to generalize the gcd property, we must make one additional assumption: that a and b are relatively prime. Then we can prove in $\mathcal{R}(a,b)$, that the gcd of F_m and F_n is F_k, where k is the gcd of m and n. The proof has two parts: We show that F_k is a divisor of both F_m and F_n, and that F_m/F_k and F_n/F_k are relatively prime. The first of these follows immediately from an observation about the skip operator already presented. The second part depends on several additional observations.

OBSERVATION 1. *F_k is a divisor of F_{nk} for all $n > 0$.*

Proof. We have already noted that $\Omega_k F = F_k \cdot F^{(a',b')}$ so every element of $\Omega_k F = F_0, F_k, F_{2k}, \ldots,$ is divisible by F_k.

OBSERVATION 2. *F_n and b are relatively prime for all $n \geq 0$.*

Proof. Suppose p is prime divisor of b. Since a and b are relatively prime, p is not a divisor of a. Modulo p, the fundamental recursion (1) becomes $F_{n+2} \equiv aF_{n+1}$, so $F_n \equiv F_1 a^{n-1}$ for $n \geq 1$. This shows that $F_n \not\equiv 0$, since p is not a divisor of a.

OBSERVATION 3. *If $A \in \mathcal{R}(a, b)$, and if p is a common prime divisor of A_k and A_{k+1}, but is not a divisor of b, then p is a divisor of A_n for all $n \geq 0$.*

Proof. If $k > 0$, $A_{k+1} = aA_k + bA_{k-1}$, so p is a divisor of A_{k-1}. By induction, p divides both A_0 and A_1, and therefore A_n for all $n \geq 0$.

OBSERVATION 4. *If positive integers h and k are relatively prime, then so are F_h and F_k.*

Proof. If p is a prime divisor of F_h and F_k, then by Observation 2, p is not a divisor of b. Since h and k are relatively prime, there exist integers r and s such that $rh + sk = 1$. Clearly r and s must differ in sign. Without loss of generality, we assume that $r < 0$, and define $t = -r$. Thus, $sk - th = 1$. Now by Observation 1, F_{sk} is divisible by F_k, and hence by p. Similarly, F_{th} is divisible by F_h, and hence, also by p. But F_{th} and F_{sk} are consecutive terms of F, so by Observation 3, p is a divisor of all F_n. That is a contradiction, and shows that F_h and F_k can have no common prime divisor.

OBSERVATION 5. *If $a' = L_k^{(a,b)}$ and $b' = -(-b)^k$, then a' and b' are relatively prime.*

Proof. Suppose, to the contrary, that p is a common prime divisor of a' and b'. Then clearly p is a divisor of b, and also a divisor of $L_k^{(a,b)}$, which equals $bF_{k-1}^{(a,b)} + F_{k+1}^{(a,b)}$ by (5). This makes p a divisor of $F_{k+1}^{(a,b)}$, which contradicts Observation 2.

With these observations, we now can prove the

THEOREM. *The gcd of F_m and F_n is F_k, where k is the gcd of m and n.*

Proof. Let $s = m/k$ and $t = n/k$, and observe that s and t are relatively prime. We consider $A = \Omega_k F = F_0, F_k, F_{2k}, \ldots$. As discussed earlier, A can also be expressed as $F_k \cdot F^{(a',b')}$ where $a' = L_k$ and $b' = -(-b)^k$. Moreover, by Observation 5, a' and b' are relatively prime. As in Observation 1, we see at once that every A_j is a multiple of F_k, so in particular, F_k is a divisor of $A_s = F_{ks} = F_m$ and $A_t = F_{kt} = F_n$. On the other hand, $F_m/F_k = F_s^{(a',b')}$ and $F_n/F_k = F_t^{(a',b')}$, are relatively prime by Observation 4. Thus, F_k is the gcd of F_m and F_n. ∎

Several remarks about this result are in order. First, in Michael [18], the corresponding result is established for the traditional Fibonacci numbers. That proof depends on the $\mathcal{R}(1, 1)$ instances of (19), Observation 1, and Observation 3, and extends to a proof for $\mathcal{R}(a, b)$ in a natural way.

Second, Holzsager [10] has described an easy construction of other sequences A_n for which $\gcd(A_n, A_m) = A_{\gcd(m,n)}$. First, for the primes p_k, define $A_{p_k} = q_k$ where the q_k are relatively prime. Then, extend A to the rest of the integers multiplicatively. That is, if $n = \prod p_i^{e_i}$ then $A_n = \prod q_i^{e_i}$. Such a sequence defines a mapping on the positive integers that carries the prime factorization of any subscript into a corresponding factorization involving the qs. This mapping apparently will commute with the gcd. By Observation 4, the terms $F_{p_k}^{(a,b)}$ are relatively prime, but since $F_2 = a$ and $F_4 = a^3 + 2ab$, the F mapping is not generally multiplicative. Thus, Holzsager's construction does not lead to examples of the form $F^{(a,b)}$.

Finally, we note that there is similar result for the (a, b)-Lucas numbers, which we omit in the interest of brevity. Both that result and the preceding Theorem also appear in Hilton and Pedersen [8]. Also, the general gcd result for F was known to Lucas, and we may conjecture that he knew the result for L as well.

In particular and in general

We have tried to show in this paper that much of the mystique of the Fibonacci numbers is misplaced. Rather than viewing F as a unique sequence with an amazing host of algebraic, combinatorial, and number theoretic properties, we ought to recognize that it is simply one example of a large class of sequences with such properties. In so arguing, we have implicitly highlighted the tension within mathematics between the particular and the general. Both have their attractions and pitfalls. On the one hand, by focusing too narrowly on a specific amazing example, we may lose sight of more general principles at work. But there is a countervailing risk that generalization may add nothing new to our understanding, and result in meaningless abstraction.

In the case at hand, the role of the skip operator should be emphasized. The proof of the gcd result, in particular, was simplified by the observation that the skip maps one $\mathcal{R}(a, b)$ to another. This observation offers a new, simple insight about the terms of Fibonacci sequences—an insight impossible to formulate without adopting the general framework of two-term recurrences.

It is not our goal here to malign the Fibonacci numbers. They constitute a fascinating example, rich with opportunities for discovery and exploration. But how much more fascinating it is that an entire world of such sequences exists. This world of the super sequences should not be overlooked.

REFERENCES

1. William Boulger, Pythagoras Meets Fibonacci, *Mathematics Teacher* **82** (1989), 277–782.
2. Carl B. Boyer, *A History of Mathematics*, 2nd ed., Revised by Uta C. Merzbach, Wiley, New York, 1991.
3. David M. Bressoud, *Factorization and Primality Testing*, Springer, New York, 1989.
4. Leonard E. Dickson, *History of the Theory of Numbers*, vol. 1: Divisibility and Primality, Chelsea, New York, 1952.
5. C. H. Edwards, Jr., *The Historical Development of the Calculus*, Springer, New York, 1979.
6. Ko Hayashi, Fibonacci numbers and the arctangent function, this MAGAZINE, **76**:3 (2003), 214–215.
7. Peter Hilton, Derek Holton, and Jean Pedersen, *Mathematical Reflections: In a Room with Many Mirrors*, Springer, New York, 1997.
8. Peter Hilton and Jean Pedersen, Fibonacci and Lucas numbers in teaching and research, *Journées Mathématiques & Informatique* **3** (1991–1992), 36–57.
9. Verner E. Hoggatt, Jr., *Fibonacci and Lucas Numbers*, Fibonacci Association, Santa Clara, CA, 1969.
10. Richard Holzsager, private correspondence, July, 2000.
11. Ross Honsberger, *Mathematical Gems III*, Mathematical Association of America, Washington, DC, 1985.
12. A. F. Horadam, A generalized Fibonacci sequence, *American Mathematical Monthly* **68** (1961), 455–459.
13. A. F. Horadam, Fibonacci number triples, *Amer. Math. Monthly* **68** (1961), 751–753.
14. A. F. Horadam, Generating functions for powers of a certain generalised sequences of numbers, *Duke Math. J.* **32** (1965), 437–446.
15. A. F. Horadam, Jacobsthal and Pell curves, *Fibonacci Quart.* **26** (1988), 77–83.
16. Thomas Koshy, *Fibonacci and Lucas Numbers with Applications*, Wiley, New York, 2001.
17. Edouard Lucas, Théorie des fonctions numériques simplement périodiques, *Amer. J. Math.* **1** (1878), 184–240, 289–321.
18. G. Michael, A new proof for an old property, *Fibonacci Quart.* **2** (1964), 57–58.
19. Charles W. Raine, Pythagorean triangles from the Fibonacci series, *Scripta Mathematica* 14 (1948), 164.
20. A. G. Shannon and A. F. Horadam, Generalized Fibonacci number triples, *Amer. Math. Monthly* **80** (1973), 187–190.
21. S. Vajda, *Fibonacci & Lucas Numbers, and the Golden Section, Theory and Applications*, Ellis Horwood Ltd., Chichester, 1989.
22. J. E. Walton and A. F. Horadam, Some further identities for the generalized Fibonacci sequence $\{H_n\}$, *Fibonacci Quart.* **12** (1974), 272–280.
23. Eric W. Weisstein, *CRC Concise Encyclopedia of Mathematics*, CRC Press, Boca Raton, FL, 1999. Also available on-line at http://mathworld.wolfram.com.

Originally appeared as:
Kalman, Dan and Robert Mena. "The Fibonacci Numbers—Exposed." *Mathematics Magazine.* vol. 76, no. 3 (June 2003), pp. 167–181.

The Fibonacci Numbers—Exposed More Discretely

*Arthur T. Benjamin and
Jennifer J. Quinn*

In the previous article, Kalman and Mena [5] propose that Fibonacci and Lucas sequences, despite the mathematical favoritism shown them for their abundant patterns, are nothing more than ordinary members of a class of super sequences. Their arguments are beautiful and convinced us to present the same material from a more *discrete* perspective. Indeed, we will present a simple combinatorial context encompassing nearly all of the properties discussed in [5].

As in the Kalman-Mena article, we generalize Fibonacci and Lucas numbers: Given nonnegative integers a and b, the *generalized Fibonacci sequence* is

$$F_0 = 0, \quad F_1 = 1, \quad \text{and for } n \geq 2, \quad F_n = aF_{n-1} + bF_{n-2}. \tag{1}$$

The *generalized Lucas sequence* is

$$L_0 = 2, \quad L_1 = a, \quad \text{and for } n \geq 2, \quad L_n = aL_{n-1} + bL_{n-2}.$$

When $a = b = 1$, these are the celebrity Fibonacci and Lucas sequences. For now, we will assume that a and b are nonnegative integers. But at the end of the article, we will see how our methods can be extended to noninteger values of a and b.

Kalman and Mena prove the following generalized Fibonacci identities

$$F_n = F_m F_{n-m+1} + bF_{m-1}F_{n-m} \tag{2}$$

$$(a+b-1)\sum_{i=1}^{n} F_i = F_{n+1} + bF_n - 1 \tag{3}$$

$$a\left(b^n F_0^2 + b^{n-1}F_1^2 + \cdots + bF_{n-1}^2 + F_n^2\right) = F_n F_{n+1} \tag{4}$$

$$F_{n-1}F_{n+1} - F_n^2 = (-1)^n b^{n-1} \tag{5}$$

$$\gcd(F_n, F_m) = F_{\gcd(n,m)} \tag{6}$$

$$L_n = aF_n + 2bF_{n-1} \tag{7}$$

$$L_n = F_{n+1} + bF_{n-1} \tag{8}$$

using the tool of difference operators acting on the real vector space of real sequences. In this paper, we offer a purely combinatorial approach to achieve the same results. We hope that examining these identities from different perspectives, the reader can more fully appreciate the unity of mathematics.

---

*Editor's Note: Readers interested in clever counting arguments will enjoy reading the authors' upcoming book, *Proofs That Really Count: The Art of Combinatorial Proof*, published by the MAA.

Fibonacci numbers—The combinatorial way

There are many combinatorial interpretations for Fibonacci and Lucas numbers [3]. We choose to generalize the "square and domino tiling" interpretation here. We show that the classic Fibonacci and Lucas identities naturally generalize to the (a, b) recurrences simply by adding a splash of color.

For nonnegative integers a, b, and n, let f_n count the number of ways to tile a $1 \times n$ board with 1×1 colored squares and 1×2 colored dominoes, where there are a color choices for squares and b color choices for dominoes. We call these objects *colored n-tilings*. For example, $f_1 = a$ since a length 1 board must be covered by a colored square; $f_2 = a^2 + b$ since a board of length 2 can be covered with two colored squares or one colored domino. Similarly, $f_3 = a^3 + 2ab$ since a board of length 3 can be covered by 3 colored squares or a colored square and a colored domino in one of 2 orders. We let $f_0 = 1$ count the empty board. Then for $n \geq 2$, f_n satisfies the generalized Fibonacci recurrence

$$f_n = af_{n-1} + bf_{n-2},$$

since a board of length n either ends in a colored square preceded by a colored $(n-1)$-tiling (tiled in af_{n-1} ways) or a colored domino preceded by a colored $(n-2)$-tiling (tiled in bf_{n-2} ways.) Since $f_0 = 1 = F_1$ and $f_1 = a = F_2$, we see that for all $n \geq 0$, $f_n = F_{n+1}$. After defining $f_{-1} = 0$, we now have a combinatorial definition for the generalized Fibonacci numbers.

THEOREM 1. *For $n \geq 0$, $F_n = f_{n-1}$ counts the number of colored $(n-1)$-tilings (of a $1 \times (n-1)$ board) with squares and dominoes where there are a colors for squares and b colors for dominoes.*

Using Theorem 1, equations (2) through (6) can be derived and appreciated combinatorially. In most of these, our combinatorial proof will simply ask a question and answer it two different ways.

For instance, if we apply Theorem 1 to equation (2) and reindex by replacing n by $n + 1$ and m by $m + 1$, we obtain

IDENTITY 1. *For $0 \leq m \leq n$,*

$$f_n = f_m f_{n-m} + bf_{m-1} f_{n-m-1}.$$

Question: How many ways can a board of length n be tiled with colored squares and dominoes?

Answer 1: By Theorem 1, there are f_n colored n-tilings.

Answer 2: Here we count how many colored n-tilings are *breakable* at the m-th cell and how many are not. To be breakable, our tiling consists of a colored m-tiling followed by a colored $(n-m)$-tiling, and there are $f_m f_{n-m}$ such tilings. To be *unbreakable* at the m-th cell, our tiling consists of a colored $(m-1)$-tiling followed by a colored domino on cells m and $m + 1$, followed by a colored $(n-m-1)$-tiling; there are $bf_{m-1} f_{n-m-1}$ such tilings. Altogether, there are $f_m f_{n-m} + bf_{m-1} f_{n-m-1}$ colored n-tilings.

Since our logic was impeccable for both answers, they must be the same. The advantage of this proof is that it makes the identity *memorable* and visualizable. See FIGURE 1 for an illustration of the last proof.

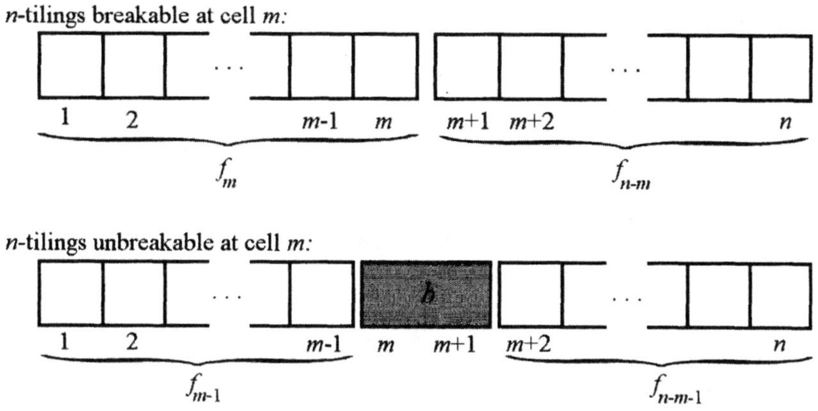

n-tilings breakable at cell *m*:

n-tilings unbreakable at cell *m*:

Figure 1 A colored *n*-tiling is either breakable or unbreakable at cell *m*

Equation (3) can be rewritten as the following identity.

IDENTITY 2. *For $n \geq 0$,*

$$f_n - 1 = (a-1)f_{n-1} + (a+b-1)[f_0 + f_1 + \cdots + f_{n-2}].$$

Question: How many colored *n*-tilings exist, excluding the tiling consisting of all white squares?

Answer 1: By definition, $f_n - 1$. (Notice how our question and answer become shorter with experience!)

Answer 2: Here we partition our tilings according to the last tile that is not a white square. Suppose the last tile that is not a white square begins on cell k. If $k = n$, that tile is a square and there are $a - 1$ choices for its color. There are f_{n-1} colored tilings that can precede it for a total of $(a-1)f_{n-1}$ tilings ending in a nonwhite square. If $1 \leq k \leq n-1$, the tile covering cell k can be a nonwhite square or a domino covering cells k and $k+1$. There are $a+b-1$ ways to pick this tile and the previous cells can be tiled f_{k-1} ways. Altogether, there are $(a-1)f_{n-1} + \sum_{k=1}^{n-1}(a+b-1)f_{k-1}$ colored *n*-tilings, as desired.

Notice how easily the argument generalizes if we partition according to the last tile that is not a square of color 1 or 2 or... or c. Then the same reasoning gives us for any $1 \leq c \leq a$,

$$f_n - c^n = (a-c)f_{n-1} + ((a-c)c + b)[f_0 c^{n-2} + f_1 c^{n-3} + \cdots + f_{n-2}]. \quad (9)$$

Likewise, by partitioning according to the last tile that is not a black domino, we get a slightly different identity, depending on whether the length of the tiling is odd or even:

$$f_{2n+1} = a(f_0 + f_2 + \cdots + f_{2n}) + (b-1)(f_1 + f_3 + \cdots + f_{2n-1}),$$
$$f_{2n} - 1 = a(f_1 + f_3 + \cdots + f_{2n-1}) + (b-1)(f_0 + f_2 + \cdots + f_{2n-2}).$$

After applying Theorem 1 to equation (4) and reindexing ($n \to n+1$) we have

IDENTITY 3. *For $n \geq 0$,*

$$a \sum_{k=0}^{n} f_k^2 b^{n-k} = f_n f_{n+1}.$$

Question: In how many ways can we create a colored n-tiling and a colored $(n+1)$-tiling?

Answer 1: $f_n f_{n+1}$.

Answer 2: For this answer, we ask for $0 \leq k \leq n$, how many colored tiling pairs exist where cell k is the last cell for which both tilings are breakable? (Equivalently, this counts the tiling pairs where the last square occurs on cell $k+1$ in exactly one tiling.) We claim this can be done $af_k^2 b^{n-k}$ ways, since to construct such a tiling pair, cells 1 through k of the tiling pair can be tiled f_k^2 ways, the colored square on cell $k+1$ can be chosen a ways (it is in the n-tiling if and only if $n-k$ is odd), and the remaining $2n-2k$ cells are covered with $n-k$ colored dominoes in b^{n-k} ways. See FIGURE 2. Altogether, there are $a \sum_{k=0}^{n} f_k^2 b^{n-k}$ tilings, as desired.

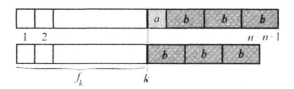

Figure 2 A tiling pair where the last mutually breakable cell occurs at cell k

The next identity uses a slightly different strategy. We hope that the reader does not *find fault* with our argument.

Consider the two colored 10-tilings offset as in FIGURE 3. The first one tiles cells 1 through 10; the second one tiles cells 2 through 11. We say that there is a *fault* at cell i, $2 \leq i \leq 10$, if both tilings are breakable at cell i. We say there is a fault at cell 1 if the first tiling is breakable at cell 1. Put another way, the pair of tilings has a fault at cell i for $1 \leq i \leq 10$ if neither tiling has a domino covering cells i and $i+1$. The pair of tilings in FIGURE 3 has faults at cells 1, 2, 5, and 7. We define the *tail* of a tiling to be the tiles that occur after the last fault. Observe that if we swap the tails of FIGURE 3 we obtain the 11-tiling and the 9-tiling in FIGURE 4 and it has the same faults.

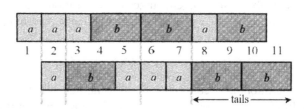

Figure 3 Two 10-tilings with their faults (indicated with gray lines) and tails

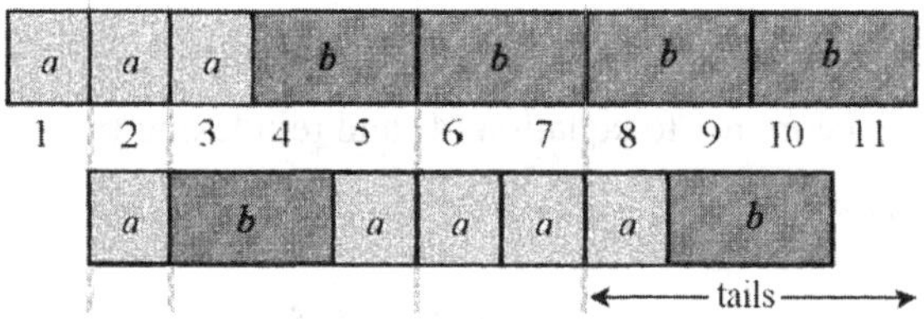

Figure 4 After tail-swapping, we have an 11-tiling and a 9-tiling with exactly the same faults

Tail swapping is the basis for the identity below, based on (5). At first glance, it may appear unsuitable for combinatorial proof due to the presence of the $(-1)^n$ term. Nonetheless, we will see that this term is merely the error term of an *almost* one-to-one correspondence between two sets whose sizes are easily counted. We use a different format for this combinatorial proof.

IDENTITY 4. $f_n^2 = f_{n+1}f_{n-1} + (-1)^n b^n$

Set 1: Tilings of two colored n-boards (a *top* board and a *bottom* board). By definition, this set has size f_n^2.

Set 2: Tilings of a colored $(n+1)$-board and a colored $(n-1)$-board. This set has size $f_{n+1}f_{n-1}$.

Correspondence: First, suppose n is odd. Then the top and bottom board must each have at least one square. Notice that a square in cell i ensures that a fault must occur at cell i or cell $i-1$. Swapping the tails of the two n-tilings produces an $(n+1)$-tiling and an $(n-1)$-tiling with the same tails. This produces a 1-to-1 correspondence between all pairs of n-tilings and all tiling pairs of sizes $n+1$ and $n-1$ that have faults. Is it possible for a tiling pair with sizes $n+1$ and $n-1$ to be fault free? Yes, with all colored dominoes in *staggered formation* as in FIGURE 5, which can occur b^n ways. Thus, when n is odd, $f_n^2 = f_{n+1}f_{n-1} - b^n$.

Similarly, when n is even, tail swapping creates a 1-to-1 correspondence between faulty tiling pairs. The only fault-free tiling pair is the all domino tiling of FIGURE 6. Hence, $f_n^2 = f_{n+1}f_{n-1} + b^n$. Considering the odd and even case together produces our identity.

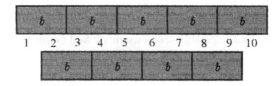

Figure 5 When n is odd, the only fault-free tiling pairs consist of all dominoes

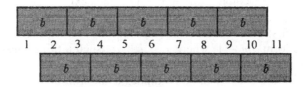

Figure 6 When n is even, the only fault-free tiling pairs consist of all dominoes

We conclude this section with a combinatorial proof of what we believe to be the most beautiful Fibonacci fact of all.

THEOREM 2. *For generalized Fibonacci numbers defined by (1) with relatively prime integers a and b,*

$$\gcd(F_n, F_m) = F_{\gcd(n,m)}. \tag{10}$$

We will need to work a little harder to prove this theorem combinatorially, but it can be done. Fortuitously, we have already combinatorially derived the identities needed to prove the following lemma.

LEMMA 1. *For generalized Fibonacci numbers defined by (1) with relatively prime integers a and b and for all* $m \geq 1$, F_m *and* bF_{m-1} *are relatively prime.*

Proof. First we claim that F_m is relatively prime to b. By conditioning on the location of the last colored domino (if any exist), equation (9) says (after letting $c = a$ and reindexing),

$$F_m = a^{m-1} + b \sum_{j=1}^{m-2} a^{j-1} F_{m-1-j}.$$

Consequently, if $d > 1$ is a divisor of F_m and b, then d must also divide a^{m-1}, which is impossible since a and b are relatively prime.

Next we claim that F_m and F_{m-1} are relatively prime. This follows from equation (5) since if $d > 1$ divides F_m and F_{m-1}, then d must divide b^{m-1}. But this is impossible since F_m and b are relatively prime.

Thus since $\gcd(F_m, b) = 1$ and $\gcd(F_m, F_{m-1}) = 1$, then $\gcd(F_m, bF_{m-1}) = 1$, as desired. ∎

To prove Theorem 2, we exploit Euclid's algorithm for computing greatest common divisors: If $n = qm + r$ where $0 \leq r < m$, then

$$\gcd(n, m) = \gcd(m, r).$$

Since the second component gets smaller at each iteration, the algorithm eventually reaches $\gcd(g, 0) = g$, where g is the greatest common divisor of n and m. The identity below shows one way that F_n can be expressed in terms of F_m and F_r. It may look formidable at first but makes perfect sense when viewed combinatorially.

IDENTITY 5. *If* $n = qm + r$, *where* $0 \leq r < m$, *then*

$$F_n = (bF_{m-1})^q F_r + F_m \sum_{j=1}^{q} (bF_{m-1})^{j-1} F_{(q-j)m+r+1}.$$

Question: How many colored $(qm + r - 1)$-tilings exist?

Answer 1: $f_{qm+r-1} = F_{qm+r} = F_n$.

Answer 2: First we count all such colored tilings that are unbreakable at every cell of the form $jm - 1$, where $1 \leq j \leq q$. Such a tiling must have a colored domino starting on cell $m - 1, 2m - 1, \ldots, qm - 1$, which can be chosen b^q ways. Before each of these dominoes is an arbitrary $(m - 2)$-tiling that can each be chosen f_{m-2} ways. Finally, cells $qm + 1, qm + 2, \ldots, qm + r - 1$ can be tiled f_{r-1} ways. See FIGURE 7. Consequently, the number of colored

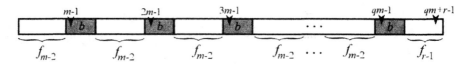

Figure 7 There are $(bF_{m-1})^q F_r$ colored $(qm + r - 1)$-tilings with no breaks at any cells of the form $jm - 1$ where $1 \leq j \leq q$

tilings with no $jm - 1$ breaks is $b^q(f_{m-2})^q f_{r-1} = (bF_{m-1})^q F_r$. Next, we partition the remaining colored tilings according to the first breakable cell of the form $jm - 1$, $1 \leq j \leq q$. By similar reasoning as before, this can be done $(bF_{m-1})^{j-1} F_m F_{(q-j)m+r+1}$ ways. (See FIGURE 8.) Altogether, the number of colored tilings is $(bF_{m-1})^q F_r + F_m \sum_{j=1}^{q}(bF_{m-1})^{j-1} F_{(q-j)m+r+1}$.

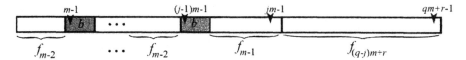

Figure 8 There are $(bF_{m-1})^{j-1} F_m F_{(q-j)m+r+1}$ colored $(qm + r - 1)$-tilings that are breakable at cell $jm - 1$, but not at cells of the form $im - 1$ where $1 \leq i < j$

The previous identity explicitly shows that F_n is an integer combination of F_m and F_r. Consequently, d is a common divisor of F_n and F_m if and only if d divides F_m and $(bF_{m-1})^q F_r$. But by Lemma 1, since F_m is relatively prime to bF_{m-1}, d must be a common divisor of F_m and F_r. Thus F_n and F_m have the same common divisors (and hence the same gcd) as F_m and F_r. In other words,

COROLLARY 1. *If $n = qm + r$, where $0 \leq r < m$, then*

$$\gcd(F_n, F_m) = \gcd(F_m, F_r).$$

But wait!! This corollary is the same as Euclid's algorithm, but with F's inserted everywhere. This proves Theorem 2 by following the same steps as Euclid's algorithm. The $\gcd(F_n, F_m)$ will eventually reduce to $\gcd(F_g, F_0) = (F_g, 0) = F_g$, where g is the greatest common divisor of m and n.

Lucas numbers—the combinatorial way

Generalized Lucas numbers are nothing more than generalized Fibonacci numbers running in circles. Specifically, for nonnegative integers a, b, and n, let ℓ_n count the number of ways to tile a circular $1 \times n$ board with slightly curved colored squares and dominoes, where there are a colors for squares and b colors for dominoes. Circular tilings of length n will be called *n-bracelets*. For example, when $a = b = 1$, $\ell_4 = 7$, as illustrated in FIGURE 9. In general, $\ell_4 = a^4 + 4a^2b + 2b^2$.

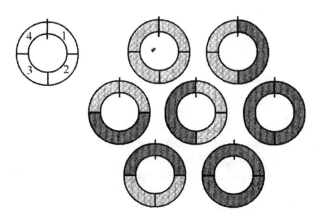

Figure 9 A circular board of length 4 and its seven 4-bracelets

From the definition of ℓ_n it follows that $\ell_n \geq f_n$ since an n-bracelet can have a domino covering cells n and 1; such a bracelet is called *out-of-phase*. Otherwise, there is a break between cells n and 1, and the bracelet is called *in-phase*. The first 5 bracelets in FIGURE 9 are in-phase and the last 2 are out-of-phase. Notice $\ell_1 = a$ and $\ell_2 = a^2 + 2b$ since a circular board of length 2 can be covered with two squares, an in-phase domino, or an out-of-phase domino. We define $\ell_0 = 2$ to allow 2 empty bracelets, one in-phase and one out-of-phase. In general for $n \geq 2$, we have

$$\ell_n = a\ell_{n-1} + b\ell_{n-2}$$

because an n-bracelet can be created from an $(n-1)$-bracelet by inserting a square to the left of the first tile or from an $(n-2)$-bracelet by inserting a domino to the left of the first tile. The *first tile* is the one covering cell 1 and it determines the phase of the bracelet; it may be a square, a domino covering cells 1 and 2, or a domino covering cells n and 1.

Since $\ell_0 = 2 = L_0$ and $\ell_1 = a = L_1$, we see that for all $n \geq 0$, $\ell_n = L_n$. This becomes our combinatorial definition for the generalized Lucas numbers.

THEOREM 3. *For all $n \geq 0$, $L_n = \ell_n$ counts the number of n-bracelets created with colored squares and dominoes where there are a colors for squares and b colors for dominoes.*

Now that we know how to combinatorially think of Lucas numbers, generalized identities are a piece of cake. Equation (7), which we rewrite as

$$L_n = af_{n-1} + 2bf_{n-2},$$

reflects the fact that an n-bracelet can begin with a square (af_{n-1} ways), an in-phase domino (bf_{n-2} ways), or an out-of-phase domino (bf_{n-2} ways). Likewise, equation (8), rewritten as

$$L_n = f_n + bf_{n-2},$$

conditions on whether or not an n-bracelet is in-phase (f_n ways) or out-of-phase (bf_{n-2} ways.)

You might even think these identities are too easy, so we include a couple more generalized Lucas identities for you to ponder along with visual hints. For more details see [4].

$$f_{n-1}L_n = f_{2n-1} \qquad \text{See FIGURE 10.}$$
$$L_n^2 = L_{2n} + 2 \cdot (-b)^n \qquad \text{See FIGURE 11.}$$

Further generalizations and applications

Up until now, all of our proofs have depended on the fact that the recurrence coefficients a and b were nonnegative integers, even though most generalized Fibonacci identities remain true when a and b are negative or irrational or even complex numbers. Additionally, our sequences have had very specific initial conditions ($F_0 = 0$, $F_1 = 1$, $L_0 = 2$, $L_1 = a$), yet many identities can be extended to handle arbitrary ones. This section illustrates how combinatorial arguments can still be used to overcome these apparent obstacles.

Arbitrary initial conditions Let a, b, A_0, and A_1 be nonnegative integers and consider the sequence A_n defined by the recurrence, for $n \geq 2$, $A_n = aA_{n-1} + bA_{n-2}$. As

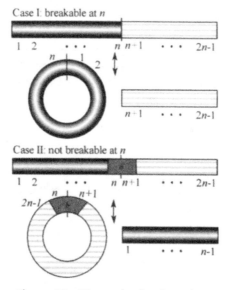

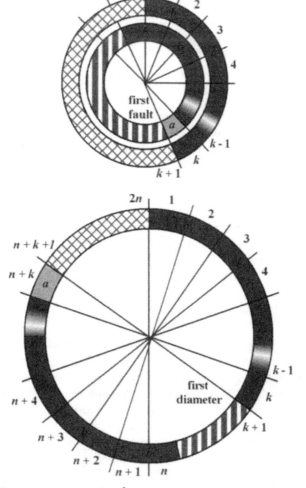

Figure 10 Picture for $f_{n-1}L_n = f_{2n-1}$

Figure 11 Picture for $L_n^2 = L_{2n} + 2 \cdot (-b)^n$ when n is even

described in [1] and Chapter 3 of [4], the *initial conditions* A_0 and A_1 determine the number of choices for the *initial tile*. Just like F_n, A_n counts the number of colored n-tilings where *except for the first tile* there are a colors for squares and b colors for dominoes. For the first tile, we allow A_1 choices for a square and bA_0 choices for a domino. So as not to be confused with the situation using ideal initial conditions, we assign the first tile a *phase* instead of a color.

For example, when $A_0 = 1$ and $A_1 = a$, the ideal initial conditions, we have a choices for the phase of an initial square and b choices for the phase of an initial domino. Since *all* squares have a choices and *all* dominoes have b choices, it follows that $A_n = f_n$. When $A_0 = 0$ and $A_1 = 1$, A_n counts the number of colored n-tilings that begin with an "uncolored" square; hence $A_n = f_{n-1} = F_n$. When $A_0 = 2$ and $A_1 = a$, A_n counts the number of colored n-tilings that begin with a square in one of a phases or a domino in one of $2b$ phases. This is equivalent to a colored n-bracelet since there are an equal number of square phases as colors and twice as many domino phases as colors (representing whether the initial domino is in-phase or out-of-phase.) Thus when $A_0 = 2$ and $A_1 = a$, we have $A_n = L_n$.

In general, there are $A_1 f_{n-1}$ colored tilings that begin with a phased square and $bA_0 f_{n-2}$ colored tilings that being with a phased domino. Hence we obtain the following identity from Kalman and Mena [5]:

$$A_n = bA_0 F_{n-1} + A_1 F_n. \tag{11}$$

Arbitrary recurrence coefficients Rather than assigning a discrete number of colors for each tile, we can assign weights. Squares have weight a and dominoes have weight b except for the initial tile, which has weight A_1 as a square and weight bA_0 as a domino. Here a, b, A_0, and A_1 do not have to be nonnegative integers, but can be chosen from the set of complex numbers (or from any commutative ring). We define the *weight of an n-tiling* to be the product of the weights of its individual tiles. For example, the 7-tiling "square-domino-domino-square-square" has weight $a^3 b^2$ with ideal initial conditions and has weight $A_1 a^2 b^2$ with arbitrary initial conditions. Inductively one can prove that for $n \geq 1$, A_n is the sum of the weights of all weighted n-tilings, which we call the *total weight* of an n-board.

If X is an m-tiling of weight w_X and Y is an n-tiling of weight w_Y, then X and Y can be glued together to create an $(m+n)$-tiling of weight $w_X w_Y$. If an m-board can be tiled s different ways and has total weight $A_m = w_1 + w_2 + \cdots + w_s$ and an n-board can be tiled t ways with total weight $A_n = x_1 + x_2 + \cdots + x_t$, then the sum of the weights of all weighted $(m+n)$-tilings breakable at cell m is

$$\sum_{i=1}^{s} \sum_{j=1}^{t} w_i x_j = (w_1 + w_2 + \cdots + w_s)(x_1 + x_2 + \cdots + x_t) = A_m A_n.$$

Now we are prepared to revisit some of our previous identities using the weighted approach. For Identity 1, we find the total weights of an n-board in two different ways. On the one hand, since the initial conditions are ideal, the total weight is $A_n = f_n$. On the other hand, the total weight is comprised of the total weight of those tilings that are breakable at cell m ($f_m f_{n-m}$) plus the total weight of those tilings that are unbreakable at cell m ($f_{m-1} b f_{n-m-1}$). Identities 2, 3, and 5 can be argued in similar fashion.

For Identity 4, we define the weight of a tiling pair to be the product of the weights of all the tiles, and define the total weight as before. Next we observe that tail swapping preserves the weight of the tiling pair since no tiles are created or destroyed in the process. Consequently, the total weight of the set of *faulty* tiling pairs (X, Y) where

X and Y are n-tilings equals the total weight of the faulty tiling pairs (X', Y'), where X' is an $(n + 1)$-tiling and Y' is an $(n - 1)$-tiling. The fault-free tiling pair, for the even and odd case, will consist of n dominoes and therefore have weight b^n. Hence identity 4 remains true even when a and b are complex numbers.

Kalman and Mena [5] prove Binet's formulas for Fibonacci numbers

$$F_n = \frac{1}{\sqrt{5}}\left[\left(\frac{1+\sqrt{5}}{2}\right)^n - \left(\frac{1-\sqrt{5}}{2}\right)^n\right], \tag{12}$$

and for more general sequences.

These can also be proved combinatorially [2]. Binet's formula follows from considering a random tiling of an infinitely long strip with cells $1, 2, 3, \ldots$, where squares and dominoes are randomly and independently inserted from left to right. The probability of inserting a square is $1/\phi$ and the probability of inserting a domino is $1/\phi^2$, where $\phi = (1 + \sqrt{5})/2$. (Conveniently, $1/\phi + 1/\phi^2 = 1$.) By computing the probability of being breakable at cell $n - 1$ in two different ways, Binet's formula instantly appears. This approach can be extended to generalized Fibonacci numbers and beyond, as described in [1].

Finally, we observe that the Pythagorean Identity presented in [5] for traditional Fibonacci numbers, which can be written as

$$(f_{n-1}f_{n+2})^2 + (2f_n f_{n+1})^2 = f_{2n+2}^2$$

can also be proved combinatorially. For details, see [4].

We hope that this paper illustrates that Fibonacci and Lucas sequences are members of a very special class of sequences satisfying beautiful properties, namely sequences defined by second order recurrence relations. But why stop there? Combinatorial interpretations can be given to sequences that satisfy higher-order recurrences. That is, if we define $a_j = 0$ for $j < 0$ and $a_0 = 1$, then for $n \geq 1$, $a_n = c_1 a_{n-1} + \cdots + c_k a_{n-k}$ counts the number of ways to tile a board of length n with colored tiles of length at most k, where each tile of length i has c_i choices of color. Again, this interpretation can be extended to handle complex values of c_i and arbitrary initial conditions. See Chapter 3 of [4]. Of course, the identities tend to be prettier for the two-term recurrences, and are usually prettiest for the traditional Fibonacci and Lucas numbers.

Acknowledgment. We thank Jeremy Rouse, Dan Kalman, and Robert Mena for their inspiration.

REFERENCES

1. A. T. Benjamin, C. R. H. Hanusa, and F. E. Su, Linear recurrences and Markov chains, *Utilitas Mathematica* **64** (2003), 3–17.
2. A. T. Benjamin, G. M. Levin, K. Mahlburg, and J. J. Quinn, Random approaches to Fibonacci identities, *Amer. Math. Monthly* **107**:6 (2000), 511–516.
3. A. T. Benjamin and J. J. Quinn, Recounting Fibonacci and Lucas Identities, *College Math. J.* **30**:5 (1999), 359–366.
4. ———, *Proofs That Really Count: The Art of Combinatorial Proof*, Mathematical Association of America, Washington, D.C., 2003.
5. D. Kalman and R. Mena, The Fibonacci Numbers—Exposed, this MAGAZINE **76**:3 (2003), 167–181.
6. T. Koshy, *Fibonacci and Lucas Numbers with Applications*, John Wiley and Sons, NY, 2001.
7. S. Vajda, *Fibonacci & Lucas Numbers, and the Golden Section: Theory and Applications*, John Wiley and Sons, New York, 1989.

Originally appeared as:

Benjamin, Arthur T. and Jennifer J. Quinn. "The Fibonacci Numbers—Exposed More Discretely." *Mathematics Magazine.* vol. 76, no. 3 (June 2003), pp. 182–192.

Part VI: Number-Theoretic Functions

In this chapter, we present Biscuits that deal with famous number-theoretic functions. One is the subject of the most famous open conjectures in mathematics and another leads to one of the most maddening unsolved problems. A third and fourth show us one of the great masters at work. The fifth is a deceptively simple function that is ubiquitous in fields from combinatorics to probability to calculus, and the sixth is the subject of nearly two hundred proofs.

The Riemann zeta function is the function $\zeta(s) = \sum_{n=1}^{\infty} n^{-s}$, for s a complex number whose real part is greater than 1. Riemann showed how to extend this function to the entire complex plane except for the point $s = 1$. The Riemann hypothesis, one of the seven Millennium Problems, states that all the complex zeroes of $\zeta(s)$ that are in the strip $0 \leq \text{Re}(s) \leq 1$ lie on the line $s = \frac{1}{2} + it$. The title of Jennifer Beineke and Chris Hughes's "Great Moments of the Riemann Zeta Function" refers both to historical events and to mathematical moments. The historical moments include Euler's proof that there are infinitely many primes, in which he proves

$$\zeta(s) = \prod_{p \text{ prime}} \left(1 - \frac{1}{p^s}\right)^{-1}$$

as well as Riemann's statement of his hypothesis and several others. Beineke and Hughes then define the moment of the modulus of the Riemann zeta function by

$$I_k(T) = \frac{1}{T} \int_0^\infty \left|\zeta\left(\frac{1}{2} + it\right)\right|^{2k} dt$$

and take us through the work of several mathematicians on properties of the second and fourth moments. They describe work by Hardy and Littlewood and by Ingham, and give a connection between the zeroes of the zeta function and random matrix theory, finishing with recent work of many researchers–including themselves. This sparkling paper appears in print for the first time in this collection ("not sold in stores"), and includes fifteen pictures.

A great part of number theory's charm is its almost endless collection of questions that are so easy to state and oh, so hard to answer. One of the more recent additions to this collection of tantalizers is the subject of our next Biscuit, Marc Chamberland's "The Collatz chameleon" (*Math Horizons*, vol. 14, no. 2 (November 2006), pp. 5–8). This is the notorious "$3x + 1$" problem, one of today's most easily-stated unsolved mathematical problems. Begin a sequence with a positive integer x and iterate the function $C(x)$, where $C(x) = 3x + 1$ if x is odd, and $C(x) = \frac{x}{2}$ if x is even. Does the set of C-iterates of x eventually include 1, for every x? In this lovely Biscuit, Chamberland tells us how this problem, like a chameleon, can appear in many different shades, and proceeds to describe the Stopping Time, Tree, Number Representation, No Adding, Chaotic, Binomial Coefficient and Functional Equation shades. Take some writing equipment and a computer, dive in, and have fun!

Another intriguing function of number theory is $p(n)$, the number of partitions of the integer n. For example, $p(5) = 7$, since 5 can be partitioned seven ways: namely $5 = 4 + 1 = 3 + 2 = 3 + 1 + 1 = 2 + 1 + 1 + 1 = 1 + 1 + 1 + 1 + 1$. Although no closed form formula for $p(n)$ is known, it does satisfy a mysterious recurrence. Specifically, $p(0) = 1$, $p(n) = 0$ for $n < 0$, and for $n \geq 1$,

$$p(n) = p(n-1) + p(n-2) - p(n-5) - p(n-7) + p(n-12) + \cdots = \sum_j (-1)^j p(n - a(j)),$$

where the sum ranges over all integers j, and $a(j) = (3j^2 + j)/2$. For example, $p(5) = p(4) + p(3) - p(0) = 5 + 3 - 1 = 7$. Although many proofs of this theorem are known, none is as short as the one given by David Bressoud and Doron Zeilberger. In "Bijecting Euler's partition recurrence" (*Amer. Math. Monthly*, vol. 92, no. 1 (January, 1985), pp. 54–55), the authors provide a one-page combinatorial proof of this remarkable identity. The proof is in the spirit of the single-sentence proof given earlier by Zagier, but the steps are easier to verify.

Incidentally, Euler was unable to prove this recurrence but discovered a beautiful consequence of this result in "Discovery of a most extraordinary law of the numbers concerning the sum of their divisors," translated by master expositor George Pólya (*Mathematics and Plausible Reasoning*, Volume I, Princeton University Press, (1954), pp. 90–98). In this paper, we are treated to Euler's thought processes as he explains why the sum of the divisors of n, denoted $\sigma(n)$, satisfies the same recurrence as $p(n)$, namely

$$\sigma(n) = \sigma(n-1) + \sigma(n-2) - \sigma(n-5) - \sigma(n-7) + \sigma(n-12) + \cdots$$

with the extra condition that if $\sigma(0)$ appears in the sum, then we assign it the value n. For example, $\sigma(5) = \sigma(4) + \sigma(3) - \sigma(0) = 7 + 4 - 5 = 6$.

The factorial function $n!$ counts the number of arrangements of n distinct objects, where n is a nonnegative integer. It makes early appearances in combinatorics, discrete probability (which some would call applied combinatorics), calculus (in Taylor and Maclaurin Series), and number theory. Euler extended the factorial function to real arguments, defining the gamma function by $\Gamma(s) = \int_0^\infty e^{-x} x^{s-1} dx$. It turns out that $\Gamma(1) = 1$ and $\Gamma(n+1) = n\Gamma(n)$; thus, if n is a positive integer, then $\Gamma(n) = (n-1)!$. But as Manjul Bhargava shows us in the Hasse Prize-winning "The factorial function and generalizations" (*Amer. Math. Monthly*, vol. 107, no. 9 (November 2000), pp. 7833–799), the gamma function was only the first of a wide variety of generalizations of $n!$. Bhargava begins with a fact from elementary number theory, namely that the product of any k consecutive integers is divisible by $k!$. He describes three other similar facts, notes their strong dependence on the structure of the integers, and wonders whether they are true in a more general framework. He then extends the definition of the factorial function to Dedekind rings, a family of rings that includes the rings of algebraic integers and polynomial rings over a finite field, and studies the questions there. This Biscuit is a bit chewier and has spent more time in the oven than most of the others in our collection; as such, it is more challenging than the rest. However, it is beautifully written and contains elegant mathematics, and working through it slowly and carefully will repay the effort. Try it and see!

Let p be an odd prime and let a be an integer relatively prime to p. The Legendre symbol $\left(\frac{a}{p}\right)$ is defined to be $+1$ or -1 according as the congruence $x^2 \equiv a \pmod{p}$ does or does not have a solution. Gauss's Quadratic Reciprocity Law (QRL) states that if p and

q are distinct odd primes, then

$$\left(\frac{p}{q}\right)\left(\frac{q}{p}\right)=(-1)^{\frac{p-1}{2}\cdot\frac{q-1}{2}}$$

Gauss gave eight different proofs of the QRL, and over the years, almost two hundred proofs have appeared. In our last functional Biscuit, "An elementary proof of the Quadratic Reciprocity Law" (*Amer. Math. Monthly*, vol. 111, no. 1 (January 2004), pp. 48–50), Sey Kim gives the QRL a proof it deserves. Elegant, highly original and easy to follow, this proof might be another one from Erdős's "Book". Work through it and you may agree that Kim's proof of the QRL should become the standard proof in elementary number theory courses from now on.

Great Moments of the Riemann zeta Function

*Jennifer Beineke and
Chris Hughes*

Abstract We all know that movies have great moments and sports have great moments, but arguably mathematics has the greatest moments of all. For those who need to be convinced of this claim, we detail below an important number-theoretic example: that of the Riemann zeta function $\zeta(s)$. Here you can learn about its memorable moments, in both a mathematical and a historical sense. Not only will you read about some current open problems involving $\zeta(s)$, but you can enjoy some stories that have arisen over the years about the mathematicians working in the area.

Great Moment I: The Discovery of the Euler Product

First, recall that the Riemann zeta function is defined by the following series:

$$\zeta(s) = \sum_{n=1}^{\infty} \frac{1}{n^s}.$$

If we consider the case where s is a real number, then a standard exercise in calculus consists of proving that the series converges when $s > 1$.

One of the more notable mathematicians who first investigated ζ for s real was Leonhard Euler, and in 1737 (89 years before Riemann was born) he determined the *Euler product* [14]:

$$\sum_{n=1}^{\infty} \frac{1}{n^s} = \prod_{p \text{ prime}} \left(1 - \frac{1}{p^s}\right)^{-1},$$

for $s > 1$. Immediately we see why the zeta function is important in analytic number theory, for the Euler product explicitly gives a connection between natural numbers and primes. In fact, Euler used it to re-prove the ancient result of Euclid that there are infinitely many prime numbers, by demonstrating that the sum of the reciprocals of the primes diverges. So why isn't the zeta function known as the "Euler zeta function"? Well, as a function of the *real* variable s, the zeta function is a very natural function to consider, but Riemann had the brilliant idea of considering s as a *complex* variable. This insight enabled him to determine how one might obtain far more information about the primes from knowledge of $\zeta(s)$, as we shall see.

Great Moment II: The Statement of the Riemann Hypothesis

In 1859, to thank the Berlin Academy of Sciences for electing him a member, Riemann wrote a ground-breaking paper [34], *Ueber die Anzahl der Primzahlen unter einer gegebenen Grösse*, (or in English: *On the number of primes less than a given magnitude*). In this

---

*Supported in part by NSF grant DMS-0502730

Figure 1. Leonhard Euler

Figure 2. G. F. B. Riemann

eight-page manuscript, his only paper in number theory, using the Euler product as a starting point, he established a connection between the number of primes less than a number x and the properties of zeros of the zeta function.

Beyond the actual results of the paper, however, perhaps the most important contributions to the field of number theory lay in the methods he employed and the use of the function $\zeta(s)$. In particular, Riemann was able to find an analytic continuation for the zeta function to the entire complex plane apart from the point $s = 1$, where the zeta function has a pole. Furthermore, using a certain contour integral, Riemann demonstrated an important symmetry of the zeta function. Namely, if we modify zeta a little bit, we can relate $\zeta(s)$ to $\zeta(1-s)$. This is called the *functional equation* for $\zeta(s)$. To obtain the exact form of the functional equation, first define the Gamma function as a function of a complex number s:

$$\Gamma(s) = \int_0^\infty e^{-x} x^{s-1} \, dx. \tag{1}$$

One can think of the Gamma function as a generalization of the factorial function, since integrating by parts $n-1$ times we see that $\Gamma(n) = (n-1)!$ for all natural numbers n. Setting

$$\zeta^*(s) = \pi^{-\frac{s}{2}} \Gamma\left(\frac{s}{2}\right) \zeta(s),$$

Riemann was able to show that $\zeta^*(s) = \zeta^*(1-s)$. Since we understand the zeta function to the right of the line $\operatorname{Re}(s) = 1$, and we understand how the Gamma function behaves, the functional equation means we now know how the zeta function behaves to the left of the line $\operatorname{Re}(s) = 0$. The remaining mystery lies in the *critical strip*, where $0 \le \operatorname{Re}(s) \le 1$.

Riemann showed that $\zeta(s)$ vanishes infinitely often in the critical strip, and went on to say that it is very probable ("sehr wahrscheinlich") that all the zeros lie on the *critical line* $\operatorname{Re}(s) = 1/2$, which is the center of the critical strip. This statement is the celebrated *Riemann Hypothesis* (RH). He goes on to say (in German)

> "Though one would wish for a strict proof here; I have meanwhile temporarily put aside the search for this after some fleeting futile attempts, as it appears unnecessary for the next objective of my investigation."

Figure 3. The first page of Riemann's handwritten manuscript *Ueber die Anzahl der Primzahlen unter eine gegebenen Grösse*. (Reproduced with permission of the SUB Göttingen, Cod. Ms. B. Riemann 3, fol. 16 r.)

The object of Riemann's investigation was the *Prime Number Theorem*, which says that the number of prime numbers less than x is asymptotic to $x/\log x$. Riemann was unable to prove this, and it had to wait until 1896 before the proof was finally completed by de la Vallée Poussin [38] and (independently) by Hadamard [17]. Their proofs relied on showing that none of the zeros of the zeta function lies on the line $\text{Re}(s) = 1$. (Remember, Riemann showed that the zeros must lie in the strip $0 \leq \text{Re}(s) \leq 1$, and conjectured that in fact they all lie on $\text{Re}(s) = 1/2$).

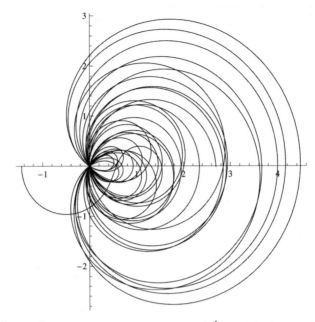

Figure 4. The real and imaginary parts of $\zeta(\tfrac{1}{2} + it)$ for $0 \leq t \leq 100$

Although no valid proof of the Riemann Hypothesis has been found, there is extensive evidence for its truth—for example, the first 10^{13} zeros are known to lie on the critical line [16], as are the 175 million zeros around zero number 10^{20} [31]. Also, at least 40% of all Riemann zeros lie on the line [5]. Many theorems in analytic number theory are conditional on the truth of RH, and its manifold ramifications make it perhaps the most important unresolved question in modern mathematics.

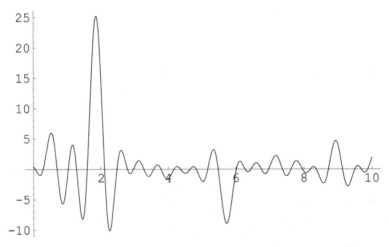

Figure 5. A plot of a variant of the zeta function, around $t = 10^{10}$

Figure 6. Charlie points out an error in Ethan's "proof" of the Riemann Hypothesis in an episode of the television show *Numb3rs*. [32] (Photo credit: CBS Paramount Television. All rights reserved.)

The increasing celebrity of the Riemann Hypothesis over the past few years is thanks in part to the Clay Mathematics Institute, a private, non-profit foundation based in Cambridge, Massachusetts. On May 24, 2000, they established the seven Millennium Prize Problems, which they view as important unsolved problems in mathematics. The solution to each problem is worth $1 million, and not surprisingly, the Riemann Hypothesis is one of the seven.

The Riemann Hypothesis even found its way into prime-time network television in 2005, on the CBS television show *Numb3rs*. The series features Don Eppes, an FBI agent, and his mathematician brother Charlie (played by Rob Morrow and David Krumholtz, respectively), who team up to solve crimes.

The episode "Prime Suspect" is summarized as follows [33]:

> When a five-year-old girl is kidnapped from her birthday party, Don and Terry lead the investigation, but must rely on Charlie's help because the girl's father, Ethan, is also a mathematician. Charlie realizes the kidnapper's motive when Ethan reveals he is close to solving Riemann's Hypothesis, a difficult math problem. If solved, the solution could not only earn him $1 million, but could break the code for internet security and unlock the world's biggest financial secret.

The Riemann Hypothesis can also be found in song, thanks to the number theorist Tom M. Apostol, a Professor Emeritus at Caltech, whose lyrics begin as follows:

Where are the zeros of zeta of s?
(Sung to the tune of *Sweet Betsy from Pike*)

Where are the zeros of zeta of s?
G. F. B. Riemann has made a good guess:
"They're all on the critical line," stated he,
"And their density's one over two pi log T."

This statement of Riemann's has been like a trigger,
And many good men, with vim and with vigor,
Have attempted to find, with mathematical rigor,
What happens to zeta as mod t gets bigger...

Figure 7. John Derbyshire in song

To hear the first verse, audio and video clips may be found on the web site of John Derbyshire [12], author of *Prime Obsession: Bernhard Riemann and the Greatest Unsolved Problem in Mathematics* [11], and first winner of the Mathematical Association of America's Euler Book Prize, in 2007.

Great Moment III: The Statement of the Lindelof Hypothesis

Besides the Riemann Hypothesis, another important conjecture involving the Riemann zeta function is the *Lindelöf Hypothesis*, dating back to 1908 [26]. This conjecture states that

$$\zeta\left(\frac{1}{2} + it\right) = O(t^\epsilon)$$

for every positive ϵ. The "Big Oh" notation is defined as follows: we say that $f(x) = O(g(x))$ as $x \to \infty$ if there is a constant K and a value x_0 such that $|f(x)| < Kg(x)$ whenever $x \geq x_0$. In other words, the Lindelöf Hypothesis is concerned with the growth of the zeta function on the line $\mathrm{Re}(s) = 1/2$, and conjectures that the modulus of $\zeta(1/2 + it)$ grows slower than any positive power of t as t tends to infinity.

We should add that the Lindelöf Hypothesis, though still highly desirable, is not as strong as the Riemann Hypothesis—that is, the Riemann Hypothesis implies the Lindelöf Hypothesis, but, as far as we know, the Lindelöf Hypothesis does not imply the Riemann Hypothesis. It was the Lindelöf Hypothesis that led to the study of moments of the Riemann zeta function.

Great Moment IV: The Second Moment

The Lindelöf Hypothesis is a conjecture on the maximum size of the zeta function. One way to understand extreme values of a function is through the study of moments. The $2k^{th}$ *moment* of the modulus of the Riemann zeta function is defined as

$$I_k(T) = \frac{1}{T} \int_0^T \left|\zeta\left(\frac{1}{2} + it\right)\right|^{2k} dt. \qquad (2)$$

The Lindelöf Hypothesis is equivalent to the statement that for any k and any positive ϵ, $I_k(T) = O(T^\epsilon)$, and thus it inspired the study of the moments of the Riemann zeta function. However, moments are now appreciated in their own right due to the intricate mathematical stucture they appear to exhibit. Furthermore, moments represent mean values of the Riemann zeta function on the critical line over a finite interval, and estimates of these average values can provide information about the zeros of $\zeta(s)$.

Figure 8. G. H. Hardy

Figure 9. J. E. Littlewood

Hardy and Littlewood were the first to obtain moments (or shall we say momentous) results. In a paper published in 1918, they [18] proved not only that the average value of $|\zeta(s)|^2$ on $\mathrm{Re}(s) = 1/2$ is infinite, but they gave an asymptotic evaluation for the second moment: as $T \to \infty$,

$$I_1(T) = \frac{1}{T}\int_0^T \left|\zeta\left(\frac{1}{2}+it\right)\right|^2 dt$$
$$\sim \log T.$$

In 1922, Hardy and Littlewood [19] found a simpler proof of this result, writing "the proof [in the earlier paper] is difficult and indirect, and the simpler proof which we give here affords an excellent illustration of the use of the approximate functional equation." The *approximate functional equation* for $\zeta(s)$ expresses the zeta function as a finite sum plus an error term, and such results have been used extensively in the study of moments since then.

Hardy and Littlewood were unaware that Riemann had already calculated something very similar. In his unpublished notes, Riemann had developed a method for accurately calculating a variant of his zeta function high up on the critical line. This result was unknown to the mathematical world until Siegel went through Riemann's *Nachlass* and was able to decipher the dense and messy handwriting. Siegel then published the result, which is now called the *Riemann-Siegel formula* [35]. This formula was one of the first uses of saddle-point approximation, and even today is one of the most complicated examples of that method.

In the meeting of the London Mathematical Society on February 10, 1921, Littlewood [27] announced several results without proof, one of which concerned the lower-order terms in the asymptotic expansion of $I_1(T)$. He showed the result to a young analyst of the time, Albert Edward Ingham, who, with Littlewood's encouragement, improved upon that result as part of his Cambridge Fellowship dissertation [22, 23].

What Littlewood and Ingham showed was that if $0 \leq \sigma \leq 1$, then

$$\int_0^T |\zeta(\sigma + it)|^2 \, dt = \int_0^T \left(\zeta(2\sigma) + \left(\frac{t}{2\pi}\right)^{1-2\sigma} \zeta(2 - 2\sigma) \right) dt + O(T^{1-\sigma}(\log T)^2). \tag{3}$$

In the case when $\sigma = 1/2$, a limit needs to be taken, but observe that the two singularities in the integrand on the right cancel. Note also that the integral contains a sum of two terms, each involving a Riemann zeta function. You will soon see why this is important.

The error term in (3) is an improvement of Ingham's over Littlewood's original result. This has been improved further since then, and Atkinson [2] found an exact expression for it in the case when $\sigma = 1/2$. A complete discussion on what is now known about the error term in the second moment will take us too far away from the main thrust of this article, but it should come as no surprise that its moments have been calculated. For more information see, for example, the book of Ivić [24].

Great Moment V: The Fourth Moment

Ingham's Cambridge Fellowship dissertation contained two great results (though he modestly called them "lemmas"). We have already seen the first, the lower-order terms of the second moment. The second result was the calculation of the asymptotic size of the fourth moment.

At the time, the approximate functional equation for $\zeta(s)$ did not seem to provide estimates with enough accuracy to obtain results for the fourth moment. In 1922, Hardy and Littlewood [19] had used the approximate functional equation to show that $I_2(T) = O((\log T)^4)$, but were not able to prove an asymptotic formula. To obtain the fourth moment asymptotics, Ingham needed to use an approximate functional equation for ζ^2, not just ζ. This approximate functional equation was due to Hardy and Littlewood, but wasn't published by them until much later, in 1929 [20].

Ingham first announced his result in a meeting of the London Mathematical Society held on April 26, 1923 [22], although the details were not published until 1926 [23]. In the meantime, Titchmarsh [37] had found an alternative proof of this result, using a smoothed type of moment. In fact, their papers are published in the same issue of the *Proceedings of the London Mathematical Society*. Ingham proved that as $T \to \infty$,

$$I_2(T) = \frac{1}{T} \int_0^T \left| \zeta\left(\frac{1}{2} + it\right) \right|^4 dt = \frac{1}{2\pi^2} \log^4 T + O(\log^3 T).$$

A little more work was done later on the fourth moment, with the lower-order terms being calculated. In the 1940s, Atkinson [1] found the lower-order terms for the smoothed moment that Titchmarsh considered. In the 1970s, Heath-Brown [21] obtained the lower-order terms for $I_2(T)$. However, the answer is quite complicated, and if it were written out in full, it would take nearly half a page to express. Finally, Motohashi [30] found an exact expression for the fourth moment, including the error term. It was the structure of that expression that motivated some exciting new work by Conrey, Farmer, Keating, Rubinstein, and Snaith on the lower-order terms of the moments, which we will discuss later in this article.

Regarding the moments problem, it took several years to proceed from the second moment to the fourth moment, and even longer to find the lower-order terms. How long would you expect for the sixth moment? And what about proving the Lindelöf Hypothesis? Unfortunately, no one has yet been able to prove results for higher moments. In fact, for a long time, no one even had a good conjecture for what the asymptotics should look like in more detail.

Great Moment VI: Conjectures!

In the early 1990s, Brian Conrey and Amit Ghosh [8] conjectured the sixth moment, although their paper wasn't published until 1998. In fact, the paper was due to be in the *Proceedings of the Amalfi Conference on Analytic Number Theory* (Maiori, 1989), but they failed to submit the final draft of their paper on time, and the conjecture was omitted from the paper. The statement of their conjecture is as follows:

$$I_3(T) = \frac{1}{T}\int_0^T \left|\zeta\left(\frac{1}{2}+it\right)\right|^6 dt \sim \frac{42}{9!}a(3)(\log T)^9, \tag{4}$$

where

$$a(k) = \prod_{p \text{ prime}} \left(1-\frac{1}{p}\right)^{k^2} \sum_{m=0}^{\infty} \left(\frac{\Gamma(m+k)}{m!\Gamma(k)}\right)^2 p^{-m}.$$

There is a folklore conjecture that the $2k^{\text{th}}$ moment of zeta should grow like

$$I_k(T) \sim c_k (\log T)^{k^2},$$

where c_k is an unknown constant. Based on the sixth-moment conjecture, and other work, Conrey and Ghosh made this conjecture much more precise by conjecturing that c_k should be decomposed as

$$c_k = \frac{g_k a(k)}{(k^2)!}, \tag{5}$$

where g_k is an integer when k is an integer. (Note that the conjecture would be vacuous without the restriction that g_k is an integer.) This work was never published, but was presented at various conferences and seminars in the early 1990s.

In the late 1990s, Brian Conrey and Steve Gonek [9] invented a new method that gave the same conjecture for the sixth moment, and also gave a conjecture for the eighth:

$$I_4(T) = \frac{1}{T}\int_0^T \left|\zeta\left(\frac{1}{2}+it\right)\right|^8 dt \sim \frac{24024}{16!}a(4)(\log T)^{16}.$$

Figure 10. Steve Gonek and Brian Conrey

At that time, therefore, the only known and conjectured values of g_k were $g_1 = 1$, $g_2 = 2$, $g_3 = 42$, and $g_4 = 24024$. This paper was published in 2001, but the results were first announced in 1998, at a conference held in Vienna.

In fact, at that same conference, Jon Keating and Nina Snaith were going to announce their conjecture for *all* the moments using random matrix theory. We therefore need to take a break in this momentous journey to explain random matrix theory, and its connections to the Riemann zeta function.

Great Moment VII: The Meeting of Montgomery and Dyson

A connection between zeros of the zeta function and the field of random matrix theory was established in the 1970s. In 1972, Hugh Montgomery, a graduate student at Cambridge, was visiting Atle Selberg at the Institute for Advanced Study at Princeton. Montgomery wanted to discuss his conjecture for how he thought the gaps between zeros of the Riemann zeta function might behave. Selberg responded that the "results were interesting."

At tea that afternoon, Montgomery was introduced to the physicist Freeman Dyson. After Montgomery told him his conjecture, Dyson quickly recognized Montgomery's results as being the same as the pair correlation of eigenvalues of random Hermitian matrices. Upon hearing Dyson's observation, Montgomery immediately recognized the importance of the connection just made. As he recalls the memorable interaction, "Just by chance this conversation took place .... This happened even before I had published the paper. I knew it was important and worth following up." Montgomery then gave his students projects to develop these ideas, but the conjecture is still unproven. However, numerical computations, heuristic calculations, and theoretical work by many people, have provided very strong supporting evidence.

What led to Dyson's work in random matrix theory? In the 1950s, a great problem faced physicists. They knew that the Schrödinger equation would yield the quantum behavior of any system, but the systems they were interested in were so large and complicated that actually solving the Schrödinger equation seemed almost impossible. The particular problem physicists were investigating was to understand the quantum mechanics of the neutrons and protons in heavy nuclei, and the permitted energy levels they could take. Just as electrons in an atom have a discrete set of energy levels, so do the particles making up the nucleus of that atom.

Figure 11. Freeman Dyson

Figure 12. Hugh Montgomery

Wigner had the insight to ask a more general question. Rather than trying to calculate the exact energy levels, he asked statistical questions about energy levels. For example, do the energy levels form clumps of levels with big gaps between the clumps (like stars clumping into galaxies)? Or are they well spaced apart, with very few large and small gaps (like trees in a forest)?

Suitably interpreted, the solutions of a Schrödinger equation are the eigenvalues of a certain Hermitian matrix, and so Wigner investigated the eigenvalues of matrices whose structure and symmetries were imposed by physical conditions. Therefore, the matrices had to be Hermitian, but all independent entries of the matrices were chosen randomly. Under these conditions, he found the probability distribution of eigenvalues, which matched up very well with experimental data.

This idea was taken up and developed by many people. In particular, in the 1960s, Dyson showed that if one replaced Hermitian matrices (which are physically appealing, since they are the matrices appearing in the Schrödinger equation), with unitary matrices (which are mathematically appealing, since they form a compact group), then the answers obtained were the same in the large-matrix-size limit. Random matrix theory is now a very broad and deep subject in mathematics. It not only applies to areas of physics, like quantum chaos and quantum field theory, but it has also appeared in combinatorics, electrical engineering, and telephony.

At first sight, random matrix theory has nothing to do with the Riemann zeta function. Certainly the connection was a surprise to Montgomery when Dyson noticed that Montgomery's result for the pair correlation function was the same as that which came from random matrix theory. However, one could argue that there is an earlier connection: The Pólya and Hilbert approach to the Riemann Hypothesis is to try to find an operator whose eigenvalues are the heights of the zeros of the zeta function, and which is also Hermitian. Such an operator is forced to have its eigenvalues lie on a straight line, and thus the zeros of the zeta function will be forced to lie on the line. In other words, if one can find a Hermitian operator associated with a matrix whose eigenvalues are the zeros of the Riemann zeta function, then one will have proved the Riemann Hypothesis. The fact that the zeros appear to have random matrix statistics suggest that this Hermitian operator must have certain other symmetries.

The meeting between Dyson and Montgomery was truly a serendipitous moment, for without this chance occurrence, who knows how long it would have been before the connection between zeros of the zeta function and random matrix theory would have been discovered? However, it still took some time for random matrix theory to be used to make predictions about the zeta function, certainly for the moments.

Great Moment VIII: The Keating-Snaith conjecture

In 1996, at a conference in Seattle organized by the American Institute of Mathematics (AIM), Peter Sarnak challenged the physicists, in particular, Jon Keating, to come up with a random matrix model for the moments, to obtain Conrey and Ghosh's "42" of the sixth moment (see equation (4)). Keating, with his graduate student Nina Snaith, set to work to try to figure it out.

They not only found a random matrix model which predicted the "42", but predicted *all* moments, and they were to announce this result in 1998, at another AIM conference held in Vienna. In fact, at the same conference, Conrey and Gonek were going to announce their result on the eighth moment, as we mentioned at the end of Section VI. Just a few minutes before he was to speak, Keating was chatting with Conrey, who mentioned that he and Steve Gonek now knew what the eighth moment should be. So for those precious few minutes, they both frantically calculated to make sure that the conjectures matched when $k = 4$, which thankfully they did. The general conjecture of Keating and Snaith takes the following form:

$$I_k(T) \sim \frac{G^2(k+1)}{G(2k+1)} a(k) (\log T)^{k^2}.$$

Here $G(z)$ is the Barnes G-function. This function is a relative of the familiar Gamma function (equation (1)), in that it satisfies a similar type of functional equation, $G(z+1) = \Gamma(z) G(z)$. Using this property, and the initial value $G(1) = 1$, one can find $G(z)$ for all integers, though the function has an analytic expression for the whole complex plane.

When the Keating-Snaith conjecture was first announced, the Barnes G-function was not used explicitly in the formula. Indeed, the constant g_k at that time, as in (5), was written as

$$g_k = (k^2)! \prod_{j=0}^{k-1} \frac{j!}{(k+j)!}.$$

Brian Conrey and David Farmer [6] checked that if k is an integer, then g_k is indeed an integer, as Conrey and Ghosh had conjectured. However, using the functional equation for the Barnes G-function, one can check that

$$\prod_{j=0}^{k-1} \frac{j!}{(k+j)!} = \frac{G^2(k+1)}{G(2k+1)}$$

where the right-hand side holds for complex values of k too, not just integers. Use of the Barnes G-function enabled several of the computations to be simplified.

Great Moment IX: The CFKRS Paper

Brian Conrey, David Farmer, Jon Keating, Mike Rubinstein, and Nina Snaith have conjectured the lower-order terms for the asymptotics for all moments [7]. Their conjecture is supported by heuristics from random matrix theory and analytic number theory, and by numerical computation. Rather than just working with the moments on the line $\mathrm{Re}(s) = 1/2$, they emphasized the importance of considering a more general integral, involving shifts off the critical line:

$$\int_0^T \left| \prod_{i=1}^k \zeta\left(\frac{1}{2} + \alpha_j + it\right) \zeta\left(\frac{1}{2} + \beta_j - it\right) \right| dt.$$

Motohashi also investigated this integral, and gave an explicit formula when $k = 2$ [30].

Using the same approximate functional equation that was so important in Hardy and Littlewood's proof of the second moment of the zeta function, Conrey, Farmer, Keating, Rubinstein, and Snaith conjectured that the main term of the formula is a sum of $\binom{2k}{k}$ terms, each involving a product of k^2 zeta functions. If you turn back to Ingham's result for the second moment (equation (3)), you will discover that you have already seen an example supporting their conjecture. Namely, the main term of the integral

$$\int_0^T |\zeta(\sigma + it)|^2 \, dt = \int_0^T \left(\zeta(2\sigma) + \left(\frac{t}{2\pi}\right)^{1-2\sigma} \zeta(2 - 2\sigma) \right) dt + O(T^{1-\sigma} (\log T)^2)$$

Figure 13. From left to right: Brian Conrey, David Farmer, Jon Keating, Mike Rubinstein, and Nina Snaith

is a sum of $\binom{2}{1}$ terms, each involving one zeta function. They also analyzed Motohashi's results on the fourth moment of ζ, confirming that their conjecture is satisfied in that case too.

The conjectures of Conrey, Farmer, Keating, Rubinstein, and Snaith have been supported by work of other number theorists. In particular, Diaconu, Goldfeld, and Hoffstein [13] came up with the same formula from a conjecture involving multiple Dirichlet series and functional equations. Other support for the conjectures comes from an automorphic function called an *Eisenstein series*. A basic example of an Eisenstein series is given by

$$E(z,s) = \frac{1}{2} y^s \sum_{(c,d)=1} |cz+d|^{-2s},$$

Figure 14. From left to right: Jeff Hoffstein, Dorian Goldfeld, and Adrian Diaconu

for Re$(s) > 1$. We will denote its normalization by

$$E^*(z,s) = \zeta^*(2s)E(z,s) = \pi^{-s}\Gamma(s)\zeta(2s)E(z,s).$$

Even this relatively simple Eisenstein series formula leads us to a connection with moments of the Riemann zeta function. To see the relationship, we can compute the constant term of the Fourier expansion of the Eisenstein series:

$$\int_0^1 E^*(x+iy,s)\,dx = \zeta^*(2s)y^s + \zeta^*(2-2s)y^{1-s}.$$

To match up with the conjecture of Conrey, Farmer, Keating, Rubinstein, and Snaith, we would like the constant term to be a sum of two terms, each involving one zeta function, and indeed, this is the case! However, there is more because the zeta functions match those of the lower-order terms in Ingham's formula. Therefore we can establish the following connections between Eisenstein series and Ingham's formula:

L-function of E^* $\qquad\qquad\qquad\qquad\qquad$ constant term of E^*

$$\int_0^T |\zeta(\sigma+it)|^2\,dt = \int_0^T \left(\zeta(2\sigma) + \left(\tfrac{t}{2\pi}\right)^{1-2\sigma}\zeta(2-2\sigma)\right)dt + O(T^{1-\sigma}(\log T)^2)$$

Beineke and Bump [3] were able to show that Eisenstein series can be constructed to generalize this phenomenon for higher values of k. That is, the constant term of the Eisenstein series is a sum of $\binom{2k}{k}$ terms, each involving the product of the appropriate k^2 zeta functions, and the L-function of the Eisenstein series matches the product of zeta functions in the moment expression. For a different approach relating Eisenstein series to moments of the Riemann zeta function, see [4].

Figure 15. Dan Bump

Great Moment X: The Moment We Have All Been Waiting For...

Approximately ninety years ago, Hardy and Littlewood published the first result on moments of $\zeta(s)$. What are some promising avenues of research in number theory and random matrix theory today?

From the paper of Conrey, Farmer, Keating, Rubinstein, and Snaith [7], we now understand the structure of the moments of zeta. Can this structure be used to find an approach to prove a result for any moment higher than the fourth? The work of Diaconu, Goldfeld, and Hoffstein [13] would lead to new rigorous results if certain mysterious hidden functional equations for their multiple Dirichlet series could be found.

Recently, Soundararajan [36] showed that, under the Riemann Hypothesis, the moments cannot grow much faster than predicted. Specifically he showed that for any $\epsilon > 0$,

$$(\log T)^{k^2} \ll I_k(t) \ll (\log T)^{k^2 + \epsilon}$$

where $I_k(T)$ is given by equation (2). So, in some sense, we are only epsilon away from the "eureka moment," but that is quite possibly a big epsilon to overcome!

However, Farmer, Gonek, and Hughes [15] have shown that the conjectures for the moments cannot remain true when k grows with T. They show that the asymptotic form of the moments must change before $k = \sqrt{(8 \log T)/(\log \log T)}$. The question of determining the structure of the moments when k is this large is still open.

The Riemann zeta function has many cousins, called L-functions, each with arithmetic significance. For each family of L-functions, similar questions regarding moments can be asked, with applications to other areas of mathematics, such as ranks of elliptic curves.

Indeed, the most recent advance has come through the simplest generalization of the Riemann zeta function: Conrey, Iwaniec, and Soundararajan [10] have calculated the sixth moment of Dirichlet L-functions, and shown that $g_3 = 42$ for that family, exactly as predicted. Thus the "42," first conjectured by Conrey and Ghosh (see Section VI) and the motivation for the use of random matrix theory in moment calculations (see Section VIII), has finally been found, albeit for Dirichlet L-functions. If this calculation could be repeated for the Riemann zeta function (that is, if equation (4) could be proven), it would be the first time since 1926 that a new moment of the zeta function had been rigorously found. Truly, a moment we are all waiting for!

If the reader would like further details on topics mentioned above relating analytic number theory to random matrix theory, an excellent resource is the volume of proceedings from the 2004 School *Random Matrix Approaches in Number Theory*, held at the Isaac Newton Institute for Mathematical Sciences [28]. The collaboration between random matrix theory and number theory has been an exciting and fruitful one, and is perhaps best exemplified by the final verse of Apostol's "Where are the zeros of zeta of s?" Rather than singing the number-theoretic version

> There's a moral to draw from this long tale of woe
> That every young genius among you must know
> If you tackle a problem and seem to get stuck,
> Just take it mod p and you'll have better luck.

one of the authors of this paper has revised the song, advising the reader to

> Just use R.M.T., and you'll have better luck.

References

[1] F. V. Atkinson, *The mean value of the zeta-function on the critical line*, Proc. London Math. Soc. [2] **47** (1941), 174–200.

[2] F.V. Atkinson, *The mean-value of the Riemann zeta function*, Acta Math. **81** (1949), 353–376.

[3] J. Beineke and D. Bump, *Moments of the Riemann zeta function and Eisenstein series I*, J. Number Theory **105** (2004), 150–174.

[4] J. Beineke and D. Bump, *Moments of the Riemann zeta function and Eisenstein series II*, J. Number Theory **105** (2004), 175–191.

[5] J. B. Conrey, *More than two fifths of the zeros of the Riemann zeta function are on the critical line*, J. Reine Angew. Math. **399** (1989), 1–26.

[6] J. B. Conrey and D. W. Farmer, *Mean values of L–functions and symmetry*, Internat. Math. Res. Notices **17** (2000), 883–908.

[7] B. Conrey, D. Farmer, J. Keating, M. Rubinstein, and N. Snaith, *Integral moments of L-functions*, Proc. London Math. Soc. [3] **91** (2005), 33–104.

[8] J. B. Conrey and A. Ghosh, *A conjecture for the sixth power moment of the Riemann zeta-function*, Internat. Math. Res. Notices **15** (1998), 775–780.

[9] J. B. Conrey and S. M. Gonek, *High moments of the Riemann zeta function*, Duke Math. J. **107** (2001), 577–604.

[10] J. B. Conrey, H. Iwaniec, and K. Soundararajan, *The sixth power moment of Dirichlet L-functions*, arXiv:0710.5176 (2007).

[11] J. Derbyshire, *Prime Obsession: Bernhard Riemann and the Greatest Unsolved Problem in Mathematics*, Joseph Henry Press (National Academies Press), 2003.

[12] J. Derbyshire, The Riemann Hypothesis in Song; http://olimu.com/Riemann/Song.htm; (accessed March 21, 2008).

[13] A. Diaconu, D. Goldfeld, and J. Hoffstein, *Multiple Dirichlet series and moments of zeta and L-functions*, Compositio Math. **139** (2003), 297–360.

[14] L. Euler, *Variae observationes circa series infinitas*, Comm. Acd. Sci. Petropolitanae **9** (1737), 222–236 (also in "Opera" (1), Vol 14, 216–244).

[15] D. W. Farmer, S. M. Gonek, and C. P. Hughes, *The maximum size of L-functions*, J. reine angew. Math. **609** (2007), 215–236.

[16] X. Gourdon and P. Demichel, *Computation of zeros of the Zeta function* http://numbers.computation.free.fr/Constants/Miscellaneous/zetazeroscompute.html; (accessed March 21, 2008).

[17] J. Hadamard, *Sur la distribution des zeros de la fonction $\zeta(s)$ et ses consequences arithmétiques*, Bull. Soc. Math. Frances **24** (1896), 199–220.

[18] G. H. Hardy and J. E. Littlewood, *Contributions to the theory of the Riemann zeta-function and the theory of the distribution of primes*, Acta Math. **41** (1918), 119–196.

[19] G. H. Hardy and J. E. Littlewood, *The approximate functional equation in the theory of the zeta-function, with applications to the divisor problem of Dirichlet and Piltz*, Proc. London Math. Soc. [2] **21** (1922), 39–74.

[20] G. H. Hardy and J. E. Littlewood, *The approximate functional equations for $\zeta(s)$ and $\zeta^2(s)$*, Proc. London Math. Soc. [2] **29** (1929), 81–87.

[21] D. R. Heath-Brown, *The fourth power moment of the Riemann zeta function*, Proc. London Math. Soc. [3] **38** (1979), 385–422.

[22] A. E. Ingham, Proc. London Math. Soc. [2] **22** (1923), xxix, (Records for April 26, 1923).

[23] A. E. Ingham, *Mean value theorems in the theory of the Riemann zeta-function*, Proc. London Math. Soc. [2] **27** (1926), 273–300.

[24] A. Ivić, *Lectures on mean values of the Riemann zeta function*, Tata Institute of Fundamental Research Lectures on Mathematics and Physics **82**, Springer-Verlag, Berlin, 1991.

[25] J. P. Keating and N. C. Snaith, *Random matrix theory and $\zeta(1/2 + it)$*, Comm. Math. Phys. **214** (2000), 57–89.

[26] E. Lindelöf, *Quelques remarques sur la croissance de la fonction $\zeta(s)$*, Bull. Sci. Math., **32** (1908), 341–356.

[27] J. E. Littlewood, *Researches in the theory of the Riemann ζ-function*, Proc. London Math. Soc. [2] **20** (1921), xxii–xxviii, (Records for Feb 10, 1921).

[28] F. Mezzadri and N. C. Snaith, eds., *Recent Perspectives in Random Matrix Theory and Number Theory*, Cambridge University Press, Cambridge, 2005.

[29] H. L. Montgomery, *The pair correlation of zeros of the zeta function*, Analytic Number Theory, Proceedings of Symposia in Pure Mathematics **24** (1973), 181–193.

[30] Y. Motohashi, Spectral Theory of the Riemann Zeta-Function, Cambridge Univ. Press, Cambridge, 1997.

[31] A. Odlyzko, *The 10^{20}-th zero of the Riemann zeta function and 175 million of its neighbors*, unpublished; http://www.dtc.umn.edu/~odlyzko/unpublished/index.html; (accessed March 21, 2008).

[32] "Prime Suspect" (*Numb3rs*), CBS (airdate February 18, 2005).

[33] "Prime Suspect" Overview (Numb3rs: Prime Suspect - TV.com); http://www.tv.com/numb3rs/prime-suspect/episode/383244/summary.html; (accessed March 21, 2008).

[34] G. F. B. Riemann, *Ueber die Anzahl der Primzahlen unter einer gegebenen Grösse*, Monatsberichte der Berliner Akademie (1859), 671–680.

[35] C. L. Siegel, *Ueber Riemanns Nachlass zur Analytischen Zahlentheorie*, Quellen und Studien zur Geschichte der Math. Astr. Phys. **2** (1932), 45–80. Reprinted in C. L. Siegel, *Gesammelte Abhandlungen*, vol. 1, Springer, 1966, pp. 275–310.

[36] K. Soundararajan, *Moments of the Riemann zeta function*, arXiv:math.NT/0612106 (2006).

[37] E. C. Titchmarsh, *The mean-value of the zeta-function on the critical line*, Proc. London Math. Soc. [2] **27** (1926), 137–150.

[38] C.-J. de la Vallée Poussin, *Recherches analytiques sur la théorie des nombres (première partie)*, Ann. Soc. Sci. Bruxelles [I] **20** (1896), 183–256.

The Collatz Chameleon

Marc Chamberland

Introduction

The $3x + 1$ Problem is today's most easily-stated unsolved mathematical problem. The origin of the problem is murky but it is largely credited to Lothar Collatz, hence it is sometimes called the Collatz Problem. Collatz considered the function

$$C(x) = \begin{cases} 3x + 1, & x \equiv 1 \pmod{2}, \\ \dfrac{x}{2}, & x \equiv 0 \pmod{2}. \end{cases}$$

The conjecture states that any positive integer eventually iterates under this map to the number one. Note that an odd number m iterates to $3m + 1$ which then iterates to $(3m + 1)/2$. One may therefore "compress" the dynamics by considering the map

$$T(x) = \begin{cases} \dfrac{3x + 1}{2}, & x \equiv 1 \pmod{2}, \\ \dfrac{x}{2}, & x \equiv 0 \pmod{2}. \end{cases}$$

For example, one has the trivial cycle consisting of $\{1, 2\}$. Most researchers work with the map T instead of C.

Despite its powerless looks, the Collatz problem has enticed many bounty hunters to pursue it with little to show at the end of the day. The famous mathematician Paul Erdős was correct when he stated, "Mathematics is not ready for such problems."

The structure of the positive integers forces any number to iterate under T to one of the following: the trivial cycle $\{1, 2\}$, a non-trivial cycle, or infinity (the orbit is divergent). The $3x + 1$ Problem claims that the first option occurs in all cases. Published results claim that this has been verified for all integers less than $100 \times 2^{50} \approx 1.12 \times 10^{17}$. This computation—completed in April 2000—was accomplished with two 133MHz and two 266MHz DEC Alpha computers and used 14.4 CPU years. An ongoing distributed-computer project by Eric Roosendaal (www.ericr.nl/wondrous) claims to have improved this to $195 \times 2^{50} \approx 2.19 \times 10^{17}$. For those who believe the second option, the existence of a nontrivial cycle, it should be mentioned that any such cycle must have a length of at least 252,000,000.

Like several mathematical conundrums, the $3x + 1$ Problem may be rephrased as other problems which bear little if any resemblance to the original. What follows are various conjectures which are equivalent to the Collatz Problem. These different "shades" of the Collatz chameleon appeal to various temperaments; share with a friend the shade which they will find the most enticing.

The "Stopping Time" Shade

There is a natural algorithm for checking that all numbers up to some N iterate to one. First, check that every positive integer up to $N - 1$ iterates to one, then consider the iterates of N.

Once the iterates go below N, you are done. For this reason, one considers the so-called *stopping time* of n, that is, the minimum number of steps (if it exists) needed to iterate below n:

$$\sigma(n) = \min\{k : T^{(k)}(n) < n\}.$$

Conjecture. *Every positive integer has a finite stopping time, that is, $\sigma(n) < \infty$ for all positive n.*

Related to this are two other auxiliary functions: the *total stopping time* (the minimum number of steps needed to iterate to 1)

$$\sigma_\infty(n) = \min\{k : T^{(k)}(n) = 1\}.$$

and the *height* (the highest point to which n iterates)

$$h(n) = \max\{T^{(k)}(n) : k \in Z^+\}.$$

Note that if n is in a divergent trajectory, then $\sigma_\infty(n) = h(n) = \infty$. These latter functions may be surprisingly large even for small values of n. For example,

$$\sigma(27) = 59, \sigma_\infty(27) = 70, h(27) = 4616.$$

The orbit of 27 is depicted in Figure 1.

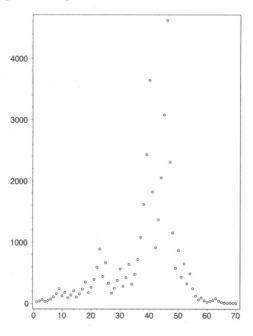

Figure 1. Orbit starting at 27.

It has been shown that almost every positive integer has finite stopping time, that is, the proportion of numbers less than N which have finite stopping time approaches 100% as N approaches infinity. It is conjectured that if one averages the stopping times of each number up to N, this average approaches a constant. Numerical evidence suggests this average is about 3.48.

The Tree Shade

The Collatz chameleon is, naturally, adept at blending in with a tree. The topic of total stopping times is intricately linked with the *Collatz tree* or *Collatz graph*, the directed graph whose vertices are predecessors of the number one via the map T. It is depicted in Figure 2.

Starting from the bottom, one may "grow" the tree as follows. If a number n is congruent to 2 mod 3, let it sprout two branches: one to $2n$ and the other to $(2n - 1)/3$. Otherwise, let it sprout only one new branch $2n$.

Conjecture. *The Collatz tree contains all the positive integers.*

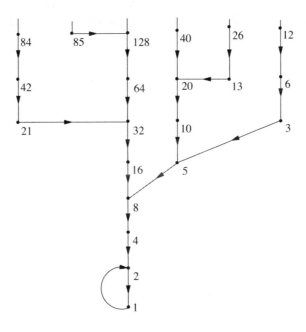

Figure 2. The Collatz graph.

The "Number Representation" Shade

The $3x + 1$ problem is ultimately dependent on how numbers may be represented with twos and threes. To this end, we say that a positive integer n has a *k-smooth representation* if and only if there exist integers $\{a_i\}$ and $\{b_i\}$ such that

$$n = \sum_{i=1}^{k} 2^{a_i} 3^{b_i}, \quad a_1 > a_2 > \cdots > a_k \geq 0, \quad 0 \leq b_1 < b_2 < \cdots < b_k$$

These numbers date back at least as early as the Indian mathematical genius Ramanujan. A 3-smooth representation of n is *special of level k* if

$$n = 3^k + 3^{k-1}2^{a_1} + \cdots + 3 \cdot 2^{a_{k-1}} + 2^{a_k}$$

in which every power of 3 appears up to 3^k. For a fixed k, each n has at most one such representation. Now we have the latest shade of the Collatz conjecture.

Conjecture. *For each positive integer m there are numbers p and q such that $2^p - 3^q m$ has a special 3-smooth representation of level $k = q - 1$. The choice of p and q is not unique, if it exists.*

The "No Adding" Shade

After investigating some calculations with the map T, one quickly finds that the "+ 1" is what complicates the analysis of the problem. Such pesky terms are not present with the daunting map R defined as

$$R(n) = \begin{cases} \frac{1}{11}n & \text{if } 11 \mid n \\ \frac{136}{15}n & \text{if } 15 \mid n \text{ and } NOTA \\ \frac{5}{17}n & \text{if } 17 \mid n \text{ and } NOTA \\ \frac{4}{5}n & \text{if } 5 \mid n \text{ and } NOTA \\ \frac{26}{21}n & \text{if } 21 \mid n \text{ and } NOTA \\ \frac{7}{13}n & \text{if } 13 \mid n \text{ and } NOTA \\ \frac{1}{7}n & \text{if } 7 \mid n \text{ and } NOTA \\ \frac{33}{4}n & \text{if } 4 \mid n \text{ and } NOTA \\ \frac{5}{2}n & \text{if } 2 \mid n \text{ and } NOTA \\ 7n & \text{otherwise} \end{cases}$$

where *NOTA* means "none of the above" conditions hold. We now have the following nonobvious shade.

Conjecture. *The R-orbit of 2^n contains 2 for each positive integer n.*

Though one need only consider the initial points 2^n, the map R is admittedly much more complicated than T. Care should be taken for those who wish to program this function. For example, the point $x = 8$ needs 50 iterates of R to reach $x = 2$.

The Chaotic Shade

A common problem-solving technique is to embed a problem into a larger class of problems and use techniques appropriate to that new space. Much work has been done along these lines for the $3x+1$ problem. An extension of T to the positive real line R^+ is interesting in that it allows tools from the study of iterating continuous maps. Consider the map f defined as

$$f(x) = x + \frac{1}{4} - \frac{2x+1}{4}\cos(\pi x) + \frac{1}{\pi}\left(\frac{1}{2} - \cos(\pi x)\right)\sin(\pi x).$$

It is a simple exercise to verify that $f(x) = T(x)$ when x is a positive integer. Numerical investigations quickly find that $\{1, 2\}$ is an attractive cycle; that is, any nearby points to $x = 1$ or $x = 2$ are drawn in to the cycle asymptotically.

Conjecture. *A randomly chosen real number greater than 1 will approach the cycle $\{1, 2\}$.*

A more technical way of stating this result is that almost every (in the sense of Lebesgue) real number greater than one will iterate as suggested. The dynamics of f admit what is called "chaotic behavior"; there are points which never settle down anywhere. This is not unique to the extension f; it holds for all continuous extensions. One way to see the complicated nature of this function is to replace x with the complex variable z. Shading points black if their iterates remained bounded produces the *filled-in Julia set* of the map. Figure 3 shows part of the Julia set for a related map.

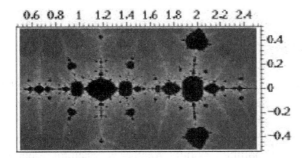

Figure 3. Julia set of an extension of T to the complex plane.

The "Binomial Coefficient" Shade

Binomial coefficients, that is, numbers of the form

$$\binom{m}{n} = \frac{m!}{(m-n)!n!},$$

occur naturally in number theory, the typical home of the Collatz chameleon. However, one is a bit surprised to find the following shade.

Conjecture. *For every positive integer a there exist numbers w, p, r, s such that a ≤ w and*

$$\left(\begin{array}{c}4(w+1)(p+r)+1\\p+r\end{array}\right)\left(\begin{array}{c}pw\\s\end{array}\right)\left(\begin{array}{c}rw\\t\end{array}\right)\times$$
$$\left(\begin{array}{c}2w+1\\w\end{array}\right)\left(\begin{array}{c}2s+2t+r+((4w+3)(p+r)+1)\\3a+(4w+4)(3t+2r+s)\end{array}\right)\times$$
$$\left(\begin{array}{c}p+r\\p\end{array}\right)\left(\begin{array}{c}3a+(4w+4)(3t+2r+s)\\2s+2t+r+((4w+3)(p+r)+1)\end{array}\right)\equiv 1 \bmod 2.$$

The "Functional Equation" Shade

Let $\omega = \exp(2\pi i/3) = \left(-1+\sqrt{3}i\right)/2$. The last shade we consider is in terms of an infinite series solution to a functional equation.

Conjecture. *The only solution of the form*

$$R(z) = \sum_{n=1}^{\infty} a_n z^n, \quad a_n \in \{0,1\}$$

to the equation

$$3z^3 R(z^3) - 3z^9 R(z^6) - R(z^2) - R(\omega z^2) - R(\omega^2 z^2) = 0$$

is the trivial solution $R(z) \equiv 0$.

Further Reading

A great place to start is the classical paper of Jeff Lagarias, The $3x + 1$ Problem and its Generalizations, *American Mathematical Monthly* 92, (1985), 1–23. More recent surveys include Günther Wirsching's *The Dynamical System Generated by the $3n + 1$ Function*, Springer, Heidelberg (1998), and the author's An Update on the $3x + 1$ Problem, (2004), www.math.grin.edu/~chamberl/3x.html. Another very helpful resource due to Lagarias is The $3x + 1$ Problem: An annotated bibliography, arxiv.org/abs/math/0309224. An experimental approach to the problem was taken by Stan Wagon, The Collatz Problem, *Mathematical Intelligencer*, 7(1), (1985), 72–76.

Originally appeared as:
Chamberland, Marc. "The Collatz Chameleon." *Math Horizons*. vol. 14, no. 2 (November 2006), pp. 5–8.

Bijecting Euler's Partition Recurrence

David Bressoud and
Doron Zeilberger

A partition of an integer n is a nonincreasing sequence of positive integers $\lambda(1) \geq \lambda(2) \geq \cdots \geq \lambda(t) > 0$, such that $\lambda(1) + \cdots + \lambda(t) = n$. The set of partitions of n is denoted $\text{Par}(n)$ and its cardinality $|\text{Par}(n)|$ is written $p(n)$. For example,

$$\text{Par}(5) = \{5; 4,1; 3,2; 3,1,1; 2,2,1; 2,1,1,1; 1,1,1,1,1\} \quad \text{and} \quad p(5) = 7.$$

There is no closed form formula for $p(n)$ but Euler ([1], p. 12) gave a very efficient way for compiling a table of $p(n)$ by proving the recurrence

$$(1) \qquad \sum_{j \text{ even}} p(n - a(j)) = \sum_{j \text{ odd}} p(n - a(j)), \quad \text{where } a(j) = (3j^2 + j)/2.$$

Euler used generating functions to prove this formula. Garsia and Milne [2] gave a very nice bijective proof of (1), utilizing their Involution Principle. We are going to give another bijective proof which does not require any iterations and is very simple. Indeed,

$$\phi: \bigcup_{j \text{ even}} \text{Par}(n - a(j)) \leftrightarrow \bigcup_{j \text{ odd}} \text{Par}(n - a(j)),$$

given below does the job.

Let $(\lambda) = (\lambda(1), \ldots, \lambda(t)) \in \text{Par}(n - a(j))$. Then define ϕ by

$$\phi((\lambda)) = \begin{cases} (t + 3j - 1, \lambda(1) - 1, \ldots, \lambda(t) - 1) \in \text{Par}(n - a(j-1)), & \text{if } t + 3j \geq \lambda(1), \\ (\lambda(2) + 1, \ldots, \lambda(t) + 1, 1, 1, \ldots, 1) \in \text{Par}(n - a(j+1)), & \text{if } t + 3j < \lambda(1), \\ \text{where there are } \lambda(1) - 3j - t - 1 \text{ 1's at the end.} \end{cases}$$

Note that applying ϕ twice yields the identity mapping, thus $\phi = \phi^{-1}$ and ϕ is invertible.

EXAMPLE. $n = 21$.

$$\phi(5,5,4,3,2) = (7,4,4,3,2,1).$$

Here $(5,5,4,3,2,) \in \text{Par}(19) = \text{Par}(n - a(1))$ so $j = 1$. The number of parts t, is 5 and we have $t + 3j \geq \lambda(1)$, since $5 + 3 \geq 5$. Now consider $\phi(7,4,4,3,2,1)$; here $j = 0$, $t = 6$, $\lambda(1) = 7$ and $6 + 0 < 7$. Also $\lambda(1) - 3j - t - 1 = 7 - 0 - 6 - 1 = 0$ so we do not add any 1's at the end and $\phi(7,4,4,3,2,1) = (5,5,4,3,2)$.

References

1. G. E. Andrews, The Theory of Partitions, Addison-Wesley, Reading, MA, 1976.
2. A. M. Garsia and S. C. Milne, A Rogers-Ramanujan bijection, J. Combin. Theory (A), 31 (1981) 289–339.

Originally appeared as:
Bressoud, David M. and Doron Zeilberger. "Bijecting Euler's Partition Recurrence." *American Mathematical Monthly*. vol. 92, no. 1 (January 1985), pp. 54–55.

Discovery of a Most Extraordinary Law of the Numbers Concerning the Sum of Their Divisors

Leonhard Euler
translated by George Pólya

> *He [Euler] preferred instructing his pupils to the little satisfaction of amazing them. He would have thought not to have done enough for science if he should have failed to add to the discoveries, with which he enriched science, the candid exposition of the ideas that led him to those discoveries.* —Condorcet

Euler

Of all mathematicians with whose work I am somewhat acquainted, Euler seems to be by far the most important for our inquiry. A master of inductive research in mathematics, he made important discoveries (on infinite series, in the Theory of Numbers, and in other branches of mathematics) by induction, that is, by observation, daring guess, and shrewd verification. In this respect, however, Euler is not unique; other mathematicians, great and small, used induction extensively in their work.

Yet Euler seems to me almost unique in one respect: he takes pains to present the relevant inductive evidence carefully, in detail, in good order. He presents it convincingly but honestly, as a genuine scientist should do. His presentation is "the candid exposition of the ideas that led him to those discoveries" and has a distinctive charm. Naturally enough, as any other author, he tries to impress his readers, but, as a really good author, he tries to impress his readers only by such things as have genuinely impressed himself.

The next section brings a sample of Euler's writing. The memoir chosen can be read with very little previous knowledge and is entirely devoted to the exposition of an inductive argument.

Euler's Memoir

Euler's memoir is given here, in English translation, *in extenso*, except for a few unessential alterations which should make it more accessible to a modern reader.

Discovery of a Most Extraordinary Law of Numbers Concerning the Sum of Their Divisors

1. Till now the mathematicians tried in vain to discover some order in the sequence of the prime numbers and we have every reason to believe that there is some mystery which the human mind shall never penetrate. To convince oneself, one has only to glance at the tables of the primes, which some people took the trouble to compute beyond a hundred thousand, and one perceives that there is no order and no rule. This is so much more surprising as the arithmetic gives us definite rules with the help of which we can continue the sequence of the primes as far as we please, without noticing, however, the least trace of order. I am myself certainly far from this goal, but I just happened to discover an extremely strange law governing the sums of the divisors of the integers which, at the first glance, appear just as irregular as the sequence of the primes, and which, in a certain sense, comprise even the latter. This law, which I shall explain in a moment, is, in my opinion, so much more remarkable as it is of such a nature that we can be assured of its truth without giving it a perfect demonstration. Nevertheless, I shall present such evidence for it as might be regarded as almost equivalent to a rigorous demonstration.

2. A prime number has no divisors except unity and itself, and this distinguishes the primes from the other numbers. Thus 7 is a prime, for it is divisible only by 1 and itself. Any other number which has, besides unity and itself, further divisors, is called composite, as for instance, the number 15, which has, besides 1 and 15, the divisors 3 and 5. Therefore, generally, if the number p is prime, it will be divisible only by 1 and p; but if p was composite, it would have, besides 1 and p, further divisors. Therefore, in the first case, the sum of its divisors will be $1 + p$, but in the latter it would exceed $1 + p$. As I shall have to consider the sum of divisors of various numbers, I shall use the sign $\sigma(n)$ to denote the sum of the divisors of the number n. Thus, $\sigma(12)$ means the sum of all the divisors of 12, which are 1, 2, 3, 4, 6 and 12; therefore, $\sigma(12) = 28$. In the same way, one can see that $\sigma(60) = 168$ and $\sigma(100) = 217$. Yet, since unity is only divisible by itself, $\sigma(1) = 1$. Now, 0 (zero) is divisible by all numbers. Therefore, $\sigma(0)$ should be properly infinite. (However, I shall assign to it later a finite value, different in different cases, and this will turn out serviceable.)

3. Having defined the meaning of the symbol $\sigma(n)$, as above, we see clearly that if p is a prime $\sigma(p) = 1 + p$. Yet $\sigma(1) = 1$ (and not $1 + 1$); hence we see that 1 should be excluded from the sequence of the primes; 1 is the beginning of the integers, neither prime nor composite. If, however, n is composite, $\sigma(n)$ is greater than $1 + n$.

In this case we can easily find $\sigma(n)$ from the factors of n. If $a, b, c, d, \ldots$ are different primes, we see easily that

$$\sigma(ab) = 1 + a + b + ab = (1+a)(1+b) = \sigma(a)\sigma(b),$$
$$\sigma(abc) = (1+a)(1+b)(1+c) = \sigma(a)\sigma(b)\sigma(c),$$
$$\sigma(abcd) = \sigma(a)\sigma(b)\sigma(c)\sigma(d)$$

and so on. We need particular rules for the powers of primes, as

$$\sigma(a^2) = 1 + a + a^2 = \frac{a^3 - 1}{a - 1}$$

$$\sigma(a^3) = 1 + a + a^2 + a^3 = \frac{a^4 - 1}{a - 1}$$

and generally,

$$\sigma(a^n) = \frac{a^{n+1} - 1}{a - 1}.$$

Using this, we can find the sum of the divisors of any number, composite in any way whatever. This we see from the formulas

$$\sigma(a^2b) = \sigma(a^2)\,\sigma(b)$$
$$\sigma(a^3b^2) = \sigma(a^3)\,\sigma(b^2)$$
$$\sigma(a^3b^4c) = \sigma(a^3)\,\sigma(b^4)\,\sigma(c)$$

and generally,

$$\sigma(a^\alpha b^\beta c^\gamma d^\delta e^\varepsilon) = \sigma(a^\alpha)\sigma(b^\beta)\sigma(c^\gamma)\sigma(d^\delta)\sigma(e^\varepsilon).$$

For instance, to find $\sigma(360)$ we set, since 360 factorized is $2^3 \cdot 3^2 \cdot 5$,

$$\sigma(360) = \sigma(2^3)\,\sigma(3^2)\,\sigma(5) = 15 \cdot 13 \cdot 6 = 1170.$$

4. In order to show the sequence of the sums of the divisors, I add the following table containing the sums of the divisors of all integers from 1 up to 99.

n	0	1	2	3	4	5	6	7	8	9
0	—	1	3	4	7	6	12	8	15	13
10	18	12	28	14	24	24	31	18	39	20
20	42	32	36	24	60	31	42	40	56	30
30	72	32	63	48	54	48	91	38	60	56
40	90	42	96	44	84	78	72	48	124	57
50	93	72	98	54	120	72	120	80	90	60
60	168	62	96	104	127	84	144	68	126	96
70	144	72	195	74	114	124	140	96	168	80
80	186	121	126	84	224	108	132	120	180	90
90	234	112	168	128	144	120	252	98	171	156

If we examine a little the sequence of these numbers, we are almost driven to despair. We cannot hope to discover the least order. The irregularity of the primes is so deeply involved in it that we must think it impossible to disentangle any law governing this sequence, unless we know the law governing the sequence of the prime itself. It could appear even that the sequence before us is still more mysterious than the sequence of the primes.

5. Nevertheless, I observed that this sequence is subject to a completely definite law and could even be regarded as a *recurring* sequence. This mathematical expression means that each term can be computed from the foregoing terms, according to an invariable rule. In fact, if we let $\sigma(n)$ denote any term of this sequence, and $\sigma(n-1)$, $\sigma(n-2)$, $\sigma(n-3)$, $\sigma(n-4)$, $\sigma(n-5)$, ... the preceding terms, I say that the value of $\sigma(n)$ can always be combined from some of the preceding as prescribed by the following formula:

$$\begin{aligned}\sigma(n) =\ & \sigma(n-1) + \sigma(n-2) - \sigma(n-5) - \sigma(n-7) \\ & + \sigma(n-12) + \sigma(n-15) - \sigma(n-22) - \sigma(n-26) \\ & + \sigma(n-35) + \sigma(n-40) - \sigma(n-51) - \sigma(n-57) \\ & + \sigma(n-70) + \sigma(n-77) - \sigma(n-92) - \sigma(n-100) \\ & + \ldots\end{aligned}$$

On this formula we must make the following remarks.

    I. In the sequence of the signs + and −, each arises twice in succession.

    II. The law of the numbers 1, 2, 5, 7, 12, 15, ... which we have to subtract from the proposed number n, will become clear if we take their differences:

Nrs.   1, 2, 5, 7, 12, 15, 22, 26, 35, 40, 51, 57, 70, 77, 92, 100, ...
Diff.   1, 3, 2, 5, 3, 7, 4, 9, 5, 11, 6, 13, 7, 15, 8, ...

In fact, we have here, alternately, all the integers 1, 2, 3, 4, 5, 6, ... and the odd numbers 3, 5, 7, 9, 11, ..., and hence we can continue the sequence of these numbers as far as we please.

    III. Although this sequence goes to infinity, we must take, in each case, only those terms for which the numbers under the sign σ are all still positive and omit the σ for negative values.

    IV. If the sign $\sigma(0)$ turns up in the formula, we must, as its value in itself is indeterminate, substitute for $\sigma(0)$ the number n proposed.

6. After these remarks it is not difficult to apply the formula to any given particular case, and so anybody can satisfy himself of its truth by as many examples as he may wish to develop. And since I must admit that I am not in a position to give it a rigorous demonstration, I will justify it by a sufficiently large number of examples.

$\sigma(1) = \sigma(0) = 1 = 1$
$\sigma(2) = \sigma(1) + \sigma(0) = 1 + 2 = 3$
$\sigma(3) = \sigma(2) + \sigma(1) = 3 + 1 = 4$
$\sigma(4) = \sigma(3) + \sigma(2) = 4 + 3 = 7$
$\sigma(5) = \sigma(4) + \sigma(3) - \sigma(0) = 7 + 4 - 5 = 6$
$\sigma(6) = \sigma(5) + \sigma(4) - \sigma(1) = 6 + 7 - 1 = 12$
$\sigma(7) = \sigma(6) + \sigma(5) - \sigma(2) - \sigma(0) = 12 + 6 - 3 - 7 = 8$
$\sigma(8) = \sigma(7) + \sigma(6) - \sigma(3) - \sigma(1) = 8 + 12 - 4 - 1 = 15$
$\sigma(9) = \sigma(8) + \sigma(7) - \sigma(4) - \sigma(2) = 15 + 8 - 7 - 3 = 13$
$\sigma(10) = \sigma(9) + \sigma(8) - \sigma(5) - \sigma(3) = 13 + 15 - 6 - 4 = 18$
$\sigma(11) = \sigma(10) + \sigma(9) - \sigma(6) - \sigma(4) = 18 + 13 - 12 - 7 = 12$
$\sigma(12) = \sigma(11) + \sigma(10) - \sigma(7) - \sigma(5) + \sigma(0) = 12 + 18 - 8 - 6 + 12 = 28$
$\sigma(13) = \sigma(12) + \sigma(11) - \sigma(8) - \sigma(6) + \sigma(1) = 28 + 12 - 15 - 12 + 1 = 14$
$\sigma(14) = \sigma(13) + \sigma(12) - \sigma(9) - \sigma(7) + \sigma(2) = 14 + 28 - 13 - 8 + 3 = 24$
$\sigma(15) = \sigma(14) + \sigma(13) - \sigma(10) - \sigma(8) + \sigma(3) + \sigma(0) = 24 + 14 - 18 - 15 + 4 + 15 = 24$
$\sigma(16) = \sigma(15) + \sigma(14) - \sigma(11) - \sigma(9) + \sigma(4) + \sigma(1) = 24 + 24 - 12 - 13 + 7 + 1 = 31$
$\sigma(17) = \sigma(16) + \sigma(15) - \sigma(12) - \sigma(10) + \sigma(5) + \sigma(2) = 31 + 24 - 28 - 18 + 6 + 3 = 18$
$\sigma(18) = \sigma(17) + \sigma(16) - \sigma(13) - \sigma(11) + \sigma(6) + \sigma(3) = 18 + 31 - 14 - 12 + 12 + 4 = 39$
$\sigma(19) = \sigma(18) + \sigma(17) - \sigma(14) - \sigma(12) + \sigma(7) + \sigma(4) = 39 + 18 - 24 - 28 + 8 + 7 = 20$
$\sigma(20) = \sigma(19) + \sigma(18) - \sigma(15) - \sigma(13) + \sigma(8) + \sigma(5) = 20 + 39 - 24 - 14 + 15 + 6 = 42$

    I think these examples are sufficient to discourage anyone from imagining that it is by mere chance that my rule is in agreement with the truth.

7. Yet somebody could still doubt whether the law of the numbers 1, 2, 5, 7, 12, 15, ... which we have to subtract is precisely that one which I have indicated, since the examples given imply only the first six of these numbers. Thus, the law could still appear as insufficiently established and, therefore, I will give some examples with larger numbers.

I. Given the number 101, find the sum of its divisors. We have

$$\begin{aligned}
\sigma(101) &= \sigma(100) + \sigma(99) - \sigma(96) - \sigma(94) \\
&+ \sigma(89) + \sigma(86) - \sigma(79) - \sigma(75) \\
&+ \sigma(66) + \sigma(61) - \sigma(50) - \sigma(44) \\
&+ \sigma(31) + \sigma(24) - \sigma(9) - \sigma(1) \\
&= 217 + 156 - 252 - 144 \\
&+ 90 + 132 - 80 - 124 \\
&+ 144 + 62 - 93 - 84 \\
&+ 32 + 60 - 13 - 1 \\
&= 893 - 791 \\
&= 102
\end{aligned}$$

and hence we could conclude, if we would not have known it before, that 101 is a prime number.

II. Given the number 301, find the sum of its divisors. We have

$$\begin{aligned}
&\text{diff.} \quad 1 \quad\quad 3 \quad\quad 2 \quad\quad 5 \\
\sigma(301) &= \sigma(300) + \sigma(299) - \sigma(296) - \sigma(294) + \\
&\quad\quad 3 \quad\quad\quad 7 \quad\quad\quad 4 \quad\quad\quad 9 \\
&+ \sigma(289) + \sigma(286) - \sigma(279) - \sigma(275) + \\
&\quad\quad 5 \quad\quad\quad 11 \quad\quad\quad 6 \quad\quad\quad 13 \\
&+ \sigma(266) + \sigma(261) - \sigma(250) - \sigma(244) + \\
&\quad\quad 7 \quad\quad\quad 15 \quad\quad\quad 8 \quad\quad\quad 17 \\
&+ \sigma(231) + \sigma(224) - \sigma(209) - \sigma(201) + \\
&\quad\quad 9 \quad\quad\quad 19 \quad\quad\quad 10 \quad\quad\quad 21 \\
&+ \sigma(184) + \sigma(175) - \sigma(156) - \sigma(146) + \\
&\quad\quad 11 \quad\quad\quad 23 \quad\quad\quad 12 \quad\quad\quad 25 \\
&+ \sigma(125) + \sigma(114) - \sigma(91) - \sigma(79) + \\
&\quad\quad 13 \quad\quad\quad 27 \quad\quad\quad 14 \\
&+ \sigma(54) + \sigma(41) - \sigma(14) - \sigma(0).
\end{aligned}$$

We see by this example how we can, using differences, continue the formula as far as is necessary in each case. Performing the computations, we find

$$\sigma(301) = 4939 - 4587 = 352.$$

We see hence that 301 is not a prime. In fact, $301 = 7 \cdot 43$ and we obtain

$$\sigma(301) = \sigma(7)\sigma(43) = 8 \cdot 44 = 352$$

as the rule has shown.

8. The examples that I have just developed will undoubtedly dispel any qualms which we might have had about the truth of my formula. Now, this beautiful property of the numbers is so much more surprising as we do not perceive any intelligible connection between the structure of my formula and the nature of the divisors with the sum of which we are here concerned. The sequence of the numbers 1, 2, 5, 7, 12, 15,... does not seem to have any relation to the matter in hand. Moreover, as the law of these numbers is "interrupted" and they are in fact a mixture of two sequences with a regular law, of 1, 5, 12, 22, 35, 51,... and 2, 7, 15, 26, 40, 57,..., we would not expect that such an irregularity can turn up in Analysis. The lack of demonstration must increase the surprise still more, since it seems wholly impossible to succeed in discovering such a property without being guided by some reliable method which could take the place of a perfect proof. I confess that I did not hit on this discovery by mere chance, but another proposition opened the path to this beautiful property–another proposition of the same nature which must be accepted as true although I am unable to prove it. And although we consider here the nature of integers to which the Infinitesimal Calculus does not seem to apply, nevertheless I reached my conclusion by differentiations and other devices. I wish that somebody would find a shorter and more natural way, in which the consideration of the path that I followed might be of some help, perhaps.

9. In considering the partitions of numbers, I examined, a long time ago, the expression

$$(1-x)(1-x^2)(1-x^3)(1-x^4)(1-x^5)(1-x^6)(1-x^7)(1-x^8)\ldots,$$

in which the product is assumed to be infinite. In order to see what kind of series will result, I multiplied actually a great number of factors and found

$$1 - x - x^2 + x^5 + x^7 - x^{12} - x^{15} + x^{22} + x^{26} - x^{35} - x^{40} + \cdots.$$

The exponents of x are the same which enter into the above formula; also the signs + and − arise twice in succession. It suffices to undertake this multiplication and to continue it as far as it is deemed proper to become convinced of the truth of this series. Yet I have no other evidence for this, except a long induction which I have carried out so far that I cannot in any way doubt the law governing the formation of these terms and their exponents. I have long searched in vain for a rigorous demonstration of the equation between the series and the above infinite product $(1-x)(1-x^2)(1-x^3)\cdots$, and I have proposed the same question to some of my friends with whose ability in these matters I am familiar, but all have agreed with me on the truth of this transformation of the product into a series, without being able to unearth any clue of a demonstration. Thus, it will be a known truth, but not yet demonstrated, that if we put

$$s = (1-x)(1-x^2)(1-x^3)(1-x^4)(1-x^5)(1-x^6)\cdots$$

the same quantity s can also be expressed as follows:

$$s = 1 - x - x^2 + x^5 + x^7 - x^{12} - x^{15} + x^{22} + x^{26} - x^{35} - x^{40} + \cdots.$$

For each of us can convince himself of this truth by performing the multiplication as far

Discovery of a Most Extraordinary Law of the Numbers... 231

as he may wish; and it seems impossible that the law which has been discovered to hold for 20 terms, for example, would not be observed in the terms that follow.

10. As we have thus discovered that those two infinite expressions are equal even though it has not been possible to demonstrate their equality, all the conclusions which may be deduced from it will be of the same nature, that is, true but not demonstrated. Or, if one of these conclusions could be demonstrated, one could reciprocally obtain a clue to the demonstration of that equation; and it was with this purpose in mind that I maneuvered those two expressions in many ways, and so I was led among other discoveries to that which I explained above; its truth, therefore, must be as certain as that of the equation between the two infinite expressions. I proceeded as follows. Being given that the two expressions

I. $s = (1-x)(1-x^2)(1-x^3)(1-x^4)(1-x^5)(1-x^6)(1-x^7)\cdots$
II. $s = 1 - x - x^2 + x^5 + x^7 - x^{12} - x^{15} + x^{22} + x^{26} - x^{35} - x^{40} + \cdots$

are equal, I got rid of the factors in the first by taking logarithms

$$\log s = \log(1-x) + \log(1-x^2) + \log(1-x^3) + \log(1-x^4) + \cdots.$$

In order to get rid of the logarithms, I differentiate and obtain the equation

$$\frac{1}{s}\frac{ds}{dx} = -\frac{1}{1-x} - \frac{2x}{1-x^2} - \frac{3x^2}{1-x^3} - \frac{4x^3}{1-x^4} - \frac{5x^4}{1-x^5} - \cdots$$

or

$$-\frac{x}{s}\frac{ds}{dx} = \frac{x}{1-x} + \frac{2x^2}{1-x^2} + \frac{3x^3}{1-x^3} + \frac{4x^4}{1-x^4} + \frac{5x^5}{1-x^5} + \cdots$$

From the second expression for s, as infinite series, we obtain another value for the same quantity

$$-\frac{x}{s}\frac{ds}{dx} = \frac{x + 2x^2 - 5x^5 - 7x^7 + 12x^{12} + 15x^{15} - 22x^{22} - 26x^{26} + \cdots}{1 - x - x^2 + x^5 + x^7 - x^{12} - x^{15} + x^{22} + x^{26} - \cdots}.$$

11. Let us put

$$-\frac{x}{s}\frac{ds}{dx} = t.$$

We have about two expressions for the quantity t. In the first expression, I expand each term into a geometric series and obtain

$$\begin{aligned}
t = x\ +\ &x^2\ +\ x^3\ +\ x^4\ +\ x^5\ +\ x^6\ +\ x^7\ +\ x^8 + \ldots \\
+\ &2x^2\quad\ \ \ +\ 2x^4\quad\ \ \ +\ 2x^6\quad\ \ \ +\ 2x^8 + \ldots \\
&+\ 3x^3\qquad\qquad +\ 3x^6\qquad\qquad +\ \ldots \\
&\quad\ \ +\ 4x^4\qquad\qquad\qquad +\ 4x^8 + \ldots \\
&\qquad\ \ +\ 5x^5\qquad\qquad\qquad\quad +\ \ldots \\
&\qquad\qquad +\ 6x^6\qquad\qquad\qquad +\ \ldots \\
&\qquad\qquad\quad\ \ +\ 7x^7\qquad\qquad +\ \ldots \\
&\qquad\qquad\qquad\qquad\quad\ \ +\ 8x^8 + \ldots.
\end{aligned}$$

Here we see easily that each power of x arises as many times as its exponent has divisors, and that each divisor arises as a coefficient of the same power of x. Therefore, if we collect the terms with like powers, the coefficient of each power of x will be the sum of the divisors of its exponent. And, therefore, using the above notation $\sigma(n)$ for the sum of the divisors of n, I obtain $t = \sigma(1)x + \sigma(2)x^2 + \sigma(3)x^3 + \sigma(4)x^4 + \sigma(5)x^5 + \ldots$.

The law of the series is manifest. And, although it might appear that some induction was involved in the determination of the coefficients, we can easily satisfy ourselves that this law is a necessary consequence.

12. By virtue of the definition of t, the last formula of No. 10 can be written as follows:

$$t(1 - x - x^2 + x^5 + x^7 - x^{12} - x^{15} + x^{22} + x^{26} - \ldots)$$
$$- x - 2x^2 + 5x^5 + 7x^7 - 12x^{12} - 15x^{15} + 22x^{22} + 26x^{26} - \ldots = 0.$$

Substituting for t the value obtained at the end of No. 11, we find

$$0 = \sigma(1)x + \sigma(2)x^2 + \sigma(3)x^3 + \sigma(4)x^4 + \sigma(5)x^5 + \sigma(6)x^6 + \ldots$$
$$- x - \sigma(1)x^2 - \sigma(2)x^3 - \sigma(3)x^4 - \sigma(4)x^5 - \sigma(5)x^6 - \ldots$$
$$- 2x^2 - \sigma(1)x^3 - \sigma(2)x^4 - \sigma(3)x^5 - \sigma(4)x^6 - \ldots$$
$$+ 5x^5 + \sigma(1)x^6 + \ldots.$$

Collecting the terms, we find the coefficient for any given power of x. This coefficient consists of several terms. First comes the sum of the divisors of the exponent of x, and then sums of divisors of some preceding numbers, obtained from that exponent by subtracting successively 1, 2, 5, 7, 12, 15, 22, 26, ... Finally, if it belongs to this sequence, the exponent itself arises. We need not explain again the signs assigned to the terms just listed. Therefore, generally, the coefficient of x^n is

$$\sigma(n) - \sigma(n-1) - \sigma(n-2) + \sigma(n-5) + \sigma(n-7) - \sigma(n-12) - \sigma(n-15) + \ldots.$$

This is continued as long as the numbers under the sign σ are not negative. Yet, if the term $\sigma(0)$ arises, we must substitute n for it.

13. Since the sum of the infinite series considered in the foregoing No. 12 is 0, whatever the value of x may be, the coefficient of each single power of x must necessarily be 0. Hence we obtain the law that I explained above in No. 5; I mean the law that governs the sum of the divisors and enables us to compute it recursively for all numbers. In the foregoing development, we may perceive some reason for the signs, some reason for the sequence of the numbers

1, 2, 5, 7, 12, 15, 22, 26, 35, 40, 51, 57, 70, 77, ...

and, especially, a reason why we should substitute for $\sigma(0)$ the number n itself, which could have appeared the strangest feature of my rule. This reasoning, although still very far from a perfect demonstration, will certainly lift some doubts about the most extraordinary law that I explained here.

Originally appeared as:

Euler, Leonard. Translated by George Pólya. "Discovery of a Most Extraordinary Law of the Numbers Concerning the Sum of their Divisors." *Mathematics and Plausible Reasoning.* vol. I. Princeton University Press, (1954), pp. 90–98.

The Factorial Function and Generalizations

Manjul Bhargava

1. INTRODUCTION. The factorial function hardly needs any introduction. Starting with its fundamental interpretation as the number of ways n people can sit in n chairs, to its occurrence in formulae for binomial coefficients, Stirling numbers, and countless other combinatorial objects, it is indeed nearly impossible to study any area of combinatorics without becoming intimately familiar with the factorial.

Perhaps it is due to this ubiquity in combinatorics that sometimes it is overlooked that the factorial also makes several important appearances in number theory! The purpose of this article is to take a closer look at some of these number-theoretic appearances, and thereby lead up to a series of generalizations of the factorial function, which recently have been applied to a variety of number-theoretic, ring-theoretic, and combinatorial problems.

The work described here began about four years ago as part of the author's thesis at Harvard University. The "generalized factorials" introduced there have since been used to give answers to some old questions; but more than that, they have given rise to several new questions that beg for answers. I hope that this expository account, together with several new results and observations, gives readers a sense of what these generalized factorials are all about, and at the same time, incites them to try their hand at some of the many very tempting questions that arise in the process!

2. THE FACTORIAL FUNCTION IN NUMBER THEORY? The most well-known instance of the factorial function arising in a number-theoretic context is probably the following divisibility result: *The product of any k consecutive integers must be divisible by $k!$*. Although admittedly a rather trivial statement, this result is more number-theoretically significant than it might first seem (indeed, we need it a bit later). The proof is of course quite simple; for we may restate this result as follows:

Theorem 1. *For any nonnegative integers k and ℓ, $(k + \ell)!$ is a multiple of $k!\,\ell!$.*

Theorem 1 is clearly equivalent to our original formulation, and moreover, its truth is clear combinatorially: the quotient $(k + \ell)!/k!\,\ell!$ is simply the binomial coefficient (and integer) $\binom{k+\ell}{k}$.

There are, however, many occurrences of the factorial function in number theory that are not quite so trivial. One beautiful such example, due to George Pólya, describes the close relationship between the factorial function and the possible sets of values taken by a polynomial.

Suppose we have a polynomial f with integer coefficients. The *fixed divisor* of f over the integers $\mathbb{Z}$, denoted by $d(\mathbb{Z}, f)$, is the greatest common divisor of all the elements in the image of f on $\mathbb{Z}$; that is,

$$d(\mathbb{Z}, f) = \gcd\{f(a) : a \in \mathbb{Z}\}.$$

For example, consider the polynomial $f(x) = x^5 + x$. If x is even, then $f(x)$ is even, and if x is odd, then again $f(x)$ is even. It follows that $d(\mathbb{Z}, f)$ must be a multiple of 2. On the other hand, we have $f(1) = 2$; hence $d(\mathbb{Z}, f)$ is exactly 2 in this case.

The question naturally arises: what are the possible values of $d(\mathbb{Z}, f)$? Can it be anything? Well, if we let $f(x) = 1000x^5 + 1000x$ (i.e., 1000 times the previous polynomial), then $d(\mathbb{Z}, f) = 2000$; that is, we can multiply an existing fixed divisor by anything we like simply by multiplying the polynomial by that amount. Therefore, we would like to answer this question for only those polynomials whose coefficients are relatively prime, i.e., for *primitive polynomials*.

In that case, our question has the following surprising answer, discovered by Pólya [27] in 1915:

Theorem 2. *Let f be a primitive polynomial of degree k, and let $d(\mathbb{Z}, f) = \gcd\{f(a) : a \in \mathbb{Z}\}$. Then $d(\mathbb{Z}, f)$ divides $k!$. (This is sharp!)*

By the phrase "this is sharp" we mean that not only is $k!$ an upper bound for the fixed divisor of a degree k polynomial, but $k!$ can actually be achieved for some primitive polynomial f; in fact, any factor of $k!$ can be achieved.

Now one may ask: what is $k!$ doing here? It's not immediately clear why the factorial function should appear in this result; yet it does. We explain this result from a more general viewpoint a little later.

But first, here is another pretty example of the factorial function arising in a number-theoretic context:

Theorem 3. *Let $a_0, a_1, \ldots, a_n \in \mathbb{Z}$ be any $n + 1$ integers. Then the product of their pairwise differences*

$$\prod_{i<j}(a_i - a_j)$$

is a multiple of $0! \, 1! \cdots n!$. (This is sharp!)

As before, "this is sharp" refers to the fact that the constant $0!1! \cdots n!$ in the statement of the theorem cannot be improved.

It is interesting to note that Theorem 3 originates in the representation theory of Lie algebras: the quotient $\prod_{i<j}(a_i - a_j)/0!1! \cdots n!$ is the dimension of a certain irreducible representation of $SU(n)$, and consequently must be an integer. This elegant result has been the subject of some recent articles (e.g., [29]), and was also once a problem posed on the Russian Mathematical Olympiad. Again, notice how the factorial function appears, and in this case how it does so numerous times.

Let us look at one more example of the factorial function for now—this one of a combinatorial nature. Recall that, for any given prime n, every function from $\mathbb{Z}/n\mathbb{Z}$ to itself can be represented by a polynomial. This is because when n is prime, $\mathbb{Z}/n\mathbb{Z}$ is a field, so one may carry out the usual Lagrange interpolation. When n is not prime, however, not every function is so representable; for when performing Lagrange interpolation, one often needs to divide, but this may not be possible in a nonfield. The question thus arises: how many functions from $\mathbb{Z}/n\mathbb{Z}$ to itself (equivalently, from $\mathbb{Z}$ to $\mathbb{Z}/n\mathbb{Z}$) are representable by a polynomial?

It so happens there is an exact formula for the number of such polynomial mappings, and it was discovered by Kempner [18] in the 1920's:

Theorem 4. *The number of polynomial functions from $\mathbb{Z}$ to $\mathbb{Z}/n\mathbb{Z}$ is given by*

$$\prod_{k=0}^{n-1} \frac{n}{\gcd(n, k!)}.$$

In particular, when n is prime, Theorem 4 tells us that there are n^n such functions; hence in this case every function from $\mathbb{Z}/n\mathbb{Z}$ to itself is polynomial, as was expected. Notice the appearance of the factorial function in the general formula.

3. THE MOTIVATING QUESTION.

To summarize, we have four—well, actually there are many more—but for now, we have four number-theoretic results in which the factorial function plays a very prominent role. But all these results involving factorials are heavily dependent on the fact that we are working in $\mathbb{Z}$—the entire set of rational integers. Indeed: in Theorem 2, we take the greatest common divisor of $f(a)$ over all a in $\mathbb{Z}$; in Theorem 3, we choose any $n + 1$ integers from $\mathbb{Z}$; in Theorem 4, we take polynomial mappings from $\mathbb{Z}$ to the integers modulo n; and so on.

What would happen if we were to change each of these occurrences of $\mathbb{Z}$ to something else? Perhaps to some subset S of $\mathbb{Z}$. Or to some other ring entirely. Or perhaps even to some subset of some other ring! Is there some other function—some generalized factorial function—that we could change each of the ordinary factorials to, so that Theorems 1–4 would still remain true?

It turns out there is such a "generalized factorial function" for *any* given subset S of $\mathbb{Z}$ that simultaneously makes Theorems 1–4 true when $\mathbb{Z}$ is replaced by S. In fact, the same holds true for any subset S of a Dedekind ring.

How can one construct these generalized factorials? Let us consider first the case when S is a subset of $\mathbb{Z}$.

4. A GAME CALLED p-ORDERING.

Let S be an arbitrary subset of $\mathbb{Z}$, and fix a prime p. A *p-ordering* of S is a sequence $\{a_i\}_{i=0}^{\infty}$ of elements of S that is formed as follows:

- Choose any element $a_0 \in S$;
- Choose an element $a_1 \in S$ that minimizes the highest power of p dividing $a_1 - a_0$;
- Choose an element $a_2 \in S$ that minimizes the highest power of p dividing $(a_2 - a_0)(a_2 - a_1)$;

and in general, at the kth step,

- Choose an element $a_k \in S$ that minimizes the highest power of p dividing $(a_k - a_0)(a_k - a_1) \cdots (a_k - a_{k-1})$.

Notice that a p-ordering of S is certainly not unique; indeed, the element a_0 is chosen arbitrarily, and at later steps there are frequently ties for which element achieves the desired minimum, and one may choose any one of these. In addition, each time one makes a choice for an a_k, it affects the choices one has in the future.

But once such a p-ordering $\{a_i\}_{i=0}^{\infty}$ has been constructed, one obtains a corresponding monotone increasing sequence $\{\nu_k(S, p)\}_{k=0}^{\infty}$ of powers of p, where the kth element $\nu_k(S, p)$ is the power of p minimized at the kth step of the

p-ordering process. More precisely, if we denote by $w_p(a)$ the highest power of p dividing a (e.g., $w_3(18) = 9$), then $\nu_k(S, p)$ is given by

$$\nu_k(S, p) = w_p((a_k - a_0) \cdots (a_k - a_{k-1})). \tag{1}$$

We refer to this sequence $\{\nu_k(S, p)\}$ as the *associated p-sequence* of S corresponding to the chosen p-ordering $\{a_i\}$ of S.

Now it would seem that since there are so many choices to be made when constructing a p-ordering, and since each choice so greatly affects all future choices, that the resulting sequence of minimal powers of p—the associated p-sequence—could be just about anything. But it turns out that:

Theorem 5. *The associated p-sequence of S is independent of the choice of p-ordering*!

Thus the associated p-sequence is intrinsic to S, dependent only on S, and we may speak of it without reference to any particular p-ordering.

Theorem 5 is not at all obvious a priori; however, by the end of the article, it should become very obvious indeed, and in many different ways!

5. THE PUNCHLINE. Let us move on to an example of a p-ordering. We start with the simplest possible case, namely the entire set of integers $\mathbb{Z}$. Then we have the following fact:

Proposition 6. *The natural ordering $0, 1, 2, \ldots$ of the nonnegative integers forms a p-ordering of $\mathbb{Z}$ for all primes p simultaneously.*

The proof is by induction: if $0, 1, 2 \ldots, k - 1$ is a p-ordering for the first $k - 1$ steps, then at the kth step we need to pick a_k to minimize the highest power of p dividing

$$(a_k - 0)(a_k - 1) \cdots (a_k - (k - 1)). \tag{2}$$

However, notice that (2) is the product of k consecutive integers; consequently it must be a multiple of $k!$. But $k!$ can actually be achieved, with the choice $a_k = k$; this value of a_k clearly minimizes the highest power of p dividing (2) for all primes p. So at the kth step we choose $a_k = k$, and the claim follows by induction. ∎

Now since any p-ordering gives the same associated p-sequence, we are in the position to calculate the associated p-sequence $\nu_k(\mathbb{Z}, p)$ of $\mathbb{Z}$. We have

$$\nu_k(\mathbb{Z}, p) = w_p((a_k - a_0) \cdots (a_k - a_{k-1}))$$
$$= w_p((k - 0) \cdots (k - (k - 1))) = w_p(k!).$$

And aha! a factorial! In fact, if we take the expression $w_p(k!)$, and multiply over all primes p, then we get exactly $k!$. So we have a definition of the factorial function purely in terms of these invariants $\nu_k(\mathbb{Z}, p)$:

$$k! = \prod_p \nu_k(\mathbb{Z}, p).$$

But by Theorem 5, $\mathbb{Z}$ is not the only set that has these invariants ν_k—any set S has these invariants! This motivates the following definition:

Definition 7. Let S be any subset of $\mathbb{Z}$. Then the *factorial function* of S, denoted $k!_S$, is defined by

$$k!_S = \prod_p \nu_k(S, p). \tag{3}$$

In particular, we have $k!_{\mathbb{Z}} = k!$.

The Factorial Function and Generalizations

It is a fundamental lemma that the number of factors not equal to one in the product (3) is necessarily finite. Hence Definition 7 makes sense for all S and k.

This seems to be a very natural definition to make—and it turns out it really is the "correct" number-theoretic generalization of the factorial, in that even when $S \neq \mathbb{Z}$, $k!_S$ still shares many important number-theoretic properties with the usual factorial.

6. SOME OLD THEOREMS REVISITED. For example, it is still true that (even for generalized factorials),

Theorem 8. *For any nonnegative integers k and ℓ, $(k + \ell)!_S$ is a multiple of $k!_S \, \ell!_S$.*

This implies, in particular, that we may associate a canonical set of binomial coefficients $\binom{n}{k}_S$ to any set $S \subseteq \mathbb{Z}$, by

$$\binom{n}{k}_S = \frac{n!_S}{k!_S (n-k)!_S}.$$

These generalized binomial coefficients surely must have many interesting properties of their own.

Theorem 8 is not quite as obvious as the result it generalizes. Indeed, as with Theorem 5, trying to prove this result directly from the definitions makes for a challenging (perhaps a bit frustrating?) exercise in combinatorics. In the next section, we give a very short proof of Theorem 8, based on our upcoming generalization of Pólya's Theorem 2.

Theorem 2 concerned the greatest common divisor $d(\mathbb{Z}, f)$ of the values of a primitive polynomial f on $\mathbb{Z}$. More generally, the *fixed divisor of f over S*, denoted by $d(S, f)$, is the greatest common divisor of the elements in the image of f on S. That is,

$$d(S, f) = \gcd\{f(a) : a \in S\}.$$

We may ask the same question about fixed divisors over S, namely: what are the possible values of $d(S, f)$ for primitive polynomials f?

Theorem 9. *Let f be a primitive polynomial of degree k, and let $d(S, f) = \gcd\{f(a) : a \in S\}$. Then $d(S, f)$ divides $k!_S$. (This is sharp!)*

As in Theorem 2, not only is $k!_S$ an upper bound on how large a fixed divisor of a primitive degree k polynomial can be on S, but $k!_S$ can actually be achieved (and as before, any factor of $k!_S$ can be achieved). Thus Theorem 9 extends Polya's result to a general setting.

The analogue of Theorem 3 can also be formulated in a similar manner. Suppose that we are required to choose our $n + 1$ integers not from $\mathbb{Z}$, but from within the set S. What then can we say about the product of their pairwise differences?

Theorem 10. *Let $a_0, a_1, \ldots, a_n \in S$ be any $n + 1$ integers. Then the product*

$$\prod_{i<j}(a_i - a_j)$$

is a multiple of $0!_S \, 1!_S \cdots n!_S$. (This is sharp!)

Again, the phrase "this is sharp" indicates that the constant $0!_S 1!_S \cdots n!_S$ cannot be improved.

As a simple example, suppose we take S to be the set of primes in $\mathbb{Z}$. Using the p-ordering algorithm, it is an easy matter to compute the first six factorials of S: $0!_S = 1$, $1!_S = 1$, $2!_S = 2$, $3!_S = 24$, $4!_S = 48$, and $5!_S = 5760$. Consequently, if $p_0, p_1, \ldots, p_5$ are any six primes, then Theorem 10 says that the product of their pairwise differences $\prod_{i<j}(p_i - p_j)$ is a multiple of 13,271,040. We may compare with the result of Theorem 3, which by itself shows only that it is a multiple of 34,560.

Finally, we consider the analogue of Kempner's result, Theorem 4. Just as before, when n is prime, any function from a subset S of $\mathbb{Z}/n\mathbb{Z}$ to $\mathbb{Z}/n\mathbb{Z}$ can be represented by a polynomial. When n is not prime, though, this is no longer the case. How many functions from S to $\mathbb{Z}/n\mathbb{Z}$ are polynomial?

Theorem 11. *The number of polynomial functions from S to $\mathbb{Z}/n\mathbb{Z}$ is given by*

$$\prod_{k=0}^{n-1} \frac{n}{\gcd(n, k!_S)}.$$

As the reader can tell, creating Theorems 8–11 was quite easy: it was done simply by changing all the previous occurrences of $\mathbb{Z}$'s in Theorems 1–4 to S's, and all the previous !'s to $!_S$'s. And remarkably, all the theorems remain true!

The same happens with many other theorems as well—we provide some further examples in Sections 11 and 13. But first, let us take a detour and indicate why all the aforementioned results are true. (If desired, the reader may skip the next section for the time being without loss of continuity.)

7. PROOFS OF THEOREMS 5 AND 8–11. The simplest proofs of Theorem 5 and Theorems 8–11 are probably via the following observations. Often, when writing polynomials, it is more convenient to use the "falling factorial" basis

$$x^{(n)} = x(x-1) \cdots (x-n+1), \quad n \geq 0,$$

rather than the more familiar basis $\{x^n : n \geq 0\}$. Indeed, much of the difference calculus is based on the important properties of these polynomials $x^{(n)}$.

It turns out that we may define an analogue of the falling factorial for any set $S \subseteq \mathbb{Z}$; namely, having fixed a p-ordering $\{a_i\}$ of S, define $x^{(n)_{S,p}}$ by

$$x^{(n)_{S,p}} = (x - a_0)(x - a_1) \cdots (x - a_{n-1}).$$

In the case $S = \mathbb{Z}$ with p-ordering $0, 1, 2, \ldots$, these polynomials coincide with the usual falling factorials $x^{(n)}$.

The generalized falling factorials $x^{(n)_{S,p}}$ can be used to develop a difference calculus for S. Although we do not need here the full details of this theory, the following result is worth mentioning:

Lemma 12. *A polynomial f over the integers, written in the form*

$$f(x) = \sum_{i=0}^{k} c_i x^{(i)_{S,p}} = \sum_{i=0}^{k} c_i (x - a_0)(x - a_1) \cdots (x - a_{i-1}), \tag{4}$$

vanishes on S modulo p^e if and only if $c_i x^{(i)_{S,p}}$ does for each $0 \leq i \leq k$.

Proof: Suppose f vanishes on S (mod p^e), but some term on the right side of (4) does not. Then let j be the smallest index for which $c_j x^{(j)s,p}$ does not vanish on S (mod p^e). Setting $x = a_j$ in (4), we find that all terms on the right side with $i > j$ vanish identically, whereas the minimality of j guarantees that all terms with $i < j$ vanish (mod p^e). It follows that $c_j a_j^{(j)s,p}$ also vanishes (mod p^e), and consequently $c_j x^{(j)s,p}$ vanishes on all of S (mod p^e), since $\{a_i\}$ is a p-ordering. This contradiction proves the lemma. ∎

We may now prove Theorems 8–11 in no time at all. We begin with our extension of Pólya's theorem, Theorem 9.

Proof of Theorem 9: For a fixed prime p, and a choice of p-ordering $\{a_i\}$ of S, write f in the form

$$f = \sum_{i=0}^{k} c_i x^{(i)s,p} = \sum_{i=0}^{k} c_i (x - a_0)(x - a_1) \cdots (x - a_{i-1}). \tag{5}$$

Since f is primitive, there is a choice of j ($0 \leq j \leq k$) such that c_j is not a multiple of p. Now by definition f vanishes on S modulo $w_p(d(S, f))$; hence Lemma 12 ensures that $c_j x^{(j)s,p}$ does also. Moreover, since c_j is relatively prime to p, it follows that $x^{(j)s,p}$ vanishes on S modulo $w_p(d(S, f))$. In particular, $w_p(d(S, f))$ divides

$$w_p\left(a_j^{(j)s,p}\right) = w_p\left((a_j - a_0)(a_j - a_1) \cdots (a_j - a_{j-1})\right) = w_p(j!_S);$$

hence $w_p(d(S, f))$ divides $w_p(k!_S)$, since $j!_S$ divides $k!_S$. Multiplying over all p, we see that $d(S, f)$ divides $k!_S$, as desired.

To see that $k!_S$ (and any factor thereof) can actually be achieved, we construct "global falling factorial" polynomials $B_{k,S}$ by setting

$$B_{k,S}(x) = (x - a_{0,k})(x - a_{1,k}) \cdots (x - a_{k-1,k}), \tag{6}$$

where $\{a_{i,k}\}_{i=0}^{\infty}$ is a sequence in $\mathbb{Z}$ that, for each prime p dividing $k!_S$, is termwise congruent modulo $v_k(S, p)$ to some p-ordering of S. Then it is clear that $d(S, B_{k,S}) = k!_S$. Furthermore, if r is any factor of $k!_S$, then $d(S, B_{k,S} + r) = r$; hence any factor of $k!_S$ can be obtained as a fixed divisor of some primitive polynomial. ∎

Theorem 9 turns out to be a very powerful tool in understanding generalized factorials. For example, it can be used to give a wonderfully quick proof of Theorem 8:

Proof of Theorem 8: By Theorem 9, there exist primitive polynomials f_k (e.g., $B_{k,S}$) and f_{n-k} (e.g., $B_{n-k,S}$) having degrees k and $n - k$ respectively, such that $d(S, f_k) = k!_S$ and $d(S, f_{n-k}) = (n - k)!_S$. By multiplication, we obtain a primitive polynomial $f = f_k f_{n-k}$ of degree n such that $k!_S(n - k)!_S$ divides $d(S, f)$. But by Theorem 9 again, we know $d(S, f)$ must divide $n!_S$. Hence $k!_S(n - k)!_S$ divides $n!_S$, as desired. ∎

Another important property of generalized factorials is given in the following lemma. Like Theorem 8, it also is not quite as innocent as it first looks, though Theorem 9 again provides the key to an easy proof.

Lemma 13. *Let $T \subseteq S$. Then $k!_S$ divides $k!_T$ for every $k \geq 0$.*

Proof: For any polynomial f, clearly $d(S, f)$ divides $d(T, f)$. Thus, in particular, $d(S, B_{k,S}) = k!_S$ divides $d(T, B_{k,S})$, and by Theorem 9, the latter must divide $k!_T$. It follows that $k!_S$ divides $k!_T$. ∎

Lemma 13 may be used to provide a quick proof of Theorem 10.

Proof of Theorem 10: For a fixed prime p, assume that $a_0, a_1, \ldots, a_n$ are the first $n + 1$ elements of a p-ordering of the set $T = \{a_0, a_1, \ldots, a_n\}$. Then since for each $0 \leq k \leq n$,

$$\nu_k(T, p) = w_p((a_k - a_0)(a_k - a_1) \cdots (a_k - a_{k-1})),$$

we find, upon taking the product over all k and then over all p, that

$$0!_T 1!_T \cdots n!_T = \pm \prod_{i<j} (a_j - a_i).$$

Now by Lemma 13, we know $k!_S$ divides $k!_T$. Therefore

$$0!_S 1!_S \cdots n!_S \Big| \prod_{i<j} (a_i - a_j),$$

proving the first assertion of the theorem.

As for the second assertion, observe that if T consists of the first $n + 1$ elements of a p-ordering of S, then

$$w_p\left(\prod_{i<j}(a_i - a_j)\right) = \nu_0(S, p)\nu_1(S, p) \cdots \nu_n(S, p) = w_p(0!_S 1!_S \cdots n!_S);$$

hence $0!_S 1!_S \cdots n!_S$ cannot be replaced by a larger constant in the statement of the theorem. ∎

For the proof of Theorem 11, we need the following refinement of Lemma 12:

Lemma 14. *A polynomial f of degree d, written in the form*

$$f(x) = \sum_{k=0}^{d} x^{(k)_{S,p}} = \sum_{k=0}^{d} b_k(x - a_0)(x - a_1) \cdots (x - a_{k-1}),$$

vanishes on S modulo p^e if and only if b_k is a multiple of $\dfrac{p^e}{\gcd(p^e, k!_S)}$ for each $0 \leq k \leq d$.

Proof: By Lemma 12, $f(x)$ vanishes on S modulo p^e if and only if $b_k x^{(k)_{S,p}}$ does for each $0 \leq k \leq d$. Now by construction of $x^{(k)_{S,p}}$, we have $w_p(d(S, x^{(k)_{S,p}})) = \nu_k(S, p)$; hence $b_k x^{(k)_{S,p}}$ vanishes on S modulo p^e if and only if b_k is a multiple of $p^e / \gcd(p^e, k!_S)$. This is the desired conclusion. ∎

Proof of Theorem 11: By the Chinese Remainder Theorem, specifying a polynomial mapping on S (modulo n) is equivalent to specifying the mapping modulo each prime power dividing n. Now it is easily seen that the formula of Theorem 11 is multiplicative; hence it suffices to verify Theorem 11 when $n = p^e$ is a prime power.

Let $\{a_i\}$ be a p-ordering of S. Then we claim that *any polynomial mapping $f: S \to \mathbb{Z}/p^e\mathbb{Z}$ can be expressed uniquely in the form*

$$f(x) = \sum_{k=0}^{\infty} c_k(x - a_0)(x - a_1)\cdots(x - a_{k-1}), \qquad (7)$$

where $0 \leq c_k < p^e/\gcd(p^e, k!_S)$ for each $k \geq 0$. Indeed, by Lemma 14, changing one of the coefficients c_k by a multiple of $p^e/\gcd(p^e, k!_S)$ in (7) does not change the function f. That is to say, the c_k are determined only modulo $p^e/\gcd(p^e, k!_S)$, so we may choose them to lie in the range $0 \leq c_k < p^e/\gcd(p^e, k!_S)$.

We now have a unique representative for each polynomial mapping from S to $\mathbb{Z}/p^e\mathbb{Z}$. Observing that there are $p^e/\gcd(p^e, k!_S)$ choices of c_k for each $k \geq 0$ yields the desired formula. ∎

Any of Theorems 9–11 may now be used to give a proof of Theorem 5.

Proof of Theorem 5: Since none of Theorems 9–11 mention p-orderings, but they do involve (and in fact define) the generalized factorials, the definition of factorials given in Section 4 could not possibly depend on any choices of p-ordering! ∎

Probably a more direct, conceptual way of seeing the truth of Theorem 5 is the following. For a positive integer d, and a large positive integer e such that $p^e > \nu_d(S, p)$, consider as an additive group the set G_d of all polynomials in $(\mathbb{Z}/p^e\mathbb{Z})[x]$ that vanish on S modulo p^e and have degree at most d. Then Lemma 12 implies that as an abelian group, G is isomorphic to

$$\bigoplus_{k=0}^{d} \mathbb{Z}/\nu_k(S, p)\mathbb{Z}.$$

Thus the numbers $\nu_k(S, p)$ (for $0 \leq k \leq d$) form the structure coefficients of this abelian group G_d; moreover, by the structure theorem for finitely generated abelian groups, these constants depend only on G_d itself, implying Theorem 5.

8. A MORE GENERAL FRAMEWORK: DEDEKIND RINGS.

Much of what we have said holds for a more general class of rings, which we may call *Dedekind rings*. A Dedekind ring is any Noetherian, locally principal ring in which all nonzero primes are maximal. This class of rings includes, for example, Dedekind domains, such as the ring of integers in an algebraic number field or a polynomial ring over a finite field. It also includes any quotients of such rings, such as $\mathbb{Z}/n\mathbb{Z}$, Galois rings, and all finite principal ideal rings.

Thus for any subset S of a Dedekind ring R, there is a corresponding sequence of factorials. However, when working in R, one constructs P-orderings using prime ideals P of R rather than prime elements p of $\mathbb{Z}$. Consequently, the factorials of a set $S \subseteq R$ in general must be considered ideals in R.

For rings R that have a canonical generator for each ideal (e.g., $\mathbb{Z}$), the factorials may also then be thought of as elements of R.

9. SOME EXAMPLES OF GENERALIZED FACTORIALS.

In this section, we take a look at some natural examples of generalized factorials. We've already seen one such:

Example 15. Let $S = \mathbb{Z}$. Then $k!_{\mathbb{Z}} = k!$.

In Example 15, there is a sequence in S that is a p-ordering of S for all primes p simultaneously (namely $0, 1, 2, \ldots$). Although such an event is rare for general sets S, there are several important sets for which it does occur. Moreover, the factorials in such cases become especially easy to compute. We state this more precisely in the following lemma (whose proof follows trivially from the definitions).

Lemma 16. *Suppose $\{a_i\}$ is a p-ordering of S for all primes p simultaneously. Then*

$$k!_S = |(a_k - a_0)(a_k - a_1) \cdots (a_k - a_{k-1})|.$$

Lemma 16 is used in our next three examples.

Example 17. Let S be the set of even integers $2\mathbb{Z}$ in $\mathbb{Z}$. Then by the same argument as in Proposition 6, we find that the natural ordering $0, 2, 4, 6, \ldots$ of $2\mathbb{Z}_{\geq 0}$ forms a p-ordering of $2\mathbb{Z}$ for all primes p. Hence, by Lemma 16, $k!_{2\mathbb{Z}} = (2k - 0)(2k - 2) \cdots (2k - (2k - 2)) = 2^k k!$. In a similar manner, we find that the set $a\mathbb{Z} + b$ of all integers that are b modulo a has factorials given by $k!_{a\mathbb{Z}+b} = a^k k!$.

Example 18. Let S be the set of powers of 2 in $\mathbb{Z}$. Then it is easy to verify that $1, 2, 4, 8, \ldots$ forms a p-ordering of S for all p; hence $k!_S = (2^k - 1)(2^k - 2) \cdots (2^k - 2^{k-1})$. More generally, suppose we take any geometric progression S in $\mathbb{Z}$ with common ratio q and first term a. Then $k!_S = a^k(q^k - 1)(q^k - q) \cdots (q^k - q^{k-1})$.

Example 19. Let S be the set of square numbers in $\mathbb{Z}$. Then one can show by analysis of quadratic residues modulo p^n for each p that $0, 1, 4, 9, \ldots$ forms a p-ordering of S for all primes p. It follows that

$$k!_S = (k^2 - 0)(k^2 - 1) \cdots \left(k^2 - (k-1)^2\right) = \frac{(2k)!}{2}.$$

Theorem 10 applied to Example 19 proves the following result, which also originates in the representation theory of Lie algebras:

Theorem 20. *Let $a_0, a_1, \ldots, a_n \in \mathbb{Z}$ be any $n + 1$ integers. Then the product of the pairwise differences of their squares*

$$\prod_{i<j} \left(a_i^2 - a_j^2\right)$$

is a multiple of $\dfrac{0! \, 2! \cdots (2n)!}{2^{n+1}}$. *(This is sharp!)*

As we mentioned, Theorem 20 arises in representation theory: the quotient $2^{n+1} \prod_{i<j}(a_i^2 - a_j^2)/0! \, 2! \cdots (2n)!$ is the dimension of a certain irreducible representation of $Sp(n)$.

Finally, we give one natural example of a subset S of $\mathbb{Z}$ that does not possess any simultaneous p-ordering; consequently, the formula for its factorials is a little more involved.

Example 21. Let S be the set of primes in $\mathbb{Z}$. Then for a fixed prime p, one can show that a p-ordering of S is given by a sequence $\{a_i\}$ having the following property: for each $e \geq 0$, the set $\{a_0, \ldots, a_{p^{e-1}(p-1)}\}$ equals $(\mathbb{Z}/p^e\mathbb{Z})^* \cup \{p\}$ when considered modulo p^e. (Such a sequence $\{a_i\}$ is guaranteed to exist for every p by Dirichlet's Theorem.) The factorials of S are therefore given by

$$k!_S = \prod_p p^{\left\lfloor \frac{k-1}{p-1} \right\rfloor + \left\lfloor \frac{k-1}{p(p-1)} \right\rfloor + \left\lfloor \frac{k-1}{p^2(p-1)} \right\rfloor + \cdots}.$$

The factorials of the set of primes seem to be strangely connected with the Bernoulli numbers. Precisely, it appears that $k!_S$ is given simply by $2^{\lfloor n/2 \rfloor}$ times the product of the denominators of the first $\left\lceil \frac{n}{2} \right\rceil$ Bernoulli numbers. This assertion may be verified using the von Staudt Theorem [24]; but is there a deeper explanation of this rather striking connection?

10. SOME SPECIAL CASES. There are many previous generalizations of factorial that can be obtained as natural special cases of the definitions we have made here.

Example 18 is reminiscent of the well-known q-factorials that arise in enumerative combinatorics. In fact, one can obtain the abstract q-factorials directly as follows: let S be the set $\{(q^k - 1)/(q - 1): k \in \mathbb{N}\}$ in the ring $\mathbb{C}[q, q^{-1}]$. (We invert q in the ring to kill the prime q.) Then we find as in Example 18 that

$$k!_S = (q - 1)^{-k}(q^k - 1)(q^{k-1} - 1) \cdots (q - 1),$$

which is indeed just the kth q-factorial.

Another natural ring on which to try out the factorial construction is $\mathbb{F}_q[t]$, the ring of polynomials over the finite field of q elements. For $S = R = \mathbb{F}_q[t]$, a t-ordering of S may be constructed as follows: let $a_0, a_1, \ldots, a_{q-1}$ be the elements of $\mathbb{F}_q$ (with $a_0 = 0$), and define a_k in general by

$$a_k = a_{c_0} + a_{c_1}t + \cdots + a_{c_h}t^h,$$

where $\sum_{i=0}^{h} c_i q^i$ is the base q expansion of k. One then easily verifies that this gives a P-ordering of $\mathbb{F}_q[t]$ not only for $P = (t)$, but for all primes $P \subset \mathbb{F}_q[t]$. It follows that

$$k!_{\mathbb{F}_q[t]} = (a_k - a_0) \cdots (a_k - a_{k-1})$$
$$= (t^{q^h} - t)^{c_h}(t^{q^{h-1}} - t)^{c_{h-1} + c_h q} \cdots (t^q - t)^{c_1 + \cdots + c_h q^{h-1}}$$

where again $\sum_{i=0}^{h} c_i q^i$ denotes the base q expansion of k. We have arrived at the well-known "Carlitz factorials." In 1938, Carlitz [9] used these factorials to construct the Carlitz module, the first example of a Drinfeld module [15].

The generalized factorials of Pólya [28], Ostrowski [26], and Gunji-McQuillan [16] can also be obtained from our construction, upon setting $S = R$ to be the ring of integers in a number field. In addition, setting $S = R$ to be the ring of integers in a function field over a finite field gives rise to what are known as the Γ-ideals of Goss [13]; they have been used by Goss as extensions of the Carlitz factorial to other function fields.

That all these classical factorials can be obtained as natural special cases of the factorials we have defined here seems to be further evidence that we have arrived at a "correct" notion of generalized factorial.

11. BASES FOR INTEGER-VALUED POLYNOMIALS. When does a polynomial take integer values on the integers? The polynomial need not have integer coefficients for this to occur, for observe that the polynomial $f(x) = x(x - 1)/2$, which has noninteger coefficients, still maps the integers to the integers. In fact, all the binomial polynomials $\binom{x}{k} = x(x-1)\cdots(x-k+1)/k!$ take integer values on the integers.

Can one classify all polynomials with this property? In 1915, Pólya gave an elegant answer to this question, by proving the following elementary but classical result:

Theorem 22 (Pólya [27]). *A polynomial is integer-valued on $\mathbb{Z}$ if and only if it can be written as a $\mathbb{Z}$-linear combination of the polynomials*

$$\binom{x}{k} = \frac{x(x-1)\cdots(x-k+1)}{k!},$$

$k = 0, 1, 2, \ldots$.

Thus the binomial polynomials, and all polynomials that can be obtained from them via addition and subtraction, are all the polynomials that are integer-valued on $\mathbb{Z}$. This result has had numerous applications, in algebraic geometry (e.g., in the theory of Hilbert polynomials), representation theory (e.g., in the theory of Chevalley groups), number theory (e.g., in the theory of Mahler expansions), as well as in combinatorics.

Subsequent to proving Theorem 22, Pólya wondered as to what generality this result could be extended. That is, for a subset S of a Dedekind domain R, when does there exist a similar *regular basis* (an R-basis consisting of one polynomial of each degree) for the set of R-valued polynomials on S? Pólya [28] answered this question when $S = R$ is the ring of integers in a quadratic number field (i.e., he characterized quadratic fields possessing such a regular basis, and gave an explicit construction of such a basis whenever it existed). Ostrowski [26] shortly thereafter extended Pólya's work to the case when $S = R$ is the ring of integers in a general number field. In the years since 1919, analogous results have been proved for various other possibilities of S and R (e.g., [7], [11], [12]), though an answer for general S and R was never obtained.

From our point of view, though, the answer to Pólya's question is easily guessed. Namely, we expect the factorials in the denominators of $\binom{x}{k}$ in Theorem 22 to be replaced by generalized factorials, and the numerators to be replaced by the generalized "falling factorials" $B_{k,S}$ of Section 7. We thereby obtain the following result:

Theorem 23. *A polynomial is integer-valued on a subset S of $\mathbb{Z}$ if and only if it can be written as a $\mathbb{Z}$-linear combination of the polynomials*

$$\frac{B_{k,S}}{k!_S} = \frac{(x-a_{0,k})(x-a_{1,k})\cdots(x-a_{k-1,k})}{k!_S},$$

$k = 0, 1, 2, \ldots$, *where the $B_{k,S}$ are the polynomials defined in* (6).

For a subset S of a general Dedekind domain R, the answer is just slightly more complicated, but again not hard to guess. When constructing our basis for the set of polynomials R-valued on S, we wish to divide our polynomials $B_{k,S}$ by the generalized factorials $k!_S$; but these factorials, being ideals in general, may not have a single generator that we can divide by! The condition, therefore, for a regular basis to exist is that every ideal $k!_S$ be principal.

Theorem 24. *The set of polynomials that are R-valued on a subset S of a Dedekind domain R has a regular basis if and only if $k!_S$ is a principal ideal for all $k \geq 0$. If this is the case, then a regular basis may be given as in Theorem 23.*

Thus this fundamental problem about integer-valued polynomials, first put forth by Pólya in 1919, is now resolved.

12. EXTENSION TO SEVERAL VARIABLES. Much of the formalism developed in the previous sections for studying polynomials in one variable can be extended to the case of several variables. Indeed, the problem is equivalent to understanding what the correct definition of "factorial" is for subsets S of $\mathbb{Z}^n$ when $n > 1$. A key observation in accomplishing this is suggested by Theorem 10, which states that choosing a_k to minimize the product (1) is equivalent to minimizing the highest power of p dividing the Vandermonde determinant

$$\begin{vmatrix} 1 & a_0 & a_0^2 & \cdots & a_0^k \\ 1 & a_1 & a_1^2 & \cdots & a_1^k \\ \vdots & \vdots & \vdots & \ddots & \vdots \\ 1 & a_k & a_k^2 & \cdots & a_k^k \end{vmatrix} = \prod_{i<j} (a_i - a_j). \tag{8}$$

In fact, we showed in [3] that $\nu_0(S, p), \nu_1(S, p), \ldots, \nu_k(S, p)$ give the p-parts of the elementary divisors of the Vandermonde matrix (8). This motivates the following more general definitions:

Definition 25. Let S be a subset of $\mathbb{Z}^n$ (or of R^n, where R is any Dedekind ring). Then for a fixed ordering $M_0, M_1, \ldots$ of the monomials of $\mathbb{Z}[x_1, \ldots, x_n]$, a *p-ordering* of S is a sequence $\mathbf{a_0}, \mathbf{a_1}, \ldots$ of elements in S inductively chosen to minimize the highest power of p dividing the determinant

$$V(\mathbf{a_0}, \mathbf{a_1}, \ldots, \mathbf{a_k}) = \begin{vmatrix} M_0(\mathbf{a_0}) & M_1(\mathbf{a_0}) & M_2(\mathbf{a_0}) & \cdots & M_k(\mathbf{a_0}) \\ M_0(\mathbf{a_1}) & M_1(\mathbf{a_1}) & M_2(\mathbf{a_1}) & \cdots & M_k(\mathbf{a_1}) \\ \vdots & \vdots & \vdots & \ddots & \vdots \\ M_0(\mathbf{a_k}) & M_1(\mathbf{a_k}) & M_2(\mathbf{a_k}) & \cdots & M_k(\mathbf{a_k}) \end{vmatrix}.$$

The *associated p-sequence* of S is then given by

$$\nu_k(S, p) = w_p\left(\frac{V(\mathbf{a_0}, \mathbf{a_1}, \ldots, \mathbf{a_n})}{V(\mathbf{a_0}, \mathbf{a_1}, \ldots, \mathbf{a_{n-1}})}\right),$$

and the *generalized factorial* $k!_S$ is

$$k!_S = \prod_p \nu_k(S, p).$$

One may verify that Definition 25, for $n = 1$ (and the usual monomial ordering $1, x, x^2, \ldots$), coincides with the notions of p-ordering, associated p-sequence, and generalized factorial given in Sections 4 and 5. Moreover, all the analogues of Theorems 5, 9–11, 23, and 24 can now be proved when S is a subset of $\mathbb{Z}^n$ (or R^n), using essentially the same techniques.

13. FURTHER APPLICATIONS. The concepts of p-ordering and generalized factorial have had some important applications to interpolation problems. We hinted at one of these problems earlier—the polynomial interpolation problem in

$\mathbb{Z}/n\mathbb{Z}$. As we have mentioned, traditional methods for performing polynomial interpolation do not work in general for subsets of $\mathbb{Z}/n\mathbb{Z}$, because these methods frequently require division operations that may not make sense in a nonfield such as $\mathbb{Z}/n\mathbb{Z}$.

But as we noted in the proof of Theorem 11, it suffices to understand the interpolation problem for $\mathbb{Z}/n\mathbb{Z}$ when $n = p^e$ is a power of a prime. In this case, it turns out that if one interpolates along a p-ordering in a certain way, then all such division-related problems can be completely avoided, and one obtains a general interpolation formula for functions on subsets of $\mathbb{Z}/n\mathbb{Z}$ (or for subsets of any finite principal ideal ring). Details may be found in [2].

Another area in which the ideas of this article have proved to be very useful is *p-adic interpolation*. A classical theorem of Mahler [21] states that that every continuous function f from the p-adic ring $\mathbb{Z}_p$ to its quotient field $\mathbb{Q}_p$ (or to any finite extension thereof) can be expressed uniquely in the form

$$f(x) = \sum_{n=0}^{\infty} c_n \binom{x}{n},$$

where the sequence c_n tends to 0 as $n \to \infty$. Do analogous series exist for other compact subsets of $\mathbb{Q}_p$, or for subsets of local fields other than $\mathbb{Q}_p$? There have been several partial results in this direction, such as the work of Amice [1], who provided answers for certain "well-distributed" sets S.

But as with Theorem 22, from our point of view it is easy to guess what the general answer should be:

Theorem 26. *Let S be any compact subset of a local field K. Then every continuous map $f : S \to K$ can be expressed uniquely in the form*

$$f(x) = \sum_{n=0}^{\infty} c_n \frac{B_{n,S}(x)}{n!_S}, \tag{9}$$

where the sequence c_n tends to 0 as $n \to \infty$.

In joint work with K. Kedlaya [5], Theorem 26 was proved, thus solving this p-adic interpolation problem for any compact subset of a local field. Moreover, by using the ideas described in Section 12, Theorem 26 has also recently been extended to the case of several variables, and in fact to arbitrary algebraic varieties over a discrete valuation domain; details may be found in [6]. For a simple treatment of locally analytic functions from this point of view, which fully extends Amice's work to general compact subsets S, see [4].

Also worth noting in this regard is the recent work of Maulik [22], who has used the ideas of p-ordering and generalized factorials to count the number of subsets of $\mathbb{Z}/n\mathbb{Z}$ that form the set of roots of some polynomial over $\mathbb{Z}/n\mathbb{Z}$.

14. SOME QUESTIONS. In the previous sections we have seen some excellent evidence that the generalized factorials defined here are the "right" number-theoretic generalizations of the factorial function to arbitrary subsets S of $\mathbb{Z}$. Are they in any sense the correct combinatorial generalization?

Question 27. *For a subset $S \subset \mathbb{Z}$, is there a natural combinatorial interpretation of $k!_S$?*

What makes an affirmative answer to Question 1 seem probable is that the generalized binomial coefficients

$$\binom{n}{k}_S = \frac{n!_S}{k!_S(n-k)!_S}$$

are always integral. Besides our tricky proof of this fact in Section 7, what is a good reason for this to be true?

Question 28. *For a subset $S \subset \mathbb{Z}$, is there a natural combinatorial interpretation for $\binom{n}{k}_S$?*

As is well-known, the factorial function has a natural extension to a continuous function on the positive reals, called the gamma function; it may be defined by

$$\Gamma(x+1) = \int_0^\infty e^{-t} t^x \, dt.$$

In addition, the gamma function may be meromorphically continued to the entire complex plane. Might there exist, for each subset S of $\mathbb{Z}$, a natural and meromorphic generalized gamma function $\Gamma_S(x)$ defined on the real/complex numbers such that $\Gamma_S(n+1) = n!_S$ for all $n \geq 0$, and $\Gamma_\mathbb{Z}(x) = \Gamma(x)$?

The factorial function also has similar "extensions" to the p-adic fields $\mathbb{Q}_p$; the most successful such interpolations are probably the well-known p-adic gamma functions $\Gamma_p(x)$ of Morita [25]. Again, it is natural to ask whether analogous p-adic interpolations $\Gamma_{S,p}$ might exist for the generalized factorials associated to other sets S.

Question 29. *For general subsets S of $\mathbb{Z}$ (or of other Dedekind domains), are there natural complex (respectively, p-adic) analytic interpolations of $k!_S$ to generalized gamma functions Γ_S (respectively, $\Gamma_{S,p}$)?*

Positive answers to Question 29 have been given for many of the special cases of generalized factorials listed in Section 10. For q-factorials, a natural complex analytic gamma function interpolating them has been given by Jackson [17], and natural p-adic extensions have been carried out by Koblitz [19]. P-adic extensions of the Carlitz factorials $k!_{\mathbb{F}_q[t]}$ were discovered by Goss [14]. For general function fields, different factorials were defined and interpolated by Thakur [30].

But as we've just seen, the special cases of our factorials for which P-adic gamma functions have been found all correspond to sets that possess simultaneous P-orderings. In fact, Dinesh Thakur has suggested the possibility that natural gamma functions interpolating these generalized factorials might exist only for sets S that have such simultaneous P-orderings. Thus the following question, interesting in its own right, may also be relevant in answering Question 29.

Question 30. *Which subsets S of $\mathbb{Z}$ (or of a Dedekind ring R) have simultaneous p-orderings for all primes p?*

An answer to Question 30 is not known even when $S = R$, where R is the ring of integers in a number field or function field. Even partial answers in these cases would be of much interest.

These were some of the questions I posed to the audience in San Diego, while giving the 1997 AMS-MAA address on which this article is based. But my audience had many other excellent questions! Here are a few of them:

Question 31. *What are analogues of Stirling's formula for generalized factorials?*

Question 32. *What is the "binomial theorem" for generalized binomial coefficients?*

Question 33. *What is the S-analogue of the exponential function*

$$e^x = \sum_{k=0}^{\infty} \frac{x^k}{k!},$$

and what properties does it have?

As is evident, there is much left to understand! We have considered here just a few of the many natural questions one may ask about generalized factorials. We expect that many of these questions have very nice answers, and hope that the examples and results included in this article will help in resolving some of these questions in the near future!

ACKNOWLEDGMENTS. I am thankful to Timothy Chow, Keith Conrad, Persi Diaconis, Joseph Gallian, Danny Goldstein, Kiran Kedlaya, Barry Mazur, Lenny Ng, Gian-Carlo Rota, Richard Stanley, and Dinesh Thakur for their encouragement and valuable conversations about generalized factorials over the past few years. Some parts of this work were done during the 1995 Summer Research Program at the University of Minnesota, Duluth, directed by Joseph Gallian and sponsored by the National Science Foundation (grant DMS-9225045) and the National Security Agency (grant number MDA 904-91-H-0036); other parts were supported by a Hertz fellowship.

REFERENCES

1. Y. Amice, Interpolation p-adique, *Bull. Soc. Math. France* **92** (1964) 117–180.
2. M. Bhargava, P-orderings and polynomial functions on arbitrary subsets of Dedekind rings, *J. reine angew. Math.* **490** (1997) 101–127.
3. M. Bhargava, Generalized factorials and fixed divisors over subsets of a Dedekind domain, *J. Number Theory* **72** (1998) 67–75.
4. M. Bhargava, Integer-valued polynomials and p-adic locally analytic functions, preprint.
5. M. Bhargava and K. S. Kedlaya, Continuous functions on compact subsets of local fields, *Acta Arith.* **91** (1999) 191–198.
6. M. Bhargava and K. S. Kedlaya, An analogue of Mahler's theorem for algebraic varieties over a discrete valuation domain, in preparation.
7. P.-J. Cahen, Polynômes à valeurs entières, *Canad. J. Math.* **24** (1972) 747–754.
8. P.-J. Cahen and J.-L. Chabert, *Integer-valued polynomials*, Mathematical Surveys and Monographs, 48, American Mathematical Society, Providence, RI, 1997.
9. L. Carlitz, A class of polynomials, *Trans. Amer. Math. Soc.* **43** (1938) 167–182.
10. L. Carlitz, Functions and polynomials (mod p^n), *Acta Arith.* **9** (1964) 67–78.
11. J.-L. Chabert, S. T. Chapman, and W. W. Smith, A basis for the ring of polynomials integer-valued on prime numbers, in *Factorization in integral domains*, Lecture Notes in Pure and Appl. Math., 189, Dekker, New York, 1997, pp. 271–284.
12. R. Gilmer, Sets that determine integer-valued polynomials, *J. Number Theory* **33** (1989) 95–100.
13. D. Goss, The Γ-ideal and special zeta-values, *Duke Math. J.* **47** (1980) 345–364.
14. D. Goss, The Γ-function in the arithmetic of function fields, *Duke Math. J.* **56** (1988) 163–191.
15. D. Goss, *Basic structures of function field arithmetic*, Ergebnisse der Mathematik und ihrer Grenzgebiete (3), 35, Springer-Verlag, Berlin, 1996.
16. H. Gunji and D. L. McQuillan, On a class of ideals in an algebraic number field, *J. Number Theory* **2** (1970) 207–222.
17. F. H. Jackson, On q-definite integrals, *Quart. J. Pure and Appl. Math.* **41** (1910) 193–203.

18. A. J. Kempner, Polynomials and their residue systems, *Trans. Amer. Math. Soc.* **22** (1921) 240–288.
19. N. Koblitz, q-Extension of the p-adic Gamma function, *Trans. Amer. Math. Soc.* **260** (1980) 449–457.
20. D. A. Lind, Which polynomials over an algebraic number field map the algebraic integers into themselves?, *Amer. Math. Monthly* **78** (1971) 179–180.
21. K. Mahler, An interpolation series for a continuous function of a p-adic variable, *J. reine angew. Math.* **199** (1958) 23–34.
22. D. Maulik, Root sets of polynomials modulo prime powers, preprint.
23. D. L. McQuillan, On a Theorem of R. Gilmer, *J. Number Theory* **39** (1991) 245–250.
24. J. W. Milnor and J. D. Stasheff, *Characteristic classes*, Annals of Mathematics Studies, No. 76, Princeton University Press, Princeton, N.J., 1974.
25. Y. Morita, A p-adic analogue of the Γ-function, *J. Fac. Sc. University Tokyo* **22** (1975) 255–266.
26. A. Ostrowski, Über ganzwertige Polynome in algebraischen Zahlkörpern, *J. reine angew. Math.* **149** (1919) 117–124.
27. G. Pólya, Über ganzwertige ganze Funktionen, *Rend. Circ. Mat. Palermo* **40** (1915) 1–16.
28. G. Pólya, Über ganzwertige Polynome in algebraischen Zahlkörpern, *J. reine angew. Math.* **149** (1919) 97–116.
29. B. Sury, An integral polynomial, *Math. Mag.* **68** (1995) 134–135.
30. D. S. Thakur, Gamma functions for function fields and Drinfeld modules, *Ann. Math.* **134** (1991) 25–64.
31. D. S. Thakur, On gamma functions for function fields, in *The arithmetic of function fields*, Ohio State Univ. Math. Res. Inst. Publ., 2, de Gruyter, Berlin, 1992, pp. 75–86.

MANJUL BHARGAVA was born in Hamilton, Ontario, Canada, but spent most of his early years in Long Island, New York. He received his A.B. summa cum laude in mathematics from Harvard University in 1996, and his Ph.D. from Princeton University in 2000. His research interests are primarily in number theory, representation theory, and combinatorics, although he also enjoys algebraic geometry, linguistics, and Indian classical music. He was the recipient of the AMS-MAA-SIAM Frank and Brennie Morgan Prize in 1997, and is a Clay Mathematics Institute Long-Term Prize Fellow and a Visiting Fellow at Princeton University.
52 Stewart Avenue, Bethpage, NY 11714-5311
bhargava@math.princeton.edu

Originally appeared as:
Bhargava, Manjul. "The Factorial Function and Generalizations." *American Mathematical Monthly.* vol. 107, no. 9 (November 2000), pp. 783–799.

An Elementary Proof of the Quadratic Reciprocity Law

Sey Y. Kim

1. INTRODUCTION. Let p be an odd prime number. For any integer a not divisible by p, the Legendre symbol $\left(\frac{a}{p}\right)$ is defined as:

$$\left(\frac{a}{p}\right) = \begin{cases} 1 & \text{if the congruence } X^2 \equiv a \pmod{p} \text{ has a solution,} \\ -1 & \text{otherwise.} \end{cases}$$

(In the first case, we say that a is a *quadratic residue* modulo p.) Euler's criterion asserts that $\left(\frac{a}{p}\right) \equiv a^{\frac{p-1}{2}} \pmod{p}$—hence that $\left(\frac{-1}{p}\right) = (-1)^{\frac{p-1}{2}}$. The celebrated quadratic reciprocity law states:

Theorem. *If p and q are distinct odd prime numbers, then*

$$\left(\frac{p}{q}\right)\left(\frac{q}{p}\right) = (-1)^{\frac{p-1}{2}\frac{q-1}{2}}.$$

The first proof of this result was discovered by C. F. Gauss (1777–1855) in 1796 by an ingenious use of mathematical induction (see [2, pp. 92–97]). Since then more than 190 proofs have been published (a list is given in [3]). They display an astonishing variety of methods derived from many branches of mathematics.

There is a standard elementary proof of the law (see [1, sec. 9.3]), due to Gauss and G. Eisenstein (1823–1852). It involves counting in two ways the lattice points inside the rectangle in the Euclidean plane whose vertices are $(0, 0)$, $(p/2, 0)$, $(0, q/2)$, and $(p/2, q/2)$. Here we provide an apparently new elementary proof following an altogether different line of reasoning.

2. THE PROOF. Define a set Φ by

$$\Phi = \left\{ a : 1 \leq a \leq \frac{pq-1}{2},\ \gcd(a, pq) = 1 \right\},$$

and then set $A = \prod_{a \in \Phi} a$.

Lemma 1. $A \equiv (-1)^{\frac{q-1}{2}} \left(\frac{q}{p}\right) \pmod{p}$ *and* $A \equiv (-1)^{\frac{p-1}{2}} \left(\frac{p}{q}\right) \pmod{q}$.

Proof. Set

$$S = \left\{ a : 1 \leq a \leq \frac{pq-1}{2},\ \gcd(a, p) = 1 \right\}, \quad T = \left\{ q \cdot 1,\ \ldots,\ q \cdot \frac{p-1}{2} \right\}.$$

Plainly, T is a subset of S. Since

$$\frac{pq-1}{2} = \frac{p-1}{2} \cdot q + \frac{q-1}{2},$$

it follows easily that $\Phi = S - T$. Hence, by Euler's criterion,

$$\prod_{a \in S} a = \prod_{a \in T} a \cdot \prod_{a \in \Phi} a = q^{\frac{p-1}{2}} \left[\frac{p-1}{2}\right]! \cdot A \equiv \left(\frac{q}{p}\right)\left[\frac{p-1}{2}\right]! \cdot A \pmod{p}.$$

On the other hand, because

$$\frac{pq-1}{2} = \frac{q-1}{2} \cdot p + \frac{p-1}{2},$$

we have

$$\prod_{a \in S} a \equiv [(p-1)!]^{\frac{q-1}{2}} \left[\frac{p-1}{2}\right]! \equiv (-1)^{\frac{q-1}{2}} \left[\frac{p-1}{2}\right]! \pmod{p},$$

where we have used Wilson's theorem: $(p-1)! \equiv -1 \pmod{p}$. We infer that

$$\left(\frac{q}{p}\right)\left[\frac{p-1}{2}\right]! \cdot A \equiv (-1)^{\frac{q-1}{2}} \left[\frac{p-1}{2}\right]! \pmod{p},$$

whence $A \equiv (-1)^{\frac{q-1}{2}} \left(\frac{q}{p}\right) \pmod{p}$. The other statement in Lemma 1 follows by symmetry. ∎

From Lemma 1 it follows immediately that $(-1)^{\frac{q-1}{2}}\left(\frac{q}{p}\right) = (-1)^{\frac{p-1}{2}}\left(\frac{p}{q}\right)$ if and only if $A \equiv 1$ or $-1 \pmod{pq}$.

Lemma 2. $A \equiv 1$ or $-1 \pmod{pq}$ if and only if $p \equiv q \equiv 1 \pmod{4}$.

Proof. We set $d = pq$. By the Chinese Remainder Theorem, the congruence $X^2 \equiv 1 \pmod{d}$ has precisely four solutions, say $X \equiv 1, -1, N, -N \pmod{d}$. The congruence $X^2 \equiv -1 \pmod{d}$ has a solution $X \equiv I \pmod{d}$ if and only if $p \equiv q \equiv 1 \pmod{4}$, in which case there are precisely four solutions, namely, $X \equiv I, -I, NI, -NI \pmod{d}$.

Now for each a in Φ there exist unique a' in Φ and δ_a in $\{-1, 1\}$ such that $a \cdot a' \equiv \delta_a \pmod{d}$. (The correspondence $a \mapsto a'$ is a permutation of Φ.) Writing

$$\Psi = \{a \in \Phi : a = a'\} = \{a \in \Phi : a^2 \equiv \pm 1 \pmod{d}\},$$

we remark that

$$A = \prod_{a \in \Phi} a \equiv \pm \prod_{a \in \Psi} a \pmod{d}.$$

If $p \equiv q \equiv 1 \pmod{4}$, we have

$$\prod_{a \in \Psi} a \equiv \pm(1 \cdot N \cdot I \cdot IN) \equiv \pm(N^2 \cdot I^2) \equiv \mp 1 \pmod{d},$$

whereas otherwise it is the case that

$$\prod_{a \in \Psi} a \equiv \pm (1 \cdot N) \not\equiv \pm 1 \pmod{d}.$$

The lemma follows immediately. ∎

Proof of theorem. Combining Lemmas 1 and 2, we conclude that

$$(-1)^{\frac{q-1}{2}} \left(\frac{q}{p}\right) = (-1)^{\frac{p-1}{2}} \left(\frac{p}{q}\right) \quad \text{if and only if} \quad p \equiv q \equiv 1 \pmod{4}.$$

The theorem now follows by considering the four cases $(p, q) \equiv (1, 1), (1, -1), (-1, 1), (-1, -1) \pmod 4$. Or, in formulas, since $p \equiv q \equiv 1 \pmod 4$ if and only if $(-1)^{\frac{p+1}{2} \frac{q+1}{2}} = -1$:

$$\left(\frac{p}{q}\right)\left(\frac{q}{p}\right) = -(-1)^{\frac{p-1}{2}}(-1)^{\frac{q-1}{2}}(-1)^{\frac{p+1}{2}\frac{q+1}{2}}.$$

But

$$\frac{p-1}{2}\frac{q-1}{2} = \frac{pq - p - q + 1}{4} = \frac{pq + p + q + 1}{4} - \frac{p+q}{2}$$

$$= \frac{p+1}{2}\frac{q+1}{2} - \left(\frac{p-1}{2} + \frac{q-1}{2} + 1\right)$$

$$\equiv \frac{p+1}{2}\frac{q+1}{2} + \left(\frac{p-1}{2} + \frac{q-1}{2} + 1\right) \pmod 2,$$

showing that $-(-1)^{\frac{p-1}{2}}(-1)^{\frac{q-1}{2}}(-1)^{\frac{p+1}{2}\frac{q+1}{2}} = (-1)^{\frac{p-1}{2}\frac{q-1}{2}}$. ∎

REFERENCES

1. D. M. Burton, *Elementary Number Theory*, 5th ed., McGraw-Hill, New York, 2002.
2. C. F. Gauss, *Disquisitiones Arithmeticae* (trans. A. A. Clarke), Yale University Press, New Haven, 1966.
3. F. Lemmermeyer, *Reciprocity Laws*, Springer-Verlag, Berlin, 2000.

Department of Mathematics & Statistics, McMaster University, Ontario L8S 4K1, Canada
kimsey@mcmaster.ca

Originally appeared as:
Kim, Sey Y. "An Elementary Proof of the Quadratic Reciprocity Law." *American Mathematical Monthly*. vol. 111, no. 1 (January 2004), pp. 48–50.

Part VII: Elliptic Curves, Cubes and Fermat's Last Theorem

The ancient Greeks had no trouble with squares, but cubes were another matter. They sought compass-and-straightedge methods of duplicating cubes and trisecting angles, but what they sought was impossible. Why? Well, the short answer is that the roots of the polynomials $x^3 - 2$ and $8x^3 - 6x + 1$ do not lie in any field that is made by successively adjoining a finite number of square roots to the rational numbers. (The long answer is Galois theory.) At any rate, this was the origin of mathematicians' fascination with cubes and higher powers, and our final chapter contains five Biscuits dealing with powers higher than two.

We begin with a Proof Without Words. It has been known for over 2000 years that the sum $1^3 + 2^3 + \cdots + n^3$ of the first n cubes is equal to $(1 + 2 + \cdots + n)^2$, the square of the sum of the first n integers. Proving this is a routine exercise in mathematical induction. This chapter's first Biscuit is J. Barry Love's Proof Without Words: cubes and squares (*Math. Magazine*, vol. 50, no. 2 (March 1977), 74; see also p. 85 of vol. I of Nelsen's book), in which he shows us a very nice pictorial proof of this old but still lively chestnut.

Our next Biscuit has its origin in a famous comment by Srinivasan Ramanujan. Replying to Hardy's comment about his taxicab's boring license number 1729, Ramanujan said that on the contrary, 1729 is the smallest positive integer that can be written as a sum of two cubes in two different ways. Joseph Silverman begins with Ramanujan's statement and takes us through quite an enjoyable journey in his Lester Ford Award-winning "Taxicabs and sums of two cubes" (*Amer. Math. Monthly*, vol. 100, no. 4 (April 1993), pp. 331–340). Included is a "Hall of Fame" of smallest positive integers of various types that can be written as sums of cubes in at least two distinct ways. As a bonus, Silverman has written a "second helping" with updates on the Hall of Fame and some remarks on the rank (as a finitely generated abelian group) of curves $x^3 + y^3 = a$, concluding with a short table of such curves.

Joseph Silverman's curves $x^3 + y^3 = a$ are cubics in two variables that have a well-defined tangent at each point. Such curves are called elliptic curves, and they are the subject of our next Biscuit, Ezra Brown's Pólya Award-winning "Three Fermat trails to elliptic curves" (*College Math. Journal*, vol. 31, no. 3 (May 2000), pp. 162–172). It so happens that Diophantus posed a problem in his *Arithmetica* that contained the first-ever appearance of elliptic curves. Brown traces the history of three problems, each associated with Fermat and each having its resolution in the world of elliptic curves. The first is to determine which positive integers n are the areas of rational-sided right triangles—the so-called Congruent Numbers Problem. The second is to determine for which values of d the equation $x^4 + dx^2y^2 + y^4 = z^2$ has nontrivial integer solutions. The third is Fermat's Last Theorem, which brings us to our final two Biscuits.

In or about 1637, Pierre de Fermat wrote the mathematical world's most famous marginal note, effectively saying that if n is an integer greater than two, then the equation

$x^n + y^n = z^n$ has no integer solutions in which $xyz \neq 0$. Then he wrote, "I have discovered a truly marvelous proof of this fact, but this margin is not big enough to contain it." This "fact" is Fermat's Last Theorem (FLT), until recently the most famous unsolved problem in mathematics.

Our next-to-last Biscuit is Willard Van Orman Quine's "Fermat's Last Theorem in combinatorial form" (*Amer. Math. Monthly*, vol. 95, no. 7. (August-September 1988), p. 636), a diabolically clever restatement of FLT as a theorem about placing rocks in boxes. Read this short but delightful piece and marvel.

We conclude with Fernando Q. Gouvêa's Lester Ford Award-winning "A Marvelous Proof" (*Amer. Math. Monthly*, vol. 101, no. 3 (March 1994), pp. 203–222). This is an eminently readable guide to the mathematics underlying Andrew Wiles' proof of FLT. Gouvêa begins with a brief history of the problem from Fermat's time to the early 1980s. He introduces us to "the main characters in the drama", namely p-adic numbers, elliptic curves, modular forms and Galois representations. He presents the Taniyama-Shimura-Weil Conjecture, namely that every elliptic curve is modular, and explains the link between elliptic curves and Wiles' approach to the proof of FLT. The paper ends with the status of Wiles' proof up in the air, but Gouvêa adds a "second helping" and brings us up to date about how Wiles brought in Richard Taylor and how the two of them finished the proof.

Proof Without Words: Cubes and Squares

J. Barry Love

$$1^3 + 2^3 + 3^3 + \cdots + n^3 =$$
$$(1 + 2 + 3 + \cdots + n)^2$$

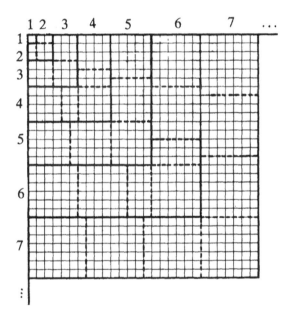

Originally appeared as:
Love, J. Barry. "Proof Without Words: Cubes and Squares." *Mathematics Magazine*. vol. 50, no. 2 (March 1977), p. 74.

Taxicabs and Sums of Two Cubes

Joseph H. Silverman

Our story begins in 1913, when the distinguished British mathematician G. H. Hardy received a bulky envelope from India full of page after page of equations. Every famous mathematician periodically receives letters from cranks who claim to have proven the most wonderous results. Sometimes the proofs, incorrect or incoherent, are included. At other times the writer solicits a reward in return for revealing his discoveries. Now this letter to Hardy, which was from a poor clerk in Madras by the name of Ramanujan, was filled with equations, all given without any sort of proof. Some of the formulas were well-known, mere exercises; while many of the others looked preposterous to Hardy's trained eye.

Who would have blamed Hardy if he had returned this missive to the sender, unread? And in fact, Ramanujan had previously sent his results to two other British mathematicians, each of whom had done just that! But instead Hardy gave some thought to these "wild theorems. Theorems such as he had never seen before, nor imagined."[1] And together, he and J. E. Littlewood, another eminent mathematician with whom Hardy often worked, succeeded in proving some of Ramanujan's amazing identities. At this point Hardy realized that this letter was from a true mathematical genius, and he became determined that Ramanujan should come to England to pursue his mathematical researches. Using travel money provided by Hardy's college, Ramanujan arrived in 1914. Over the next several years he continued to produce and publish highly original material, and he also collaborated with Hardy on a number of outstanding papers.

In 1918, at the age of 30, Ramanujan was elected a Fellow of the Royal Society and also of Trinity College, both signal honors which he richly deserved. Unfortunately, in the colder climate of England he contracted tuberculosis. He returned to his native Madras and died, in 1920, at the age of 33.

During all of Ramanujan's life, he considered numbers to be his personal friends. To illustrate, Hardy tells the story of how one day he visited Ramanujan in the hospital. At a loss for something to say, Hardy remarked that he had arrived at the hospital in taxicab number 1729. "It seemed to me," he continued, "a rather dull number." To which Ramanujan replied "No, Hardy! It is a very interesting number. It is the smallest number expressible as a sum of two cubes in two different ways:"[2]

$$1729 = 1^3 + 12^3 = 9^3 + 10^3.$$

---

*This article is an expanded version of talks given at M.I.T. and Brown University
[1] [10], page 32
[2] [10], page 37

Hardy then asked for the smallest number which is a sum of two *fourth* powers in two different ways, but Ramanujan did not happen to know.[3]

Rather than studying sums of higher powers, we will instead concern ourselves with another question that Hardy could just as easily have asked, namely for the smallest number which is a sum of two cubes in three (or more) distinct ways. In this case the answer is given in [3],

$$87{,}539{,}319 = 436^3 + 167^3 = 423^3 + 228^3 = 414^3 + 255^3,$$

although if one is willing to allow both positive and negative integers there is the much smaller solution

$$4104 = 16^3 + 2^3 = 15^3 + 9^3 = (-12)^3 + 18^3.$$

In this note we will be taking a leisurely number-theoretic stroll centered around the problem of writing numbers as sums of two cubes, specifically the search for integers with many such representations. It will be some time before we actually return to this specific question, but along the way we will view some beautiful mathematics which illustrates some of the interplay that can occur between geometry, algebra, and number theory.

We will thus be looking at solutions of the equation

$$X^3 + Y^3 = A.$$

What can be said about such solutions? First we have the elementary result that if A is a non-zero integer, then there are only finitely many solutions in integers X and Y. Of course, if X and Y are restricted to be positive integers, then this is obvious. But in any case we can use the factorization

$$A = X^3 + Y^3 = (X + Y)(X^2 - XY + Y^2)$$

to see that A must factor as $A = BC$ in such a way that

$$B = X + Y \quad \text{and} \quad C = X^2 - XY + Y^2.$$

Now there are only finitely many ways of factoring A as $A = BC$, and for each such factorization we substitute the first equation (i.e. $Y = B - X$) into the second to obtain

$$X^2 - X(B - X) + (B - X)^2 = C.$$

Thus each factorization of A yields at most two values for X, each of which gives one value for $Y = B - X$. Therefore there are only finitely many integer solutions (X, Y).[4]

[Aside: This proof that the equation $X^3 + Y^3 = A$ has only finitely many solutions in integers depends heavily on the factorization of the polynomial $X^3 + Y^3$. It is similarly true, but quite difficult to prove, that an equation like $X^3 + 2Y^3 = A$ has only finitely many integer solutions. This was first proven by A. Thue in 1909 [11]. By way of contrast, note that the equation $X^2 - 2Y^2 = 1$ has infinitely many solutions.]

---

[3] The answer, $635{,}318{,}657 = 59^4 + 158^4 = 133^4 + 134^4$, appears to have been discovered by Euler. And it does not seem to be known if there are any numbers which are a sum of two fifth powers in two different ways.

[4] Exercise: Show that any solution in integers satisfies $\max\{|X|, |Y|\} \leq 2\sqrt{|A|/3}$.

As is generally the case in mathematics, when a person wants to work on a difficult mathematical problem, it is nearly always best to start with a related easier problem. Then, step by step, one approaches the original goal. In our case we are confronted with the equation

$$X^3 + Y^3 = A,$$

and a natural first question is to ask for the solutions in real numbers. In other words, what does the graph of this curve look like? Since $dY/dX = -(X/Y)^2$, the graph is always falling. Further, it is symmetric about the line $Y = X$ and has the line $Y = -X$ as an asymptote. With this information it is easy to sketch the graph, which is illustrated in Figure 1. We will denote the resulting curve by C.

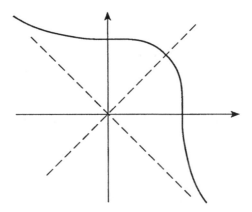

Figure 1. The curve C: $X^3 + Y^3 = A$

If (X, Y) is a point on this curve C, then so is the reflected point (Y, X). But there is another, less obvious, way to produce new points on the curve C which will be very important for the sequel. Thus suppose that $P = (X_1, Y_1)$ and $Q = (X_2, Y_2)$ are two (distinct) points on C, and let L be the line connecting P to Q. Then L will (usually) intersect C at exactly one other point. To see this, suppose that L is given by the equation $Y = mX + b$. Then substituting the equation of L into the equation of C gives the cubic equation

$$X^3 + (mX + b)^3 = A.$$

By assumption, this equation already has the solutions X_1 and X_2, since P and Q lie on the intersection $C \cap L$, so the cubic equation has exactly one other solution X_3. (We run into a problem if $m = -1$, but we'll deal with that later.) Then letting $Y_3 = mX_3 + b$, we have produced a new point $R = (X_3, Y_3)$ on C. Further, even if $P = Q$, the same procedure will work provided we take L to be the tangent line to C at P. Thus given any two points P and Q on C, we have produced a third point R. Finally we define an "addition law" on C by setting

$$P + Q = (\text{reflection of } R \text{ about the line } Y = X).$$

In other words, if $R = (X_3, Y_3)$, then $P + Q = (Y_3, X_3)$. (See Figure 2.) The reason we define $P + Q$ with this reflection will soon become clear, but first we have to deal with the pesky problem that sometimes our procedure fails to yield a third point.

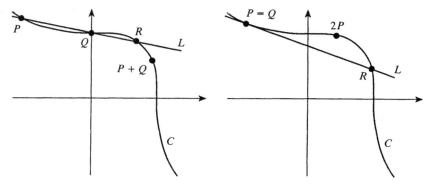

Figure 2. The addition law on the curve $C: X^3 + Y^3 = A$

This problem arises whenever the line connecting P and Q has slope -1, and thus is parallel to the asymptote $Y = -X$. We solve the problem in cavalier fashion; since there is no actual third intersection point we create one by fiat. To be precise, we take the XY-plane and add an extra point, which we will denote by $\mathscr{O}$. This point has the property that the lines going through $\mathscr{O}$ are exactly the lines with slope -1. Further, if P is a point in the XY-plane, then the unique line through P and $\mathscr{O}$ is the line through P having slope -1. Having done this, we are now in the happy position of being able to assert that given any two points P and Q on C, the above procedure will yield a unique third point R on C, so we are able to define $P + Q$ for any P and Q. (By definition, we set $\mathscr{O} + \mathscr{O} = \mathscr{O}$.)

[*Aside.* Some readers will doubtless recognize that what we have done is start the construction of the real projective plane. The projective plane consists of the usual XY-plane together with one point for each direction. In our case, we only needed the direction with slope -1. The projective plane has the agreeable property that any two lines, even "parallel" ones, intersect at exactly one point. This is true because if two lines are parallel, then they intersect at the point corresponding to their common direction.]

The justification for our use of the symbol "$+$" is now this:

The addition law $P + Q$ described above makes the points of C into an abelian group.

To be more precise, if $P = (X, Y)$ is a point on C, we define its inverse to be its reflection, $-P = (Y, X)$. Then for all points P, Q, R on C we have

$$P + \mathscr{O} = \mathscr{O} + P = P \quad \text{(identity)}$$

$$P + (-P) = \mathscr{O} \quad \text{(inverse)}$$

$$P + Q = Q + P \quad \text{(commutativity)}$$

$$(P + Q) + R = P + (Q + R) \quad \text{(associativity)}.$$

All of these properties are quite easy to check except for the associativity, which we will return to in a moment.

Note that the addition procedure outlined above is entirely mechanical. We can write down explicit formulas for the sum of two points, although there are a number of special cases. For example, if $P = (X, Y)$, then

$$P + P = 2P = \left(\frac{Y(X^3 + A)}{Y^3 - X^3}, \frac{X(Y^3 + A)}{X^3 - Y^3} \right).$$

The formula for the sum of two distinct points $P_1 = (X_1, Y_1)$ and $P_2 = (X_2, Y_2)$ is somewhat more complicated:

$P_1 + P_2$

$$= \left(\frac{A(Y_1 - Y_2) - X_1 X_2 (Y_1 X_2 - Y_2 X_1)}{X_1 X_2 (X_1 - X_2) + Y_1 Y_2 (Y_1 - Y_2)}, \frac{A(X_1 - X_2) - Y_1 Y_2 (X_1 Y_2 - X_2 Y_1)}{X_1 X_2 (X_1 - X_2) + Y_1 Y_2 (Y_1 - Y_2)} \right).$$

And now that we are in possession of these formulas, verification of the associative law is a tedious, but straightforward, task.

Let us now look at an example, and what better choice than Ramanujan's equation

$$X^3 + Y^3 = 1729.$$

We already know two interesting points on this curve, namely

$$P = (1, 12) \quad \text{and} \quad Q = (9, 10).$$

Using the addition law, we can easily compute some more points, such as

$$P + Q = \left(\frac{46}{3}, \frac{-37}{3} \right), \quad P - Q = \left(\frac{453}{56}, \frac{-397}{56} \right), \quad 2P = \left(\frac{20760}{1727}, \frac{-3457}{1727} \right),$$

$$2Q = \left(\frac{24580}{271}, \frac{-24561}{271} \right), \quad 3P = \left(\frac{-5150812031}{107557668}, \frac{5177701439}{107557668} \right).$$

As the discerning reader will notice, the numbers produced seem to grow with frightening rapidity. But of far more importance is the fact that although we have not produced any new integer solutions, all of the new solutions are at least in rational numbers.

This is a consequence of the fact that the addition law on C is given by rational functions. That is, the coordinates of $P + Q$ are given by quotients of polynomials in the coordinates of P and Q. Thus if the coordinates of P and Q are rational numbers, then so are the coordinates of $P + Q$. (If this is not clear, look for example at the formula for $2P$ given above. If A, X, and Y are all rational numbers, then the formula shows that the coordinates of $2P$ are also rational.) Let us define

$$C(\mathbb{Q}) = \{(X, Y): X \text{ and } Y \text{ are rational numbers and } X^3 + Y^3 = A\} \cup \{\mathscr{O}\}.$$

Here $\mathbb{Q}$ is the usual symbol for the field of rational numbers. Then the remarks given above provide a proof of the following fact, first noted by Poincaré around 1900 [6]:

The set $C(\mathbb{Q})$ is a subgroup of C. In other words, $C(\mathbb{Q})$ supplied with the addition law from C becomes a group in its own right.

The sum of two cubes "Hall of Fame"

The following material gives the smallest integer I know with the indicated property. It has been collected from various sources, including [3] and [12]. Note that we are counting $X^3 + Y^3$ and $Y^3 + X^3$ as being the same representation.

Positive integers, 2 representations, cube-free
$$1729 = 1^3 + 12^3 = 9^3 + 10^3 = 7 \cdot 13 \cdot 19$$

Any integers, 3 representations, not cube-free
$$4104 = 16^3 + 2^3 = 15^3 + 9^3 = (-12)^3 + 18^3 = 2^3 \cdot 3^3 \cdot 19$$

Any integers, 3 representations, cube-free
$$3{,}242{,}197 = 141^3 + 76^3 = 138^3 + 85^3 = (-171)^3 + 202^3 = 7 \cdot 31 \cdot 67 \cdot 223$$

Positive integers, 3 representations, not cube-free
$$87{,}539{,}319 = 436^3 + 167^3 = 423^3 + 228^3 = 414^3 + 255^3$$
$$= 3^3 \cdot 7 \cdot 31 \cdot 67 \cdot 223$$

Positive integers, 3 representations, cube-free
$$15{,}170{,}835{,}645 = 517^3 + 2468^3 = 709^3 + 2456^3 = 1733^3 + 2152^3$$
$$= 3^2 \cdot 5 \cdot 7 \cdot 31 \cdot 37 \cdot 199 \cdot 211$$

Any integers, 4 representations, not cube-free
$$42{,}549{,}416 = 348^3 + 74^3 = 282^3 + 272^3$$
$$= (-2662)^3 + 2664^3 = (-475)^3 + 531^3$$
$$= 2^3 \cdot 7 \cdot 13 \cdot 211 \cdot 277$$

Positive integers, 4 representations, not cube-free
$$26{,}059{,}452{,}841{,}000 = 29620^3 + 4170^3 = 28810^3 + 12900^3$$
$$= 28423^3 + 14577^3 = 24940^3 + 21930^3$$
$$= 2^3 \cdot 5^3 \cdot 31 \cdot 43^3 \cdot 97 \cdot 109$$

Any integers, 4 representations, cube-free
Unknown!

Any integers, 5 representations, not cube-free
$$1{,}148{,}834{,}232 = 1044^3 + 222^3 = 920^3 + 718^3$$
$$= 846^3 + 816^3 = (-7986)^3 + 7992^3 = (-1425)^3 + 1593^3$$
$$= 2^3 \cdot 3^3 \cdot 7 \cdot 13 \cdot 211 \cdot 277$$

We have come a long way in our study of the solutions of the equation $X^3 + Y^3 = A$. What we have found is that the set of solutions in rational numbers becomes, in a very natural way, an abelian group. So if we can say something significant about this group, then we might feel that at last we have some understanding about this set of rational solutions. An answer to this problem is provided by one of the most celebrated theorems of the twentieth century. Aside from its intrinsic interest, this result has been the starting point for much of the study of Diophantine equations over the past 70 years. The theorem, as we state it, was first proven by L. J. Mordell in 1922 [5]. It was subsequently vastly generalized by A. Weil in his 1928 thesis, and so is usually called the *Mordell-Weil Theorem*:

There exists a finite set of points $P_1, \ldots, P_r$ in $C(\mathbb{Q})$ so that every point in $C(\mathbb{Q})$ can be obtained from $P_1, \ldots, P_r$ by addition and subtraction. In fancier language, the group $C(\mathbb{Q})$ is finitely generated. The points $P_1, \ldots, P_r$ are called generators for $C(\mathbb{Q})$.

This seems fairly satisfactory. Even though our equation may have infinitely many rational solutions, they can all be obtained by starting with some finite subset and applying a completely mechanical procedure. For example, on the curve

$$X^3 + Y^3 = 7,$$

every rational point is a multiple of the single generating point $(2, -1)$. Similarly, it is probably true that every rational point on Ramanujan's curve has the form $nP + mQ$, where $P = (1, 12)$, $Q = (9, 10)$, and n and m are allowed to range over all integers, although I am not aware that anyone has verified this probable fact.

It may come as a surprise, then, to learn that the Mordell-Weil Theorem is far less satisfactory than it appears. This is due to the fact that it is not *effective*. What this means is that currently we do not have an algorithm which will determine, for every value of A, a set of generators $P_1, \ldots, P_r$ for $C(\mathbb{Q})$. In fact, there is not even a procedure for determining exactly how many generators are needed, although it is possible to give a rather coarse upper bound. This problem of making the Mordell-Weil Theorem effective is one of the major outstanding problems in the subject.

Another open problem concerns the number of generators needed. As mentioned above, the curve $X^3 + Y^3 = 7$ requires only one generator, while Ramanujan's curve $X^3 + Y^3 = 1729$ needs at least two. The question is whether there are curves which require a large number of generators. More precisely, is it true that for every integer r there is some value of A so that the rational points on the curve $X^3 + Y^3 = A$ require at least r generators?

We now return to our original problem, namely the study of *integer* solutions to the equation $X^3 + Y^3 = A$. Specifically, we seek values of A for which this equation has many solutions. We have observed that the equation

$$X^3 + Y^3 = 7$$

has the solution $P = (2, -1)$, and by taking multiples of P we can produce a sequence of points

$$P = (2, -1), \quad 2P = \left(\frac{5}{3}, \frac{4}{3}\right), \quad 3P = \left(\frac{-17}{38}, \frac{73}{38}\right), \quad 4P = \left(\frac{-1256}{183}, \frac{1265}{183}\right), \ldots.$$

Further, it is true (but moderately difficult to prove) that this sequence of points $P, 2P, 3P, \ldots$ never repeats. Each point in the sequence has rational coordinates, so we can write nP in the form

$$nP = \overbrace{P + P + \cdots + P}^{n\text{-terms}} = \left(\frac{a_n}{d_n}, \frac{b_n}{d_n}\right),$$

where a_n, b_n, and d_n are integers. We are going to take the first N of these rational solutions and multiply our original equation by a large integer so as to clear the denominators of all of them. Thus let

$$B = d_1 d_2 \cdots d_N.$$

Then the equation
$$X^3 + Y^3 = 7B^3$$
has at least N solutions in integers, namely
$$\left(\frac{a_n B}{d_n}, \frac{b_n B}{d_n}\right), \quad 1 \le n \le N.$$

(Actually $2N$ solutions, since we can always switch X and Y, but for simplicity we will generally count pairs of solutions (X, Y) and (Y, X).) We have thus found the following answer to our original problem:

> Given any integer N, there exists a positive integer A for which the equation $X^3 + Y^3 = A$ has at least N solutions in integers.

Of course, Ramanujan and Hardy were probably talking about solutions in positive integers; but with a bit more work one can show that infinitely many of the points $P, 2P, 3P, \ldots$ have positive coordinates, so we even get positive solutions.

Naturally, it is of some interest to make this result quantitative, that is, to describe how large A must be for a given value of N. The following estimate is essentially due to K. Mahler [4], with the improved exponent appearing in [8]:

> There is a constant $c > 0$ such that for infinitely many positive integers A, the number of positive integer solutions to the equation $X^3 + Y^3 = A$ exceeds $c\sqrt[3]{\log A}$.

In some sense we have now answered our original question. There are indeed integers which are expressible as a sum of two cubes in many different ways. But a nagging disquiet remains. We have not really produced a large number of intrinsically integral solutions. Rather, we have cleared the denominators from a lot of rational solutions. In the solutions produced above, the X and Y coordinates will generally have a large common factor whose cube will divide A. What happens if we rule out this situation? The simplest way to do so is to restrict attention to integers A that are *cube-free*; that is, A should not be divisible by the cube of any integer greater than 1. This is a reasonable restriction since if D^3 divides A, then an integer solution (X, Y) to $X^3 + Y^3 = A$ really arises from the rational solution $(X/D, Y/D)$ to the "smaller" equation $X^3 + Y^3 = A/D^3$.

Notice Ramanujan's example
$$1729 = 1^3 + 12^3 = 9^3 + 10^3 = 7 \cdot 13 \cdot 19$$
is cube-free. But the example with three representations given earlier,
$$87{,}539{,}319 = 436^3 + 167^3 = 423^3 + 228^3 = 414^3 + 255^3 = 3^3 \cdot 7 \cdot 31 \cdot 67 \cdot 223,$$
is not cube-free. As far as I have been able to determine, the smallest cube-free integer which can be expressed as a sum of two positive cubes in three distinct ways was unearthed by P. Vojta in 1983 [12]:
$$15{,}170{,}835{,}645 = 517^3 + 2468^3 = 709^3 + 2456^3 = 1733^3 + 2152^3$$
$$= 3^2 \cdot 5 \cdot 7 \cdot 31 \cdot 37 \cdot 199 \cdot 211.$$

And now our problem has become so difficult that Vojta's number holds the current record! There is no cube-free number known today which can be written as a sum of two positive cubes in four or more distinct ways.

We are going to conclude by describing a relationship between the two problems that we have been studying. The first problem was that of expressing a number as a sum of two *rational cubes*, and in that case we saw that all solutions arise from a finite generating set and speculated as to how large this generating set might be. The second problem was that of writing a cube-free number as a sum of two *integral cubes*, and here we saw that there are only finitely many solutions and we wondered how many solutions there could be. It is not a priori clear that these two problems are related, beyond the obvious fact that an integral solution is also a rational solution. In 1974 V. A. Dem'janenko stated that if there are a large number of integer solutions, then any generating set for the group of rational points must also be large, but his proof was incomplete. (See [2, page 140] for Lang's commentary on and generalization of Dem'janenko's conjecture.) The following more precise version of the conjecture was proven in 1982 [8]:

For each integer A, let $N(A)$ be the number of solutions in integers to the equation $X^3 + Y^3 = A$, and let $r(A)$ be the minimal number of rational points on this curve needed to generate the complete group of rational points (as in the Mordell-Weil Theorem). There is a constant $c > 1$ such that for every cube-free integer A,

$$N(A) \leq c^{r(A)}.$$

Note that the requirement that A be cube-free is essential. For as we saw above, we can make $N(7B^3)$ as large as we want by choosing an appropriate value of B. On the other hand, $r(7B^3) = r(7)$ for every value of B, so an inequality of the form $N(A) \leq c^{r(A)}$ cannot be true if we allow arbitrary values of A. (To see that $r(7B^3) = r(7)$, note that the groups of rational points on the curves $X^3 + Y^3 = 7$ and $X^3 + Y^3 = 7B^3$ are the same via the map $(X, Y) \to (BX, BY)$.)

Final Remark. The cubic curve $X^3 + Y^3 = A$ is an example of what is called an *elliptic curve*. Those readers interested in learning more about the geometry, algebra, and number theory of elliptic curves might begin with [9] and continue with the references listed there. Elliptic curves and the related theory of elliptic functions appear frequently in areas as diverse as number theory, physics, computer science, and cryptography.

ACKNOWLEDGMENTS. I would like to thank Jeff Achter and Greg Call for their helpful suggestions.

REFERENCES

1. Dem'janenko, V. A.: On Tate height and the representation of a number by binary forms, *Math. USSR Isv.* **8**, 463–476 (1974).
2. Lang, S.: *Elliptic Curves: Diophantine Analysis.* Berlin: Springer-Verlag, 1978
3. Leech, J.: Some solutions of Diophantine equations. *Proc. Camb. Philos. Soc.* **53**, 778–780 (1957).
4. Mahler, K.: On the lattice points on curves of genus 1. *Proc. London Math. Soc.* **39**, 431–466 (1935).
5. Mordell, L. J.: On the rational solutions of the indeterminate equation of the third and fourth degree. *Proc. Camb. Math. Soc.* **21**, 431–466 (1922).
6. Poincaré, H.: Sur les propriétés arithmétiques des courbes algébriques. *J. de Liouville* **7**, 161–233 (1901).
7. Silverman, J. H.: Integer points and the rank of Thue elliptic curves. *Invent. Math.* **66**, 395–404 (1982).
8. Silverman, J. H.: Integer points on curves of genus 1. *J. London Math. Soc.* **28**, 1–7 (1983).

9. Silverman, J. H., Tate, J.: *Rational Points on Elliptic Curves*. New York: Springer-Verlag, 1992.
10. Snow, C. P., Foreword to: *A Mathematician's Apology*. by G. H. Hardy (1940), Cambridge: Cambridge University Press, 1967.
11. Thue, A.: Über Annäherungswerte algebraischer Zahlen. *J. reine ang. Math.* **135**, 284–305 (1909).
12. Vojta, P.: private communication, 1983.

Added in proof: Richard Guy has pointed out to the author that Rosenstiel, Dardis and Rosenstiel have recently found the non-cube-free example

$$6963472309248 = 2421^3 + 19083^3 = 5436^3 + 18948^3 = 10200^3 + 18072^3$$
$$= 13322^3 + 16630^3$$

which is smaller than the example in the above table, and which is in fact the smallest such example. See *Bull. Inst. Math. Appl.* 27(1991), 155–157.

Mathematics Department
Brown University
Providence, RI 02912 USA
jhs@gauss.math.brown.edu

Originally appeared as:
Silverman, Joseph H. "Taxicabs and Sums of Two Cubes." *American Mathematical Monthly.* vol. 100, no. 4 (April 1993), pp. 331–340.

Second Helping

Joseph H. Silverman

The nth *Taxicab Number*, denoted Taxicab(n), is the smallest number that can be expressed as a sum of two positive cubes in at least n ways (where $X^3 + Y^3$ and $Y^3 + X^3$ count as a single representation). Our "clearing denominators" trick shows that Taxicab(n) exists, but even the most efficient search method known, which is due to Dan Bernstein [1], is only able to compute the first few taxicab numbers. Table 1 gives the known values up to Taxicab(5) and the conjectured value of Taxicab(6). (See [3, A011541] or [5].) For example, Taxicab(5) is

$$48988659276962496 = 2^6 \cdot 3^3 \cdot 7^4 \cdot 13 \cdot 19 \cdot 43 \cdot 73 \cdot 97 \cdot 157$$
$$= 365757^3 + 38787^3 = 362753^3 + 107839^3 = 342952^3 + 205292^3$$
$$= 336588^3 + 221424^3 = 331954^3 + 231518^3$$

It's amusing to observe that if we allow X or Y to be negative, then Taxicab(5) can be written as a sum of two cubes in two more ways,

$$48988659276962496 = 714700^3 - 681184^3 = 622316^3 - 576920^3.$$

We also note that the values for Taxicab(3) and Taxicab(4) beat the values given in our "Hall of Fame".

> Taxicab(1) = 2
> Taxicab(2) = 1729
> Taxicab(3) = 87539319
> Taxicab(4) = 6963472309248
> Taxicab(5) = 48988659276962496
> Taxicab(6) $\stackrel{?}{=}$ 24153319581254312065344

TABLE 1. The first few taxicab numbers

We have seen that it is easier to get many representations of A as a sum of two cubes if we allow X or Y to be negative. Current practice is to define the nth *Cabtaxi Number*, denoted Cabtaxi(n), to be the smallest number that can be expressed as a sum of two cubes (positive, negative, or zero) in at least n ways. For example,

$$\text{Cabtaxi}(3) = 728 = 2^3 \cdot 7 \cdot 13 = 12^3 - 10^3 = 9^3 - 1^3 = 8^3 + 6^3.$$

Notice that the 3rd, 4th and 5th Cabtaxi numbers are considerably smaller than the examples given in the "Hall of Fame". The first 9 cabtaxi numbers are known and are listed in Table 2. (See [3, A047696] or [4].)

Cabtaxi(1) = 1
Cabtaxi(2) = 91
Cabtaxi(3) = 728
Cabtaxi(4) = 2741256
Cabtaxi(5) = 6017193
Cabtaxi(6) = 1412774811
Cabtaxi(7) = 11302198488
Cabtaxi(8) = 137513849003496
Cabtaxi(9) = 424910390480793000

TABLE 2. The first few cabtaxi numbers

Allowing negative values for X and Y makes the taxicab problem easier. On the other hand, if we require that X and Y have no common factors, then the taxicab problem becomes immensely more difficult. If we further insist that X and Y be positive, then Vojta's example is the only number known with more than two representations! However, if we allow X and Y to be negative, then there is an example with five representations [3, A047697]:

$$506433677359393 = 7 \cdot 19 \cdot 37 \cdot 43 \cdot 67 \cdot 79 \cdot 139 \cdot 3253$$
$$= 139264^3 - 129951^3 = 95866^3 - 72087^3 = 80025^3 - 18218^3$$
$$= 78532^3 + 28065^3 = 77074^3 + 36489^3.$$

Let C be the taxicab curve $X^3 + Y^3 = A$. The Mordell-Weil theorem says that every point in $C(\mathbb{Q})$ can be obtained from a finite generating set $P_1, \ldots, P_r$ through repeated addition and subtraction. We write $r(A)$ for the minimal number of points needed to generate $C(\mathbb{Q})$. This number is called the *rank of* $C(\mathbb{Q})$. It is interesting to pose the following fundamental question.

> *Is there an absolute upper bound for the rank $r(A)$, or can we make $r(A)$ arbitrarily large by choosing larger and larger values for A?*

The evidence either way is less than overwhelming. Up until 2004, the largest known value of $r(A)$ was 7. At that time Elkies and Rogers [2] combined some ingeneous algebraic manipulations with an intelligently designed search algorithm to find examples having $r(A)$ equal to 8, 9, 10, and 11. They also say that "it is widely believed" that $r(A)$ can be arbitrarily large. And that is where the matter stands today.

Table 3 (see [2] or [3, A060748]) lists the smallest known values of A giving each rank up to 11. Not surprisingly, the values of A increase very rapidly as the rank increases. The values of A in Table 3 have many prime factors, for example the last entry factors as

$$1329399805658495217 4157235 = 3 \cdot 5 \cdot 7 \cdot 13 \cdot 19 \cdot 23 \cdot 31 \cdot 43 \cdot 59 \cdot$$
$$61 \cdot 73 \cdot 79 \cdot 103 \cdot 109 \cdot 157 \cdot 457.$$

Somewhat surprisingly, these curves of high rank have very few integer points. Indeed, $A = 19$ and $A = 657$ have only one representation as a sum of two cubes, and all of the other values of A in Table 3 have no representations at all.

$r(A)$	A
1	6
2	19
3	657
4	21691
5	489489
6	9902523
7	1144421889
8	1683200989470
9	349043376293530
10	137006962414679910
11	1329399805658495217 4157235

TABLE 3. Curves $C : X^3 + Y^3 = A$ whose rank is large

References

[1] D. J. Bernstein. Enumerating solutions to $p(a) + q(b) = r(c) + s(d)$. *Math. Comp.*, 70(233):389–394, 2001.

[2] N. D. Elkies and N. F. Rogers. Elliptic curves $x^3 + y^3 = k$ of high rank. In *Algorithmic number theory*, volume 3076 of *Lecture Notes in Comput. Sci.*, pages 184–193. Springer, Berlin, 2004. http://arxiv.org/abs/math.NT/0403116.

[3] N. J. A. Sloane. The On-Line Encyclopedia of Integer Sequences. www.research.att.com/~njas/sequences/.

[4] Wikipedia. Cabtaxi number — Wikipedia, The Free Encyclopedia, 2006. www.wikipedia.org, accessed 30-April-2007.

[5] Wikipedia. Taxicab number — Wikipedia, The Free Encyclopedia, 2007. www.wikipedia.org, accessed 30-April-2007.

E-mail address: jhs@math.brown.edu

MATHEMATICS DEPARTMENT, BOX 1917 BROWN UNIVERSITY, PROVIDENCE, RI 02912 USA

Three Fermat Trails to Elliptic Curves

Ezra Brown

1   Mysterious Curves and Distinguished Visitors

You may have wondered, as I once did, what lies beyond quadratic equations within mathematics. In geometry, we meet the Pythagorean Theorem; in algebra, the quadratic formula; in analytic geometry, the conic sections—ellipses, parabolas, and hyperbolas; and eventually, in multivariable calculus, the quadric surfaces—ellipsoids, hyperboloids, paraboloids, hyperbolic paraboloids. All of these involve quadratic equations in one or more variables.

Beyond quadratics... what? In the first semester of calculus, we find extreme values of cubics and polynomials of higher degree, but no special names are attached to those curves. A course in number theory will include mention of Fermat's Last Theorem, and if you're lucky, you might learn about Tartaglia, Fior, Cardano, and Ferrari, the sixteenth-century Italians who figured out the third- and fourth-degree analogs of the quadratic formula. But what then?

One day in graduate school, I found out "what then" in a book lying open on a desk in the coffee room. I picked it up and saw the two graphs in Figure 1.

I was struck by their interesting shapes and began to read. These were, I learned, the graphs of the equations $y^2 = x^3 - 7x + 6$ and $y^2 = x^3 - 2x + 4$, respectively. Both their pictures and their names intrigued me. For these were, I learned, two

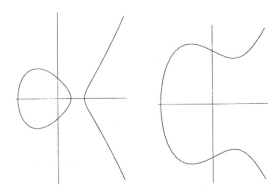

Figure 1

examples of elliptic curves—curves that were graphs of cubic equations of the form $y^2 = ax^3 + bx^2 + cx + d$, in which the right-hand side has three distinct complex roots. I wondered about their names... they certainly didn't look like ellipses.

My next encounter with elliptic curves was in 1968. Picture the scene: a visiting speaker was talking with some graduate students after his lecture, and the topic of doctoral dissertations arose. The speaker reeled off a list of fields that were not currently fashionable in mathematics and, in his opinion, that graduate students should avoid as dissertation topics. At one point, he mentioned elliptic curves—and I pounced.

"What are elliptic curves? Anything like ellipses?"

The speaker became thoughtful. "Oh, they're part of algebraic geometry... not too much in vogue these days... mostly studied at really high-powered institutions, especially in Japan and Britain... questions are hard... not much future in studying them... better do something within your grasp. Oh, the name? Something to do with elliptic integrals—that's where the name comes from. Still interested? Tell me your background in complex analysis, algebraic geometry and analytic number theory. Oh... in that case, you'd better stick to your quadratic forms, young man."

So I did. But I was curious enough to read up on elliptic integrals. Here's what I found:

- Elliptic integrals are definite integrals of several different types, including ones of the form

$$w = w(v) = \int_0^v \frac{1}{\sqrt{(1-u^2)(1-k^2u^2)}}\, du. \qquad (1)$$

- Such great mathematicians as Euler and Legendre (eighteenth century) and Abel and Jacobi (1820s and 1830s) made major studies of elliptic integrals. Just why these integrals were important, the source did not say.
- Abel transformed the study of elliptic integrals into the study of their inverse functions—that is, the functions you get from the integral in (1) by viewing v as a function of w—that he called elliptic functions. Just why he did this, the source did not say.
- An elliptic function, according to one source, is a function f of a complex variable z that is doubly periodic—that is, there exist complex numbers α and β such that for all complex z, $f(z) = f(z+\alpha) = f(z+\beta)$ and such that the ratio α/β is not a real number. Just what they had to do with ellipses, curves, or elliptic integrals, the source did not say.

Not terribly enlightening, I thought, and not much of a source, either. Not only that, but the source was totally silent regarding the mysterious elliptic curves. Oh, well... I tucked the subject of elliptic curves quietly away and "stuck to my quadratic forms." But I wondered....

2 Reviews and Rejections

One day in the mid 1980s, after discovering that a particular reference located in a certain reviewing journal was not quite what I needed, I hid my disappointment by idly leafing through the volume, hoping to turn up something of interest. Did I ever!

What "turned up" was a review of an article [16] about an old problem in elementary number theory called Euler's Congruent Numbers problem. Briefly, Euler asked for a characterization of congruent numbers, those rational numbers that are the areas of right triangles with rational sides. For example, 6 is a congruent number, since 6 is the area of the 3-4-5 right triangle. What caught my eye was a statement by the reviewer that the author, Jerrold Tunnell, had recast the entire congruent numbers problem into a problem involving *elliptic curves*—and then proceeded, essentially, to solve it.

Elliptic curves, again.

Maybe I should look at this paper... but the next day I got an inspiration for a problem I was working on, and once again elliptic curves slid quietly back into their customary comfortable corner... waiting.

They did not have long to wait. Several years later, a thick package from a journal proved to contain a rejection of a paper I had submitted some months earlier. The referee said that the problem I was studying could easily be solved, because "...a simple change of variables transforms any question about solutions of the equations at hand into a problem about *elliptic curves* [my emphasis]..."—at which point I dropped the letter on the floor.

Not only had elliptic curves resurfaced, but they had jumped out of the water and swatted me in the face. Now I was truly curious about these creatures, and this curiosity heightened over the next few years. For these innocent-looking cubic polynomials have turned up in a powerful method, due to Lenstra, for factoring large integers [9], in the Goldwasser-Kilian Primality Test [4], in public key cryptography [8], [10], in Tunnell's resolution of the congruent numbers problem [16], and finally, in Gerhard Frey's transformation of Fermat's Last Theorem into a problem about elliptic curves [3], ultimately solved in spectacular fashion (with an assist from Richard Taylor [15]) by Andrew Wiles [17]. In a word, elliptic curves are the latest silver bullets in the world of mathematics.

I'd like to take you on a brief tour of the world of elliptic curves, so that you can learn what they are, where they came from and how they got their name. Then, we'll look at three problems that illustrate their attraction. These three have an added attraction: they're all associated with Fermat.

Oh, yes: about that rejected paper—tell you later.

3 The Arc Length of an Ellipse

The story of how elliptic curves got their name begins with the work of G. C. Fagnano (1682–1766) who showed that computing the arc length of an ellipse leads to the integals mentioned in the previous section. (The story of elliptic curves begins, as do many mathematical stories, with the ancient Greeks and Alexandrians —but we'll get to that later.)

An ellipse centered at the origin and having the ends of its major and minor axes at $(\pm a, 0)$ and $(0, \pm b)$, respectively, is the graph of the equation

$$\frac{x^2}{a^2} + \frac{y^2}{b^2} = 1.$$

Fagngno knew that to get a very nice parametrization of this ellipse, just set $x = x(t) = a\cos t$ and $y = y(t) = b\sin t$; as t varies from 0 to 2π, the point $(x(t), y(t))$ traces out the entire ellipse.

For a curve parametrized by functions $x(t)$ and $y(t)$ for t between t_0 and t_1, the arc length of the curve is given by the integral

$$L = \int_{t_0}^{t_1} \sqrt{\left(\frac{dx}{dt}\right)^2 + \left(\frac{dy}{dt}\right)^2}\, dt.$$

Because of symmetry, we obtain the arc length of the entire ellipse by calculating the arc length in the first quadrant and multiplying by 4. You can check that this means t varies from 0 to $\pi/2$. Hence,

$$L = 4\int_0^{\pi/2} \sqrt{a^2 \cos^2 t + b^2 \sin^2 t}\, dt = 4b\int_0^{\pi/2} \sqrt{1 - k^2 \sin^2 t}\, dt,$$

where k is some constant involving a and b.

Now set $\sin t = u$, and so

$$\cos t = \frac{1}{\sqrt{1-u^2}}, \quad \text{and} \quad du = \cos t \, dt.$$

Thus,

$$L = 4b \int_0^1 \frac{\sqrt{1-k^2 u^2}}{\sqrt{1-u^2}} \, du = 4b \int_0^1 \frac{1-k^2 u^2}{\sqrt{(1-u^2)(1-k^2 u^2)}} \, du.$$

In order to find the arc length of the ellipse, we must evaluate

$$I(u) = \int \frac{du}{\sqrt{(1-u^2)(1-k^2 u^2)}},$$

which is known as an elliptic integral.

If the change of variables

$$v^2 = g(u) = (1-u^2)(1-k^2 u^2) = 1 - (1+k^2)u^2 + k^2 u^4 \tag{2}$$

would allow us to evaluate the integral, we'd do it. Now that's a quartic polynomial, and elliptic curves are of the form $y^2 = f(x)$, where f is a cubic polynomial—and elliptic curves are supposed to have something to do with elliptic integrals. Notice that since $k \neq \pm 1$, g has the four distinct roots ± 1 and $\pm k$. This allows us to transform $v^2 = g(u)$ to $y^2 = f(x)$, where f is a cubic polynomial, as follows:

If g is a quartic polynomial with four distinct roots α, β, γ, and δ, then we have that

$$v^2 = (u-\alpha)(u-\beta)(u-\gamma)(u-\delta).$$

Dividing both sides by $(u-\alpha)^4$ and doing a little algebra leads to

$$\left(\frac{v}{(u-\alpha)^2}\right)^2 = \left(1 + (\alpha-\beta)\frac{1}{u-\alpha}\right)\left(1 + (\alpha-\gamma)\frac{1}{u-\alpha}\right)\left(1 + (\alpha-\delta)\frac{1}{u-\alpha}\right).$$

Since the roots are distinct, none of the factors on the right collapses to 1. It follows that if we set

$$x = \frac{1}{u-\alpha}, \quad y = \frac{v}{(u-\alpha)^2},$$

then we have constructed a birational transformation between $v^2 = g(u)$ and

$$y^2 = f(x) = (1 + (\alpha-\beta)x)(1 + (\alpha-\gamma)x)(1 + (\alpha-\delta)x)$$

where $f(x)$ is a cubic polynomial in x—with three distinct roots, by the way. Thus, to find the arc length of an ellipse, we must evaluate

$$I(x) = \int \frac{dx}{\sqrt{x^3 + ax^2 + bx + c}},$$

and that is why $y^2 = x^3 + ax^2 + bx + c$ is called an elliptic curve.

Unfortunately, the change of variables (2) does not lead to an evaluation of the integral. You can't have everything.

4 Chord and Tangent Addition

Before we get to the problems, there's one more thing you need to know about elliptic curves—something that explains how Diophantus, Bachet, and Fermat were able to produce rational points (that is, points with rational coordinates) on curves, seemingly out of thin air.

It took the work of several giants of mathematics to show that the air wasn't as thin as all that. In the late seventeenth century, Newton (see [6, pp. 9–12] for details) deduced that each new point lay on the intersection of the given curve with a line that was either (a) tangent to the curve at some other point, or (b) a chord joining two points already on the curve (see Figure 2). In the nineteenth century, Jacobi tied the chord-and-tangent method in with elliptic integrals, and Weierstrass made a beautiful connection between an addition formula for elliptic functions and this chord-and-tangent way of producing new points from old on an elliptic curve. Finally, in 1901, Poincaré combined all of these ideas in a landmark paper on the arithmetical properties of algebraic curves.

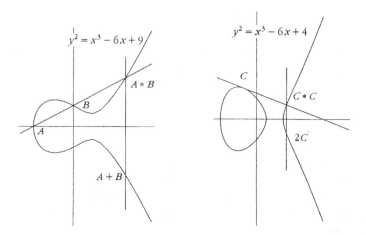

Figure 2

What these outstanding mathematicians showed was that the chord-and-tangent construction can be used to define a way of "adding" points on an elliptic curve. That is, given two points A and B on an elliptic curve, there's a third point on the curve called $A + B$, and this "addition" satisfies the group laws of closure, associativity, existence of an identity element, and existence of inverses.

Here are two examples. In Figure 2, the curve on the left, $y^2 = x^3 - 6x + 9$, contains the points $A = (-3, 0)$ and $B = (0, 3)$, and the chord joining A and B meets the curve at a third point $A * B = (4, 7)$. Instead of this third point being $A + B$, it turns out that $A + B$ is the reflection $(4, -7)$ of $A * B$ in the x-axis. Similarly, the curve on the right, $y^2 = x^2 - 6x + 4$, contains the point $C = (-1, 3)$, and the tangent line at C meets the curve at the point $C * C = (9/4, 21/8)$. We call the reflection $(9/4, -21/8)$ of $C * C$ in the x-axis the double $2C = C + C$ of C. (Just why the operation is defined this way is another story.)

The upshot is that the set of all points on an elliptic curve forms a group under this addition, and the rational points on the curve form a subgroup. It is this subgroup of rational points that has been extensively studied, and whose properties can be brought to bear on many ancient problems. And that is perhaps the most magical aspect of this area of mathematics: we can do arithmetic with the points on an elliptic curve, and that this arithmetic comes directly out of the geometry of the curve!

And now, as promised, for three of those problems.

5 Congruent Numbers and Elliptic Curves

A right triangle with rational sides is called a rational right triangle; the area of such a triangle, which equals half the product of the legs, is clearly rational. But suppose we specified the area in advance; is there, for example, a rational right triangle with area 1? with area 5? with area, say, 157?

Commenting on an old problem of Diophantus, Pierre de Fermat proved, in a marginal note—not *the* marginal note—in his copy of Bachet's edition of Diophantus' *Arithmetica*, that the difference of two fourth powers is never a square. Now, if x, y, and z are the sides of a rational right triangle area $xy/2 = w^2$, a little algebra shows that

$$(x^2 - y^2)^2 = z^4 - (2w)^4,$$

contrary to Fermat's result. Hence, the area of a rational right triangle is never a square; in particular, no rational right triangle exists with area 1. This was the first theorem on what became known as the Congruent Numbers problem.

Leonhard Euler defined a congruent number to be a rational number that is the area of some right triangle with rational sides. As I mentioned earlier, the area of the 3-4-5 right triangle is 6; therefore, 6 is a congruent number. The 5-12-13 and 7-24-25 triangles have areas 30 and 84, respectively, so 30 and 84 are also congruent numbers. You should be able to find a rational right triangle with area 5 (start with the 9-40-41 triangle) and, with a bit more work, one with area 7. Euler conjectured, but could not prove, that if n is a squarefree integer of the form $8k + 5$, $8k + 6$, or $8k + 7$, then n is a congruent number. If he was right, then $157 = 8 \cdot 19 + 5$ should be a congruent number—and it is. More about that later.

There's another characterization of congruent numbers, which turns out to be useful:

Theorem. *Let n be a positive rational number. Then n is a congruent number if and only if there exist three rational squares in an arithmetic progression with common difference n.*

Proof. If n is the area of the rational right triangle with legs X and Y and hypotenuse Z, then we have $n = XY/2$; since $X^2 + Y^2 = Z^2$, a little algebra shows that

$$\left(\frac{X+Y}{2}\right)^2 = \frac{X^2 = Y^2}{4} + \frac{XY}{2} = \left(\frac{Z}{2}\right)^2 + n, \text{ and similarly}$$

$$\left(\frac{X-Y}{2}\right)^2 = \left(\frac{Z}{2}\right)^2 - n.$$

If we let $u = Z/2$, then $u^2 - n$, u^2 and $u^2 + n$ are three rational squares (rational since Z is rational) in an arithmetic progression (AP) with common difference n.

Conversely, if we have three such squares $u^2 - n, u^2$ and $u^2 + n$, just set $X = \sqrt{u^2 + n} - \sqrt{u^2 - n}$, $Y = \sqrt{u^2 + n} + \sqrt{u^2 - n}$, and $Z = 2u$. Again, a little algebra shows that $X^2 + Y^2 = Z^2$, so that we do have a right triangle (rational, since $u^2 - n, u^2$ and $u^2 + n$ are rational squares); a little more algebra shows that this triangle has area n, and so n is a congruent number. We're done.

For example, $\{1/4, 25/4, 49/4\}$ is an AP of rational squares with common difference $n = 6$, and the $(3, 4, 5)$ right triangle has area 6. On the other hand, the $(9, 40, 41)$ right triangle has area 180, and so the $(9/6, 40/6, 41/6)$ right triangle has area 5. Sure enough, $\{(31/12)^2, (41/12)^2, (49/12)^2\}$ is an AP of rational squares with common difference $n = 5$.

Euclid characterized all integral right triangles (X, Y, Z) with $\gcd(x, y, z) = 1$ as being of the form

$$X = r^2 - s^2, Y = 2rs, Z = r^2 + s^2$$

where r and s are relatively prime integers of opposite parity. Noting that the area of this triangle is $rs(r^2 - s^2)$, a systematic way to generate squarefree congruent numbers goes something like this. For each odd k, list all products $rs(r^2 - s^2)$ with $r + s = k$. If $rs(r^2 - s^2)$ is divisible by a square m^2, so are both X and Y (try it!); so if we write $rs(r^2 - s^2) = m^2 n$ with n squarefree, then

$$\{(r^2 - s^2)/m, 2rs/m, (r^2 + s^2)/m\}$$

is a rational right triangle with area n. Here's a short list:

Pythagorean Triple	Area	Squarefree Congruent Number
(3, 4, 5)	6	6
(5, 12, 13)	30	30
(7, 24, 25)	84	21
(8, 15, 17)	60	15
(9, 40, 41)	180	5
(25, 312, 313)	3900	39

Well, this is fine, but where do elliptic curves come in?

Right here. If we take the product of three rational squares $u^2 - n, u^2$ and $u^2 + n$ in an arithmetic progression, this product is certainly a rational square v^2—but of a very particular form, namely

$$v^2 = (u^2 - n)u^2(u^2 + n) = u^6 - n^2 u^2 = (u^2)^3 - n^2(u^2).$$

That is, if n is a congruent number, then the point $(x = u^2, y = v)$ is a rational point on the elliptic curve $y^2 = x^3 - n^2 x$. From our example with $n = 6$, we saw that $u = 5/2$, and a little algebra again shows that $(25/4, 35/8)$ is a point on the elliptic curve $y^2 = x^3 - 36x$. We observe that $25/4$ is a rational square with even denominator, which turns out to be the key to the whole business, though we won't prove the

Key Lemma. *n is a congruent number if and only if there exists a rational point (x, y) on the elliptic curve $y^2 = x^3 - n^2 x$ such that x is a rational square with even denominator.*

So, even though $(25, 120)$ is a point on $y^2 = x^3 - 49x$ with x a rational square, its denominator is odd, and that's not enough to prove that 7 is a congruent number.

But 7 *is* a congruent number, and the procedure outlined above should not take too long to produce both the relevant triangle and the rational point on $y^2 = x^3 - 49x$. Try it!

The Congruent Numbers trail wound from Diophantus through Fermat and Euler and came to (pretty much) an end in 1983 with the work of Jerrold Tunnell [16]. It was Tunnell who finally resolved the Congruent Numbers problem by tackling the equivalent problem involving the group of rational points on the elliptic curves $y^2 = x^3 - n^2 x$. In part, he found the following necessary condition for n to be a congruent number—and it's fairly easy to check:

Tunnell's Criterion. Suppose that n is a square-free positive integer which is a congruent number. (a) If n is odd, then the number of integer triples (x, y, z) satisfying $n = 2x^2 + y^2 + 8z^2$ is just twice the number of integer triples (x, y, z) satisfying $n = 2x^2 + y^2 + 32z^2$. (b) If n is even, then the number of integer triples (x, y, z) satisfying $\frac{n}{2} = 4x^2 + y^2 + 8z^2$ is just twice the number of integer triples (x, y, z) satisfying $\frac{n}{2} = 4x^2 + y^2 + 32z^2$.

You can use this criterion to verify that a particular number is not a congruent number—try it on 11, 26, and 43, for example. It's almost true that if the criterion holds, then n is a congruent number. (The "almost" part is deep; if you want to pursue the matter, Koblitz' book [7] contains a wealth of information, is very well written, and will take you as far as you'd care to go.)

Oh, yes; 157 is a congruent number, but the denominator of a side of a right triangle with rational sides and area 157 has 22 digits. The two legs of this triangle are

$$x = \frac{6803298487826435051217540}{411340519227716149383203} \quad \text{and} \quad y = \frac{411340519227716149383203}{21666555693714761309610}.$$

I'll leave it as an exercise for you to find the hypotenuse!

6 A Truly Marvelous Proof—via Elliptic Curves

Fermat's Last Theorem states that if n is an integer greater than 2, then the equation $x^n + y^n = z^n$ has no solutions in which x, y, and z are all nonzero integers. This statement is one of the major milestones on the longest trail in mathematics. This trail begins almost 4000 years ago in ancient Mesopotamia, with a clay tablet known as Plimpton 322 (see [13, p. 3]), containing a table of integer triples (x, y, z) for which $x^2 + y^2 = z^2$, whose authors are unknown. It ends in 1995 in Princeton, with a mathematics journal known as the *Annals of Mathematics*, containing two papers [17], [15] which provide a proof of Fermat's Last Theorem, whose authors are Andrew Wiles and Richard Taylor. Many singular and notable events stand along that trail, especially an obscure day, probably in 1637, when Fermat wrote down "that theorem" in the margin of his copy of Bachet's translation of Diophantus' *Arithmetica*, and that heart-stopping moment in June 1993 when Andrew Wiles announced a proof. Many mathematicians, including Euler, Legendre, Dirichlet, Germain, and Kummer, labored in vain to that end, and a good bit of modern number theory was developed in these attempts. Singh's book [12] is a highly readable and mathematically unsophisticated account of this most famous of mathematical puzzles from its genesis to its resolution. For a mathematically more technical account, a good place to begin is with David Cox's article [2].

What is interesting for us is that the home stretch of the trail passes right through the world of elliptic curves, and that Fermat's Last Theorem (FLT, for short) was finally proved by, in essence, grafting the whole problem onto an elliptic curve.

Previous attempts at a proof of FLT usually begin by assuming, to the contrary, that nonzero integers a, b, c exist for which $a^n + b^n = c^n$ with $n > 2$ an integer. The attempt would proceed by analyzing the curve $x^n + y^n = z^n$ directly. One example was Gabriel Lamé's work in the 1840s; he began by writing

$$y^n = z^n - x^n = \prod_{k=0}^{n-1} (z - \zeta^k x),$$

where $\zeta = e^{2\pi i/n}$ satisfies $\zeta^n = 1$. He thought that he had proved that each factor on the right is a perfect nth power, and this led him to a desired contradiction. But Ernst Kummer pointed out a flaw in Lamé's argument, which he, Kummer, did his best to patch. The patch didn't always work, however, and the problem remained unsolved.

It was Gerhard Frey [3] who completely transformed FLT into a problem about elliptic curves. In essence, Frey said this: if I have a solution $a^n + b^n = c^n$ to the Fermat equation for some exponent $n > 2$, then I'll use it to construct the following elliptic curve:

$$\mathbf{E}: y^2 = x(x - a^n)(x + b^n) = g(x).$$

Now if f is a polynomial of degree k and if $r_1, r_2, \ldots, r_k$ are all of its roots, then the discriminant $\Delta(f)$ of f is defined by

$$\Delta(f) = \prod_{1 \leq i < j \leq k} (r_i - r_j)^2.$$

If f is monic with integer coefficients, it turns out that $\Delta(f)$ is an integer. The three roots of the polynomial $g(x)$ on the right-hand side of the Frey curve are 0, a^n, and $-b^n$; using the fact that $a^n - (-b^n) = a^n + b^n = c^n$ and a little algebra, we find that $\Delta(g) = (abc)^{2n}$.

Frey said that an elliptic curve with such a discriminant must be really strange. In particular, such a curve cannot possibly be what is called modular (never mind what that means). Now here's a thought, said he; what if you could manage to prove two things: first, that a large class of elliptic curves *is* modular, and second, that the Frey curve is always a member of that class of curves? Why, you'd have a contradiction —from which you could conclude that there is no such curve. That is, there is no such solution to the Fermat equation...that there is no counterexample to Fermat's Last Theorem...and so Fermat's Last Theorem is true.

And *that* is exactly what Andrew Wiles [17]—with a last-minute assist from Richard Taylor [15]—did.

7 $x^4 + dx^2y^2 + y^4 = z^2$ and Elliptic Curves

A solution of

$$x^4 + dx^2 y^2 + y^4 = z^2 \tag{3}$$

with x, y, and z nonnegative integers is called trivial either if $xy = 0$ or if $d = n^2 - 2$ and $x = y = 1$. The first mention of this equation was made by Fermat, who proved that if $d = 0$, then (3) has only trivial solutions. His proof appeared in —yes—1637 as a marginal note in his copy of—you guessed it—Bachet's edition of Diophantus' *Arithmetica*. Over the years, many mathematicians tackled this problem, including Leibniz, who showed that the case $d = 6$ has only trivial solutions, and Euler, who gave several elegant methods for generating nontrivial solutions of (3).

As a result of reading a review in *Mathematical Reviews*, I got interested in this problem. In particular, it seemed curious that there were 23 values of d between 0 and 100 about which it was unknown whether nontrivial solutions to (3) exist. I was able to show that in 22 of these cases, there are indeed only trivial solutions to (3). The exceptional case is $d = 85$—apparently a solution Euler missed, since it is a solution he might have found, namely

$$1287^4 + 85 \cdot 1287^2 \cdot 4340^2 + 4340^4 = 54858119^2.$$

I wrote up what I'd done and sent it off to a journal, only to receive the aforementioned bad news that the entire problem could be transformed into a problem about elliptic curves, whose solution, I was assured by the referee, was routine.

Here's how the transformation works. Suppose that $X^4 + dX^2Y^2 + Y^4 = Z^2$; if we multiply both sides of this equation by X^2/Y^6, we are led to

$$\left(\frac{X}{Y}\right)^6 + d\left(\frac{X}{Y}\right)^4 + \left(\frac{X}{Y}\right)^2 = \left(\frac{ZX^3}{Y^3}\right)^2.$$

If we now let $y = ZX^3/Y^3$ and $x = (X/Y)^2$, we obtain

$$y^2 = x^3 + dx^2 + x; \tag{4}$$

since the roots of the right hand side are distinct if $d \neq \pm 2$, (4) is the equation of an elliptic curve.

Hence, if there is a nontrivial solution to (3), then the elliptic curve (4) will have a rational point (x, y) such that x is a perfect square. If this sounds familiar, it should; it's almost the same thing that happens when the Congruent Numbers problem is transformed to the world of elliptic curves. To take another example, since $(25, 245)$ is a rational point on the elliptic curve $y^2 = x^3 + 71x^2 + x$ in which the x-coordinate is a perfect square, it follows that $X^4 + 71X^2Y^2 + Y^4 = Z^2$ has a nontrivial solution—namely, $(X, Y, Z) = (5, 1, 49)$.

Resolving the problem at hand, however, while not nearly as hard or deep as the resolution of the Congruent Numbers problem, is still not entirely as routine as all that; furthermore, at the time, my knowledge of elliptic curves was practically nil.

Then I got an idea. I sent the paper off to a different journal, explaining in the cover letter that my approach to the problem was completely elementary and avoided the complicated machinery of elliptic curves. They accepted the paper [1].

8 Now What?

The future looks bright for the world of elliptic curves. Research in the area is booming, and there are many old problems in number theory that have been around for a long time and just might yield to the elliptic curve approach. In fact, ...oh, yes, of course you have questions.

- *You've said that elliptic curves are cubics of the form $y^2 = x^3 + ax^2 + bx + c$, in which the cubic polynomial in x has distinct roots. What do you get if the* *polynomial has repeated roots, such as x^3 or $x^3 + x^2$?* Those curves are called singular cubics, and they aren't studied as much as elliptic curves are, mainly because they don't have anything comparable to the chord-and-tangent addition of points on an elliptic curve.

- *Is there a special name for curves that involve polynomials of degree greater than 3?* For some of them, yes. In Section 3, we saw that an equation of the form $y^2 = f(x)$ with f a polynomial of degree 4 can be transformed into an elliptic curve, provided f has distinct roots. Curves of the form $y^2 = g(x)$, with g a polynomial of degree ≥ 5, are called *hyperelliptic curves*, and a goodly bit is known about them. But that's another story. Maybe some other time.

- *You said something back in Section 4 about addition of points on an elliptic curve being defined in a peculiar way for a particular reason. What's the reason?* Now *that's* a story I'd love to tell—but the margin of this paper is not large enough to contain it. As my grandmother used to say, "Tell you tomorrow!"

References

1. Ezra Brown, $x^4 + dx^2y^2 + y^4 = z^2$: Some cases with only trivial solutions—and a solution Euler missed, *Glasgow Math. J.* 31 (1989), 297–307.
2. David A. Cox, Introduction to Fermat's Last Theorem, *Amer. Math. Monthly* 101 (1994), 3–14.
3. Gerhard Frey, Links between stable elliptic curves and certain Diophantine equations, *Ann. Univ. Saraviensis*, Series Mathematicae 1 (1986), 1–40.
4. Shafi Goldwasser and J. Kilian, Almost all primes can be quickly certified, *Proc. 18th Annual ACM Symposium on Theory of Computing* (1986), 316–329.
5. Thomas L. Heath, *Diophantus of Alexandria*, Cambridge University Press, 1910.
6. Anthony W. Knapp, *Elliptic Curves*, Princeton University Press, 1992.
7. Neal Koblitz, *Introduction to Elliptic Curves and Modular Forms*, Springer-Verlag, 1984.
8. Neal Koblitz, Elliptic curve cryptosystems, *Math. Comp.* 48 (1987), 203–209.
9. Hendrik W. Lenstra, Factoring integers with elliptic curves, *Annals of Mathematics* 126 (1987), 649–673.
10. Victor S. Miller, Use of elliptic curves in cryptography, *Advances in Cryptology—CRYPTO '85*, Lecture Notes in Computer Science 218 (1986), 417–426.
11. Joseph H. Silverman and John Tate, *Rational Points on Elliptic Curves*, Springer-Verlag, 1992.
12. Simon Singh, *Fermat's Enigma*, Walker Publishing Co., 1997.
13. John Stillwell, *Mathematics and its History*, Springer-Verlag, 1989.
14. John Stillwell, The evolution of elliptic curves, *Amer. Math. Monthly* 102 (1995), 831–837.
15. Richard Taylor and Andrew Wiles, Ring-theoretic aspects of certain Hecke algebras, *Annals of Mathematics* 142 (1995), 553–572.
16. Jerrold Tunnell, A classical Diophantine problem and modular forms of weight 3/2, *Inventiones Math.* 72 (1983), 323–334.
17. Andrew Wiles, Modular elliptic curves and Fermat's Last Theorem, *Annals of Mathematics* 142 (1995), 443–551.

Originally appeared as:

Brown, Ezra. "Three Fermat Trails to Elliptic Curves." *The College Mathematics Journal*. vol. 31, no. 3 (May 2000), pp. 162–172.

Fermat's Last Theorem in Combinatorial Form

W.V. Quine

Fermat's Last Theorem can be vividly stated in terms of sorting objects into a row of bins, some of which are red, some blue, and the rest unpainted. The theorem amounts to saying that when there are more than two objects, the following statement is never true:

Statement. *The number of ways of sorting them that shun both colors is equal to the number of ways that shun neither.*

I shall show that this statement is equivalent to Fermat's equation $x^n + y^n = z^n$, when n is the number of objects, z is the number of bins, x is the number of bins that are not red, and y is the number of bins that are not blue. There are z^n ways of sorting the objects into bins; x^n of these ways shun red and y^n of them shun blue. So where A is the number of ways that shun both colors, B is the number of ways that shun red but not blue, C is the number of ways that shun blue but not red, and D is the number of ways that shun neither, we have

$$x^n = A + B, \qquad y^n = A + C, \qquad z^n = A + B + C + D.$$

These equations give $x^n + y^n = z^n$ if and only if $A = D$, which is to say, if and only if the statement above holds.

Originally appeared as:
Quine, W. V. "Fermat's Last Theorem in Combinatorial Form." *American Mathematical Monthly*. vol. 95, no. 7 (August-September 1988), p. 636.

"A Marvelous Proof"

Fernando Q. Gouvêa

No one really knows when it was that the story of what came to be known as "Fermat's Last Theorem" really started. Presumably it was sometime in the late 1630s that Pierre de Fermat made that famous inscription in the margin of Diophantus' *Arithmetic* claiming to have found "a marvelous proof". It seems now, however, that the story may be coming close to an end. In June, 1993, Andrew Wiles announced that he could prove Fermat's assertion. Since then, difficulties seem to have arisen, but Wiles' strategy is fundamentally sound and may yet succeed.

The argument sketched by Wiles is an artful blend of various topics that have been, for years now, the focus of intensive research in number theory: elliptic curves, modular forms, and Galois representations. The goal of this article is to give mathematicians who are not specialists in the subject access to a general outline of the strategy proposed by Wiles. Of necessity, we concentrate largely on background material giving first a brief description of the relevant topics, and only afterwards describe how they come together and relate to Fermat's assertion. Readers who are mainly interested in the structure of the argument and who do not need or want too many details about the background concepts may want to skim through Section 2, then concentrate on Section 3. Our discussion includes a few historical remarks, but history is not our main intention, and therefore we only touch on a few highlights that are relevant to our goal of describing the main ideas in Wiles' attack on the problem.

Thanks are due to Barry Mazur, Kenneth Ribet, Serge Lang, Noriko Yui, George Elliot, Keith Devlin, and Lynette Millett for their help and comments.

1 PRELIMINARIES. We all know the basic statement that Fermat wrote in his margin. The claim is that for any exponent $n \geq 3$ there are no non-trivial integer solutions of the equation $x^n + y^n = z^n$. (Here, "non-trivial" will just mean that none of the integers x, y, and z is to be equal to zero.) Fermat claims, in his marginal note, to have found "a marvelous proof" of this fact, which unfortunately would not fit in the margin.

This statement became known as "Fermat's Last Theorem," not, apparently, due to any belief that the "theorem" was the last one found by Fermat, but rather due to the fact that by the 1800s all of the other assertions made by Fermat had been either proved or refuted. This one was the last one left open, whence the name. In what follows, we will adopt the abbreviation FLT for Fermat's statement, and we will refer to $x^n + y^n = z^n$ as the "Fermat equation."

The first important results relating to FLT were theorems that showed that Fermat's claim was true for specific values of n. The first of these is due to Fermat himself: very few of his proofs were ever made public, but in one that was he shows that the equation

$$x^4 + y^4 = z^2$$

has no non-trivial integer solutions. Since any solution of the Fermat equation with exponent 4 gives a solution of the equation also, it follows that Fermat's claim is true for $n = 4$.

Once that is done, it is easy to see that we can restrict our attention to the case in which n is a prime number. To see this, notice that any number greater than 2 is either divisible by 4 or by an odd prime, and then notice that we can rewrite an equation

$$x^{mk} + y^{mk} = z^{mk}$$

as

$$(x^m)^k + (y^m)^k = (z^m)^k,$$

so that any solution for $n = mk$ yields at once a solution for $n = k$. If n is not prime, we can always choose k to be either 4 or an odd prime, so that the problem reduces to these two cases.

In the 1750s, Euler became interested in Fermat's work on number theory, and began a systematic investigation of the subject. In particular, he considered the Fermat equation for $n = 3$ and $n = 4$, and once again proved that there were no solutions. (Euler's proof for $n = 3$ depends on studying the "numbers" one gets by adjoining $\sqrt{-3}$ to the rationals, one of the first instances where one meets "algebraic numbers.") A good historical account of Euler's work is to be found in [**Wei83**]. In the following years, several other mathematicians extended this step by step to $n = 5, 7, \ldots$. A general account of the fortunes of FLT during this time can be found in [**Rib79**].

Since then, ways for testing Fermat's assertion for any specific value of n have been developed, and the range of exponents for which the result was known to be true kept getting pushed up. As of 1992, one knew that FLT was true for exponents up to 4 000 000 (by work of J. Buhler).

It is clear, however, that to get general results one needs a general method, i.e., a way to connect the Fermat equation (for any n) with some mathematical context which would allow for its analysis. Over the centuries, there have been many attempts at doing this; we mention only the two biggest successes (omitting quite a lot of very good work, for which see, for example, [**Rib79**].

The first of these is the work of E. Kummer, who, in the mid-nineteenth century, established a link between FLT and the theory of cyclotomic fields. This link allowed Kummer to prove Fermat's assertion when the exponent was a prime that had a particularly nice property (Kummer named such primes "regular"). The proof is an impressive bit of work, and was the first general result about the Fermat equation. Unfortunately, while in numerical tests a good percentage of primes seem to turn out to be regular, no one has yet managed to prove even that there are infinitely many regular primes. (And, ironically, we do have a proof that there are infinitely many primes that are *not* regular.) A discussion of Kummer's approach can be found in [**Rib79**]; for more detailed information on the cyclotomic theory, one could start with [**Was82**].

The second accomplishment we should mention is that of G. Faltings, who, in the early 1980s, proved Mordell's conjecture about rational solutions to certain kinds of polynomial equations. Applying this to the Fermat equations, one sees that for any $n \geq 4$ one can have only a *finite* number of non-trivial solutions. Once again, this is an impressive result, but its impact on FLT itself turns out to be minor because we have not yet found a way to actually determine how many solutions should exist. For an introduction to Faltings' work, check [**CS86**], which contains an English translation of the original paper.

Wiles' attack on the problem turns on another such linkage, also developed in the early 1980s by G. Frey, J.-P. Serre, and K. A. Ribet. This one connects FLT with the theory of elliptic curves, which has been much studied during all of this century, and thereby to all the machinery of modular forms and Galois representations that is the central theme of Wiles' work. The main goal of this paper is to describe this connection and then to explain how Wiles attempts to use it to prove FLT.

Notation. We will use the usual symbols $\mathbb{Q}$ for the rational numbers and $\mathbb{Z}$ for the integers. The integers modulo m will be written[1] as $\mathbb{Z}/m\mathbb{Z}$; we will most often need them when m is a power of a prime number p. If p is prime, then $\mathbb{Z}/p\mathbb{Z}$ is a field, and we commemorate that fact by using an alternative notation: $\mathbb{F}_p = \mathbb{Z}/p\mathbb{Z}$.

2 THE ACTORS. We begin by introducing the main actors in the drama. First, we briefly (and very informally) introduce the p-adic numbers. These are not so much actors in the play as they are part of the stage set: tools to allow the actors to do their job. Then we give brief and impressionistic outlines of the theories of Elliptic Curves, Modular Forms, and Galois Representations.

2.1 p-adic Numbers. The p-adic numbers are an extension of the field of rational numbers which are, in many ways, analogous to the real numbers. Like the real numbers, they can be obtained by defining a notion of distance between rational numbers, and then passing to the completion with respect to that distance. For our purposes, we do not really need to know much about them. The crucial facts are:

1. For each prime number p there exists a field $\mathbb{Q}_p$ which is complete with respect to a certain notion of distance and contains the rational numbers as a dense subfield.
2. Proximity in the p-adic metric is closely related to congruence properties modulo powers of p. For example, two integers whose difference is divisible by p^n are "close" in the p-adic world (the bigger the n, the closer they are).
3. As a consequence, one can think of the p-adics as encoding congruence information: whenever one knows something modulo p^n for every n, one can translate this into p-adic information, and vice-versa.
4. The field $\mathbb{Q}_p$ contains a subring $\mathbb{Z}_p$, which is called the *ring of p-adic integers*. (In fact, $\mathbb{Z}_p$ is the closure of $\mathbb{Z}$ in $\mathbb{Q}_p$.)

There is, of course, a lot more to say, and the reader will find it said in many references, such as [**Kob84**], [**Cas86**], [**Ami75**], and even [**Gou93**]. The p-adic numbers were introduced by K. Hensel (a student of Kummer), and many of the basic ideas seem to appear, in veiled form, in Kummer's work; since then, they have become a fundamental tool in number theory.

---

[1]Many elementary texts like to use $\mathbb{Z}_m$ as the notation for the integers modulo m; for us (and for serious number theory in general), this notation is inconvenient because it collides with the notation for the p-adic integers described below.

2.2 Elliptic Curves. Elliptic curves are a special kind[2] of algebraic curves which have a very rich arithmetical structure. There are several fancy ways of defining them, but for our purposes we can just define them as the set of points satisfying a polynomial equation of a certain form.

To be specific, consider an equation of the form

$$y^2 + a_1 xy + a_3 y = x^3 + a_2 x^2 + a_4 x + a_6,$$

where the a_i are integers (there is a reason for the strange choice of indices on the a_i, but we won't go into it here). We want to consider the set of points (x, y) which satisfy this equation. Since we are doing number theory, we don't want to tie ourselves down too seriously as to what sort of numbers x and y are: it makes sense to take them in the real numbers, in the complex numbers, in the rational numbers, and even, for any prime number p, in $\mathbb{F}_p$ (in which case we think of the equation as a congruence modulo p). We will describe the situation by saying that there is an underlying object which we call *the curve E* and, for each one of the possible fields of definition for points (x, y), we call the set of possible solutions the "points of E" over that field. So, if we consider all possible complex solutions, we get the set $E(\mathbb{C})$ of the complex points of E. Similarly, we can consider the real points $E(\mathbb{R})$, the rational points $E(\mathbb{Q})$, and even the $\mathbb{F}_p$-points $E(\mathbb{F}_p)$.

We haven't yet said when it is that such equations define elliptic curves. The condition is simply that the curve be *smooth*. If we consider the real or complex points, this means exactly what one would expect: the set of points contains no "singular" points, that is, at every point there is a well-defined tangent line. We know, from elementary analysis, that an equation $f(x, y) = 0$ defines a smooth curve exactly when there are no points on the curve at which both partial derivatives of f vanish. In other words, the curve will be smooth if there are no common solutions of the equations

$$f(x, y) = 0 \qquad \frac{\partial f}{\partial x}(x, y) = 0 \qquad \frac{\partial f}{\partial y}(x, y) = 0.$$

Notice, though, that this condition is really algebraic (the derivatives are derivatives of polynomials, and hence can be taken formally). In fact, we can boil it down to a (complicated) polynomial condition in the a_i. There is a polynomial $\Delta(E) = \Delta(a_1, a_2, a_3, a_4, a_6)$ in the a_i such that E is smooth if and only if $\Delta(E) \neq 0$. This gives us the means to give a completely formal definition (which makes sense even over $\mathbb{F}_p$). The number $\Delta(E)$ is called the *discriminant* of the curve E.

Definition 1. *Let K be a field. An elliptic curve over K is an algebraic curve determined by an equation of the form*

$$y^2 + a_1 xy + a_3 y = x^3 + a_2 x^2 + a_4 x + a_6,$$

where each of the a_i belongs to K and such that $\Delta(a_1, a_2, a_3, a_4, a_6) \neq 0$.

Specialists would want to rephrase that definition to allow other equations, provided that a well-chosen change of variables could transform them into equations of this form.

---

[2] Perhaps it's best to dispel the obvious confusion right up front: ellipses are not elliptic curves. In fact, the connection between elliptic curves and ellipses is a rather subtle one. What happens is that elliptic curves (over the complex numbers) are the "natural habitat" of the elliptic integrals which arise, among other places, when one attempts to compute the arc length of an ellipse. For us, this connection will be of very little importance.

It's about time to give some examples. To make things easier, let us focus on the special case in which the equation is of the form $y^2 = g(x)$, with $g(x)$ a cubic polynomial (in other words, we're assuming $a_1 = a_3 = 0$). In this case, it's very easy to determine when there can be singular points, and even what sort of singular points they will be. If we put $f(x, y) = y^2 - g(x)$, then we have

$$\frac{\partial f}{\partial x}(x, y) = -g'(x) \quad \text{and} \quad \frac{\partial f}{\partial y}(x, y) = 2y,$$

and the condition for a point to be "bad" becomes

$$y^2 = g(x) \quad -g'(x) = 0 \quad 2y = 0,$$

which boils down to $y = g(x) = g'(x) = 0$. In other words, a point will be bad exactly when its y-coordinate is zero and its x-coordinate is a *double root* of the polynomial $g(x)$. Since $g(x)$ is of degree 3, this gives us only three possibilities:

- $g(x)$ has no multiple roots, and the equation defines an elliptic curve;
- $g(x)$ has a double root;
- $g(x)$ has a triple root.

Let's look at one example of each case, and graph the real points of the corresponding curve.

For the first case, consider the curve given by $y^2 = x^3 - x$. Its graph is in figure 1(a) (to be precise, this is the graph of its real points). A different example of the same case is given by $y^2 = x^3 + x$; see figure 1(b). (The reason these look so

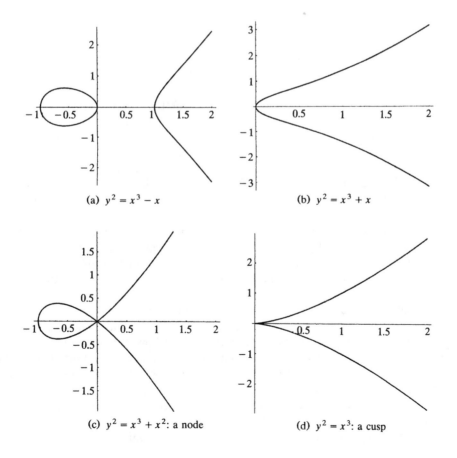

(a) $y^2 = x^3 - x$

(b) $y^2 = x^3 + x$

(c) $y^2 = x^3 + x^2$: a node

(d) $y^2 = x^3$: a cusp

different is that we are only looking at the real points of the curve; in fact, over the complex numbers these two curves are isomorphic.)

When there are "bad" points, what has happened is that either two roots of $g(x)$ have "come together" or all three roots have done so. In the first case, we get a loop. At the crossing point, which is usually called a "node," the curve has two different tangent lines. See Figure 1(c), where we have the graph of the equation $y^2 = x^3 + x^2$ (double root at zero).

In the final case, not only have all three roots of $g(x)$ come together, but also the two tangents in the node have come together to form a sort of "double tangent" (this can be made precise with some easy algebra of polynomials, but it's more fun to think of it geometrically). The graph now looks like Figure 1(d), and we call this kind of singular point a "cusp".

How does all this relate to the discriminant Δ we mentioned above? Well, if r_1, r_2 and r_3 are the roots of the polynomial $g(x)$, the discriminant for the equation $y^2 = g(x)$ turns out to be

$$\Delta = K(r_1 - r_2)^2(r_1 - r_3)^2(r_2 - r_3)^2,$$

where K is a constant. This does just what we want: if two of the roots are equal, it is zero, and if not, not. Furthermore, it is not too hard to see that Δ is actually a polynomial in the coefficients of $g(x)$, which is what we claimed. In other words, all that the discriminant is doing for us is giving a direct algebraic procedure for determining whether there are singular points.

While this analysis applies specifically to curves of the form $y^2 = g(x)$, it actually extends to all equations of the sort we are considering: there is at most one singular point, and it is either a node or a cusp.

One final geometric point: as one can see from the graphs, these curves are not closed. It is often convenient to "close them up." This is done by adding one more point to the curve, usually referred to as "the point at infinity." This can be done in a precise way by embedding the curve into the projective plane, and then taking the closure. For us, however, the only important thing is to remember that we actually have one extra point on our curves. (One should imagine it to be "infinitely far up the y-axis," but keep in mind that there is only one "point at infinity" on the y-axis, so that it is *also* "infinitely far down.")

With some examples in hand, we can proceed to deeper waters. In order to understand the connection we are going to establish between elliptic curves and FLT, we need to review quite a large portion of what is known about the rich arithmetic structure of these curves.

The first thing to note is that one can define an operation on the set of points of an elliptic curve that makes it, in a natural way, an abelian group. The operation is usually referred to as "addition." The identity element of this group turns out to be the point at infinity (it would be more honest to say that we *choose* the point at infinity for this role).

We won't enter into the details of how one adds points on an elliptic curve. In fact, there are several equivalent definitions, each of which has its advantages! The reader should see the references for more details of how it is done (and the proof that one does get a group). The main thing to know about the definition, for now, is that it preserves the field of definition of the points: adding two rational points gives a rational point, and so on.

What this means is that for every choice of a base field, we can get a group of points on the curve with coordinates in that field, so that in fact an elliptic curve gives us a whole bunch of groups, which are, of course, all related (though sometimes related in a mysterious way). So, given an E, we can look at the complex points $E(\mathbb{C})$, which form a complex Lie group which is topologically a torus, or we can look at the real Lie group $E(\mathbb{R})$, which turns out to be either isomorphic to the circle S^1 or to the direct product $\mathbb{Z}/2\mathbb{Z} \times S^1$. (Look back at the examples above; can you see which is which?)

From an arithmetical point of view, however, the most interesting of these groups is the group of rational points, $E(\mathbb{Q})$. A point $P \in E(\mathbb{Q})$ gives a solution in rational numbers of our cubic equation, and looking for such solutions is, of course, an example of solving a diophantine equation, a sort of problem that is quite important in number theory. What is especially nice about $E(\mathbb{Q})$ is the fact, proved by L. Mordell (and extended by A. Weil) in the 1920s, that it is a *finitely generated* abelian group. What this means is just the following: there is a finite list of rational points on the curve (or, if one prefers, of rational solutions to the equation) such that every other rational solution is obtained by combining (using the addition law) these points with one another. These points are called the *generators* of the group $E(\mathbb{Q})$, which is usually called the *Mordell-Weil group* of E.

The curves we considered earlier have very simple Mordell-Weil groups. For the curve given by $y^2 = x^3 - x$ (figure 1a), it has four points; and for $y^2 = x^3 + x$ (figure 1b) it has two. It is easy, though, to give more interesting examples. Here is one, chosen at random from [**Cre92**]: if E is the curve defined by $y^2 + y = x^3 - x^2 - 2x + 2$, the Mordell-Weil group $E(\mathbb{Q})$ is an infinite cyclic group, generated by the point $(2, 1)$.

Of course, knowing that we have a finitely generated group raises the obvious question of estimating or computing the number of generators needed and of how one might go about actually finding these generating points. Both of these questions are still open, even though there are rather precise conjectures about what their answers should be. For many specific curves, both the number and the generators themselves have been completely worked out (see, for example, the tables in [**Cre92**]), but the general problem still seems quite difficult.

A fundamental component of the conjectural plan for determining the generators is considering, for each prime number p, the reduction of our curve modulo p. The basic idea is quite simple: since our equation has integer coefficients, we can reduce it modulo p" and look for solutions in the field $\mathbb{F}_p$ of integers modulo p. This should give a finite[3] group $E(\mathbb{F}_p)$, whose structure should be easier to analyse than that of the big group $E(\mathbb{Q})$. It's a rather simple idea, but several complications[4] arise.

The main thing that can go wrong is that the reduction modulo p may fail to be an elliptic curve. That is actually very easy to see. To tell whether the curve is elliptic (that is, if it has no singular points), one needs to look at Δ. It is perfectly possible for Δ to be nonzero (so the curve over $\mathbb{Q}$ is elliptic) while being at the same time congruent to zero modulo p (so that the curve over $\mathbb{F}_p$ is singular). This phenomenon is called *bad reduction*, and it is easy to come up with examples. One might take $p = 5$, and look at the curve $y^2 = x^3 - 5$. This turns out to be an elliptic curve over $\mathbb{Q}$, but its reduction modulo 5 is going to have a cusp. One says, then, that the curve has bad reduction at 5. In fact, the discriminant turns out to be

---

[3] It is finite because, apart from the point at infinity, there are only p^2 possible points. In fact, the maximum possible number of points is smaller than that, but that fact takes some proving.

[4] It may seem a bit perverse to dwell on the nature of these complications, but it will turn out that we need to have at least some understanding of how this goes later on.

$\Delta = -10800$, which is clearly divisible by 2, 3, and 5, so that the curve has bad reduction at each of these. (In each case, it's easy to verify that the reduced curve has a cusp.)

We want to classify the possible types of reduction, but there is one further glitch that we have to deal with before we can do so. To see what it is, consider the curve $y^2 = x^3 - 625x$. At first glance, it seems even worse than the first, and the discriminant, which turns out to be $\Delta = -15625000000$, looks *very* divisible by 5. But look what we can do: let's change variables by setting $x = 25u = 5^2 u$ and $y = 125v = 5^3 v$. Then our equation becomes

$$(5^3 v)^2 = (5^2 u)^3 - 625(5^2 u),$$

which simplifies to

$$5^6 v^2 = 5^6 u^3 - 5^6 u,$$

and hence to

$$v^2 = u^3 - u,$$

which is not only a nice elliptic curve, but has good reduction at 5. In other words, this example shows that *curves which are isomorphic over $\mathbb{Q}$ can have very different reductions modulo p*.

It turns out that among all the possible equations for our curve, one can choose an equation that is *minimal*, in the sense that its discriminant will be divisible by fewer primes than the discriminant for other equations. Since the primes that divide the discriminant are the primes of bad reduction, a minimal equation will have reduction properties that are as good as possible. When studying the reduction properties of the curve, then, one must also pass to such a minimal equation (and there are algorithms to do this).

Well, then, suppose we have done so, and have an elliptic curve E given by a minimal equation. Then we can classify all prime numbers into three groups:

- *Primes of good reduction*: those which do not divide the discriminant of the minimal equation. The curve modulo p is an elliptic curve, and we have a group $E(\mathbb{F}_p)$.
- *Primes of multiplicative reduction*: those for which the curve modulo p has a node. If the singular point is (x_0, y_0), it turns out that the set $E(\mathbb{F}_p) - \{(x_0, y_0)\}$ has a group structure, and is isomorphic to the multiplicative group $\mathbb{F}_p - \{0\}$.
- *Primes of additive reduction*: those for which the curve modulo p has a cusp. If the singular point is (x_0, y_0), the set $E(\mathbb{F}_p) - \{(x_0, y_0)\}$ once again has a group structure, and is isomorphic to the additive group $\mathbb{F}_p$.

No curve can have good reduction everywhere, so there will always be some bad primes, but the feeling one should get is that multiplicative reduction is somehow not as bad as additive reduction. There are various technical reasons for this, which we don't really need to go into. Instead, we codify the information about the reduction types of the curve into a number, called the *conductor* of the curve. We define the conductor to be a product $N = \prod p^{n(p)}$, where

$$n(p) = \begin{cases} 0 & \text{if } E \text{ has good reduction at } p \\ 1 & \text{if } E \text{ has multiplicative reduction at } p \\ \geq 2 & \text{if } E \text{ has additive reduction at } p \end{cases}$$

(The exact value of $n(p)$ for the case of additive reduction depends on some rather subtle properties of the reduction modulo such primes; most of the time, the exponent is 2.) The result is that one can tell, by looking at the conductor, exactly what the reduction type of E at each prime is.

The elliptic curves we will want to consider are those whose reduction properties are as good as possible. Since good reduction at all primes is not possible, we opt for the next best thing: good reduction at almost all primes, multiplicative reduction at the others. Such curves are called *semistable*:

Definition 2. *An elliptic curve is called semistable if all of its reductions are either good or multiplicative. Equivalently, a curve is semistable if its conductor is square-free.*

A crucial step in the application of Wiles' theorem to FLT will be verifying that a certain curve is semistable. Just to give us some reference points, let's look at a few examples.

1. Let E_1 be the curve $y^2 = x^3 - 5$, which we considered above. One checks that this equation is minimal, and that the curve has additive reduction at 2, 3, and 5, so that it is not semistable. The conductor turns out to be equal to 10800 (essentially, the same as the discriminant!).
2. Let E_2 be the curve $y^2 + y = x^3 + x$. This has multiplicative reduction at 7 and 13 (checking this makes a nice exercise) and good reduction at all other primes. Hence, E_2 is semistable and its conductor is 91.
3. Let E_3 be the curve $y^2 = x^3 + x^2 + 2x + 2$ (which is minimal). This has discriminant $\Delta = -1152 = -2^7 \cdot 3^2$, so that the bad primes are 2 and 3. It turns out that the reduction is multiplicative at 3 and additive at 2, and the conductor is 384; the curve is not semistable.
4. *The main example for the purpose at hand*: Let $a, b,$ and c be relatively prime integers such that $a + b + c = 0$. Consider the curve E_{abc} whose equation[5] is $y^2 = x(x - a)(x + b)$. Depending on what $a, b,$ and c are, this equation may or may not be minimal, so let's make the additional assumptions that $a \equiv -1 \pmod 4$ and that $b \equiv 0 \pmod{32}$. In this case, the equation is *not* minimal. A minimal equation for this curve turns out to be

$$y^2 + xy = x^3 + \frac{b-a-1}{4}x^2 - \frac{ab}{16},$$

which we get by the change in variables $x \to 4x, y \to 8y + 4x$. One can then compute that the discriminant is $\Delta = a^2b^2c^2/256$ (not surprising: a constant times the product of the squares of the differences of the roots of the original cubic), and that the curve is semistable. The primes of bad reduction are those that divide abc (this would be easy to see directly from the equation, by checking when there is a multiple root modulo p), and therefore the conductor is equal to the product of the primes that divide abc:

$$N = \prod_{p \mid abc} p$$

---

[5] It may strike the reader as funny that c is absent from the equation. Keep in mind, however, that c is completely determined by a and b, so that it is really not as absent as all that. The crucial point is that the roots of the cubic on the right hand side are 0, a, and $-b$, so that the differences of the roots are (up to sign) exactly $a, b,$ and c.

(this number is sometimes called the *radical* of *abc*). We will be using curves of the form E_{abc} (for very special a, b, and c) when we make the link with FLT.

We need a final bit of elliptic curve theory. It is interesting to look at the number of points in the groups $E(\mathbb{F}_p)$ as p ranges through the primes of good reduction for E. Part of the motivation for this is the reasoning that if the group $E(\mathbb{Q})$ is large (i.e. there are many rational solutions), one would expect that for many choices of the prime p many of the points in $E(\mathbb{Q})$ would survive reduction modulo p, so that the group $E(\mathbb{F}_p)$ would be large. Therefore, one would like to make some sort of conjecture that said that if the $E(\mathbb{F}_p)$ are very large for many primes p, then the group $E(\mathbb{Q})$ will be large.

Elaborating and refining this idea leads to the conjecture of Birch and Swinnerton-Dyer, which we won't get into here. But even this coarse version suggests that the variation of the size of $E(\mathbb{F}_p)$ as p runs through the primes should tell us something about the arithmetic on the curve. To "encode" this variation, we start by observing that the (projective) line over $\mathbb{F}_p$ has exactly $p + 1$ points (the p elements of $\mathbb{F}_p$, plus the point at infinity). We take this as the "standard" number of points for a curve over $\mathbb{F}_p$, and, when we look at $E(\mathbb{F}_p)$, record how far from the standard we are. To be precise, given an elliptic curve E and a prime number p at which E has good reduction, we define a number a_p by the equation

$$\#E(\mathbb{F}_p) = p + 1 - a_p.$$

For primes of bad reduction, we extend the definition in a convenient way; it turns out that we get $a_p = \pm 1$ when the reduction is multiplicative (with a precise rule to decide which) and $a_p = 0$ when it is additive.

The usual way to "record" the sequence of the a_p is to use them to build a complex analytic function called the *L-function* of the curve E. It then is natural to conjecture that this L-function has properties similar to those of other L-functions that arise in number theory, and that one can read off properties of E from properties of its L-function. This is a huge story which we cannot tell in this article, but which is really very close to some of the issues which we do discuss later on. Suffice it to say, for now, that we get a function

$$L(E, s) = \sum_{n=1}^{\infty} \frac{a_n}{n^s},$$

where the a_p are exactly the same as the ones we just introduced, the a_n are determined from the a_p by "Euler product" expansion for the L-function, and the series can be shown to converge when $\text{Re}(s) > 3/2$. The L-function is conjectured to have an analytic continuation to the whole complex plane and to satisfy a certain functional equation.

It is time to introduce the other actors in the play and to explain how they relate to elliptic curves. The reader who would like to delve further into this theory has a lot to choose from. As an informal introduction, one could look at J. Silverman's article [**Sil93**], which relates elliptic curves to "sums of two cubes" and Ramanujan's taxicab number. Various introductory texts are available, including [**Cas91**], [**Hus87**], [**Kna92**], [**Sil86**], and [**ST92**]. Each of these has particular strengths; the last is intended as an undergraduate text. In addition to these and other texts, the interested reader might enjoy looking at symbolic manipulation software that will handle elliptic curves well. Such capabilities are built into GP-PARI and SIMATH, and can be added to *Mathematica* by using Silverman's *EllipticCurveCalc* package (which is what we used for most of the computations in this paper), and to *Maple* by using Connell's *Apecs* package. See[C⁺], [Z⁺], [SvM], [Con].

2.3 Modular Forms. Modular forms start their lives as analytic objects (or, to be more honest, as objects of group representation theory), but end up playing a very intriguing role in number theory. In this section, we will *very* briefly sketch out their definition and explain their relation to elliptic curves.

Let $\mathfrak{h} = \{x + iy | y > 0\}$ be the complex upper half-plane. As is well known (and, in any case, easy to check), matrices in $SL_2(\mathbb{Z})$ act on $\mathfrak{h}$ in the following way. If $\gamma \in SL_2(\mathbb{Z})$ is the matrix

$$\gamma = \begin{pmatrix} a & b \\ c & d \end{pmatrix},$$

(so that $a, b, c,$ and d are integers and $ad - bc = 1$), and $z \in \mathfrak{h}$, we define

$$\gamma \cdot z = \frac{az + b}{cz + d}.$$

It is easy to check that if $z \in \mathfrak{h}$ then $\gamma \cdot z \in \mathfrak{h}$, and that $\gamma_1 \cdot (\gamma_2 \cdot z) = (\gamma_1 \gamma_2) \cdot z$.

We want to consider functions on the upper half-plane which are "as invariant as possible" under this action, perhaps when restricted to a smaller group. The subgroups we will need to consider are the "congruence subgroups" which we get by adding a congruence condition to the entries of the matrix. Thus, for any positive integer N, we want to look at the group

$$\Gamma_0(N) = \left\{ \gamma = \begin{pmatrix} a & b \\ c & d \end{pmatrix} \in SL_2(\mathbb{Z}) | c \equiv 0 \pmod{N} \right\}.$$

We are now ready to begin defining modular forms. They will be functions $f: \mathfrak{h} \to \mathbb{C}$, holomorphic, which "transform well" under one of the subgroups $\Gamma_0(N)$. To be specific, we require that there exist an integer k such that

$$f\left(\frac{az + b}{cz + d}\right) = (cz + d)^k f(z).$$

Applying this formula to the special case in which the matrix is

$$\begin{pmatrix} 1 & 1 \\ 0 & 1 \end{pmatrix}$$

shows that any such function must satisfy $f(z + 1) = f(z)$, and hence must have a Fourier expansion

$$f(z) = \sum_{n=-\infty}^{\infty} a_n q^n \quad \text{where } q = e^{2\pi i z}.$$

We require that this expression in fact only involve non-negative powers of q (and in fact we extend that requirement to a finite number of other, similar, expansions, which the experts call the "Fourier expansions at the other cusps"). A function satisfying all of these constraints is called *a modular form of weight k on $\Gamma_0(N)$*. The number N is usually called the *level* of the modular form f.

We will need to consider one special subspace of the space of modular forms of a given weight and level. Rather than having a Fourier expansion with non-negative powers only, we might require *positive* powers only (in the main expansion and in the ones "at the other cusps"). We call such modular forms *cusp forms*; they turn out to be the more interesting part of the space of modular forms.

Finally, one must make a remark on the relation between the theory at various levels: if N divides M, then every form of level N (and weight k) gives rise to (a number of) forms of level M (and the same weight). The subspace generated by all forms of level M and weight k which arise in this manner (from the various divisors of M) is called the space of *old forms* of level M. With respect to a natural inner product structure on the space of modular forms, one can then take the orthogonal complement of the space of old forms. This complement is called the space of *new forms*, which are the ones we will be most interested in.

What really makes the theory of modular forms interesting for arithmetic is the existence of a family of commuting operators on each space of modular forms, called the Hecke operators. We will not go into the definition of these operators (they are quite natural from the point of view of representation theory); for us the crucial things will be:

- For each positive integer n relatively prime to the level N, there is a Hecke operator T_n acting on the space of modular forms of fixed weight and level N.
- The Hecke operators commute with each other.
- If m and n are relatively prime, then $T_{nm} = T_n T_m$.

We will be especially interested in modular forms which are eigenvectors for the action of all the Hecke operators, i.e., forms for which there exist numbers λ_n such that $T_n(f) = \lambda_n f$ for each n which is relatively prime to the level. We will call such forms *eigenforms*.

This is all quite strange and complicated, so let's immediately point out one connection between modular forms and elliptic curves. Suppose one has a modular form which is

- of weight 2 and level N,
- a cusp form,
- new,
- an eigenform.

If that is the case, one can normalize the form so that its Fourier expansion looks like

$$f(z) = \sum_{n=1}^{\infty} a_n q^n \text{ with } a_1 = 1.$$

Suppose that, once we have done the normalization,

- all of the Fourier coefficients a_n are integers.

Then there exists an elliptic curve whose equation has integer coefficients, whose conductor is N, and whose a_n are exactly the ones that appear in the Fourier expansion of f. In particular, the L-function of E can be expressed in terms of f (as a Mellin transform), and the nice analytic properties of f then allow us to prove that the L-function does have an analytic continuation and does satisfy a functional equation.

This connection between forms and elliptic curves is so powerful that it led people to investigate the matter further. The first one to suggest that *every* elliptic curve should come about in this manner was Y. Taniyama, in the mid-fifties. The suggestion only penetrated the mathematical culture much later, largely due to the work of G. Shimura, and it was made more precise by A. Weil's work pinning down the role of the conductor. We now call this the "Shimura-Taniyama-Weil

Conjecture." Here it is:

Conjecture 1 (Shimura-Taniyama-Weil). *Let E be an elliptic curve whose equation has integer coefficients. Let N be the conductor of E, and for each n let a_n be the number appearing in the L-function of E. Then there exists a modular form of weight 2, new of level N, an eigenform under the Hecke operators, and (when normalized) with Fourier expansion equal to $\Sigma a_n q^n$.*

For any specific curve, it is not too hard to check that this is true. One takes E, determines the conductor and the a_n for a range of n. Since the space of modular forms of weight 2 and level N is finite-dimensional, knowing enough of the a_n must determine the form, and we can go and look if it is there. (In general, given a list of a_n, it is not at all easy to determine whether $\Sigma a_n q^n$ is the Fourier expansion of a modular form, so we need to do it the other way: we generate a basis of the space of modular forms, then try to find our putative form as a linear combination of the basis.) If we find a form with the right (initial chunk of) Fourier expansion, this gives prima facie evidence that the curve satisfies the STW conjecture. To clinch the matter, one can use a form of the Čebotarev density theorem to show that if *enough* (in an explicit sense) of the a_n are right, then they all are.

This method has been used to verify the STW conjecture for any number of specific curves (see, for example, [**Cre92**]). The conjecture has a really crucial role in the theory of elliptic curves; in fact, curves that satisfy the conjecture are known as "modular elliptic curves," and many of the fundamental new results in the theory have only been proved for curves that have this property.

As our final remark on modular forms, we point out that it is possible, for any given N, to determine (essentially using the Riemann-Roch theorem) the exact dimension of the space of cusp forms of weight 2 and level N. This gives us a very good handle on what curves of that conductor should exist (if the STW conjecture is true).

For more information on modular forms, one might look at [**Lan76**], [**Miy89**], or [**Shi71**]. There is an intriguing account of the Shimura-Taniyama-Weil conjecture, in a very different spirit, in Mazur's article [**Maz91**], and a useful survey in [**Lan91**].

2.4 Galois Representations. The final actors in our play are Galois representations. One starts with the Galois group of an extension of the field of rational numbers. To understand this Galois group, one can try to "represent" the elements of the group as matrices. In other words, one can try to find a vector space on which our Galois group acts, which gives a way to associate a matrix to each element of the group. This in fact gives a group homomorphism from the Galois group to a group of matrices (this need not be injective; when it is, one calls the representation "faithful").

Rather than work with specific finite extensions of $\mathbb{Q}$, we work with the Galois group $G = \text{Gal}(\overline{\mathbb{Q}}/\mathbb{Q})$ of the algebraic closure of $\mathbb{Q}$. This is a huge group (which one makes more manageable by giving it a topology) that hides within itself an enormous amount of arithmetic information. The representations we will be considering will be into 2×2 matrices over various fields and rings, and they will (for the most part) be obtained from elliptic curves and from modular forms.

To see how to get Galois representations from an elliptic curve, let's start with an elliptic curve E, whose equation has coefficients in $\mathbb{Z}$. Choose a prime p. Since the (complex, say) points of E form a group, one can look in this group for points which are of order p (that is, for points (x, y) such that adding them to themselves

p times gives the identity). It turns out that (over $\mathbb{C}$) there are p^2 such points, and they form a subgroup which we donate by $E[p]$. In fact, this group is isomorphic to the product of two copies of $\mathbb{F}_p$:

$$E[p] \cong \mathbb{F}_p \times \mathbb{F}_p.$$

Now, the points in $E[p]$ are a priori complex, but on closer look one sees that in fact they are all defined over some extension of $\mathbb{Q}$, and in particular that transforming the coefficients of a point of order p by the Galois group G yields another point of order p. In fact, it's even better than that: since the rule for adding points is defined in rational terms, the whole group structure is preserved. Since $E[p]$ looks like a vector space of dimension 2 over $\mathbb{F}_p$, this means that each element of G acts as a linear transformation on this space, and hence that we get a representation

$$\bar{\rho}_{E,p}: G \to \mathrm{GL}_2(\mathbb{F}_p).$$

(We use a bar to remind ourselves that this is a representation "modulo p.")

Now, $\mathrm{GL}_2(\mathbb{F}_p)$ is a finite group, and G is very infinite, so this representation, while it tells us a lot, can't be the whole story. It turns out, however, that we can use p-adic numbers to get a whole lot more. Instead of considering only the points of order p, we can consider points of order p^n for each n. This gives a whole bunch of subgroups

$$E[p] \subset E[p^2] \subset E[p^3] \subset \ldots$$

and a whole bunch of representations, into $\mathrm{GL}_2(\mathbb{F}_p)$, then into $\mathrm{GL}_2(\mathbb{Z}/p^2\mathbb{Z})$, then into $\mathrm{GL}_2(\mathbb{Z}/p^3\mathbb{Z})\ldots$. Putting all of these together ends up by giving us a p-adic representation

$$\rho_{E,p}: G \to \mathrm{GL}_2(\mathbb{Q}_p)$$

which hides within itself all of the others. The representations $\rho_{E,p}$ contain a lot of arithmetic information about the curve E.

And how does it look on the modular forms side? Well, it follows from the work of several mathematicians (M. Eichler, G. Shimura, P. Deligne, and J.-P. Serre) that, whenever we have a modular form f (of any weight) which is an eigenform for the action of the Hecke operators and whose Fourier coefficients (after normalization) are integers, we can construct a representation

$$\rho_{f,p}: G \to \mathrm{GL}_2(\mathbb{Q}_p)$$

which is attached to f in a precise sense which is too technical to explain here. (The construction of the representation is quite difficult, and in fact no satisfactory expository account is yet available.)

The crucial thing to know, for our purposes, is that *when an elliptic curve E arises from a modular form f, then the representations $\rho_{E,p}$ and $\rho_{f,p}$ are the same.* In fact, a converse is also true: given a curve E, if one can find a modular form f such that $\rho_{E,p}$ is the same as $\rho_{f,p}$ then E will be modular.

3 THE PLAY. We are now ready to take the plunge and try to see how all of this theory relates to Fermat's Last Theorem. The idea is to assume that FLT is false, and then, using this assumption, to construct an elliptic curve that contradicts just about every conjecture under the sun.

3.1 Linking FLT to Elliptic Curves. So let's start by assuming FLT is false, i.e., that there exist three non-zero integers $u, v,$ and w such that $u^p + v^p + w^p = 0$ (as we know, we only need to consider the case of prime exponent p, which is therefore odd, so that we can recast a solution in Fermat's form to be in the form above). Since we know that the theorem is true for $p = 3$, we might as well assume that $p \geq 5$. We may clearly assume that $u, v,$ and w are relatively prime, which means that precisely one of them must be even. Let's say v is even. Since p is bigger than two, we can see, by looking at the equation modulo 4, that one of u and w must be congruent to -1 modulo 4, and the other must be congruent to 1. Let's say $u \equiv -1 \pmod{4}$.

Let's use this data to build an elliptic curve, following an idea due to G. Frey (see [**Fre86**], [**Fre87a**], [**Fre87b**]). We consider the curve

$$y^2 = x(x - u^p)(x + v^p).$$

This is usually known as the Frey curve. Following our discussion, above, of the curve E_{abc}, we already know quite a bit about the Frey curve. Here's a summary:

1. Since v is even and $p \geq 5$, we know that we have $v^p \equiv 0 \pmod{32}$. We also know that $u^p \equiv -1 \pmod{4}$. This puts us in the right position to use what we know about curves E_{abc}.
2. The minimal discriminant of the Frey curve is

$$\Delta = \frac{(uvw)^{2p}}{256}.$$

3. The conductor of the Frey curve is the product of all the primes dividing $u^p v^p w^p$, which is, of course, the same as the product of all the primes dividing uvw.
4. The Frey curve is semistable.

Now, as Frey observed in the mid-1980s, this curve seems much too strange to exist. For one thing, its conductor is extremely small when compared to its discriminant (because of that exponent of $2p$). For another, its Galois representations are pretty weird. Very soon, people were pointing out that there were several conjectures that would rule out the existence of Frey's curve, and therefore would prove that Fermat was correct in saying that his equation had no solutions.

3.2 FLT follows from the Shimura-Taniyama-Weil Conjecture. It was already clear to Frey that it was likely that the existence of his curve would contradict the Shimura-Taniyama-Weil conjecture, but he was unable to give a solid proof of this. A few months after Frey's work, Serre pinpointed, in a letter to J.-F. Mestre, exactly what one would need to prove to establish the link. In this letter (published as [**Ser87a**]), Serre describes the situation with the phrase "STW + ε implies Fermat." Because of this, the missing theorem became known, for a while, as "conjecture epsilon." This conjecture was proved by K. A. Ribet in [**Rib90**] about a year later, and this established the link. A survey of these results can be found in [**Lan91**].

What Serre noticed was that the representation modulo p

$$\bar{\rho}_{E,p} : G \to \mathrm{GL}_2(\mathbb{F}_p)$$

obtained from the Frey curve was rather strange. It looked like the sort of representation one would get from a modular form of weight 2, but if one applied the "usual recipe" for guessing the level of that modular form, the answer came out to be $N = 2$. He also showed that the modular form must be a cusp form. The problem is that *there are no cusp forms of weight 2 and level 2!*

So suppose there is a solution of the Fermat equation for some prime p, and use this solution to build a Frey curve E. Let N be the conductor of E (which we determined above). Suppose, also, that STW holds for E, so that there exists a modular form of weight 2 and level N whose Galois representation is the same as the one for E. Then we have the following curious situation: we have a representation $\bar{\rho}$ which we know comes from a modular form of weight 2 and level N, but which *looks* as if it should come from a modular form of smaller level.

Here is where Ribet's theorem comes in: he proves that (under certain hypotheses which will hold in our case) whenever this happens the modular form of smaller level must actually exist! Notice that this doesn't mean that the original modular form came from lower level; what it means is that there is a form of lower level whose representation reduces modulo p to the same representation.

The upshot of Ribet's theorem is the following:

Theorem 1 (Ribet). *Suppose STW holds for all semistable elliptic curves. Then FLT is true.*

This is true because if FLT were false, one could choose a solution of the Fermat equation and use it to construct a Frey curve, which would be a semistable elliptic curve. By STW, this curve would be attached to a modular form, so that its Galois representation is attached to a modular form. By Ribet's theorem, there must exist a modular form of weight 2 and level 2 which gives the same representation modulo p. Just a little more work allows one to check that this modular form must be a cusp form. But this is a contradiction, because there are *no* cusp forms of weight 2 and level 2.

3.3 Deforming Galois Representations. It is now that we come to Wiles' work. His idea was that one can attack the problem of proving STW by using the Galois representations, and in particular by thinking of "deformations" of Galois representations. The idea is to consider not only a representation modulo p, but also *all* the possible p-adic representations attached to it (one speaks of "all the possible lifts" of the representation modulo p). These can be thought of as "deformations" because, from the p-adic point of view, they are "close" to the original representation.

This sort of idea had been introduced by B. Mazur in [**Maz89**]. Mazur showed that one could often obtain a "universal lift," i.e., a representation into GL_2 of a big ring such that all possible lifts were "hidden" in this representation. If one knew that the representation modulo p were modular, then one could make another big ring "containing" all the lifts which are attached to modular forms. The abstract deformation theory then provides us with a homomorphism between these two rings, and one can try to prove that this is an isomorphism. If so, it follows that all lifts are modular.

What Wiles proposes to do is very much in this spirit, except that he restricts himself to lifts that have especially nice properties. He starts with a representation modulo p, and supposes that it is modular and that it satisfies certain technical assumptions. Then he considers all possible deformations which "look like they could be attached to forms of weight 2," and gets a deformation ring. Considering all deformations which are attached to modular forms of weight 2 gives a second ring (which is closely related to the algebra generated by the Hecke operators, in fact). Wiles then attempts to prove, using a vast array of recent results, including ideas of Mazur, Ribet, Faltings, V. Kolyvagin, and M. Flach, that these two rings are the same.

It is not hard to see that the homomorphism between the two rings we want to consider is surjective. The difficulty is to prove it is also injective. Wiles reduces this question to bounding the size of a certain cohomology group. It is here that the brilliant ideas of Kolyvagin and of Flach come in. About five years ago, Kolyvagin came up with a very powerful method for controlling the size of certain cohomology groups, using what he calls "Euler systems" (see [**Kol91**] and the survey of the method in [**Maz93**]). This method seems to be adaptable to any number of situations, and has been used to prove several important recent results. The initial breakthrough showing how one could begin to use Kolyvagin's method in our context is due to Flach (see [**Fla92**]), who found a way to construct something that can be thought of as the beginning of an Euler system applicable to our situation. Wiles called on all these ideas to construct a "geometric Euler system" which plays a central role in the argument. (*It is at this point that the current difficulty lies.*)

From the bound on the cohomology group one will get a proof that the two rings are in fact isomorphic. Translated back to the language of representations, this means that if one starts with a representation modulo p which satisfies Wiles' technical assumptions (and is modular), then any lift of the kind Wiles considers is also modular.

3.4 Put it all together.... Assume, then, that one can prove that all lifts of a modular representation are still modular. Now suppose we have an elliptic curve E whose representation modulo p we can prove (by some means) to be modular. Suppose also that this representation satisfies Wiles' technical assumptions. Then any lift of this representation is modular. But the p-adic representation $\rho_{E,p}$ attached to E is one such lift! It follows that this representation is modular, and hence that E is modular.

All we need, now, is to prime the pump: we must find a way to decide that the representation modulo p is modular, and then use that to clinch the issue. What Wiles does is quite beautiful.

First of all, he takes a semistable elliptic curve, and looks at the Galois representation modulo 3 attached to this curve. At this point, there are two possibilities. The representation, as we pointed out above, amounts to an action of the Galois group on the vector space $\mathbb{F}_3 \times \mathbb{F}_3$. Now, it may happen that there is a subspace of that vector space which is invariant under every element of the Galois group. In that case, one says that the representation is *reducible*. If not, it is *irreducible*.

One has to be just a little more careful, Just as it sometimes happens that a real matrix has complex eigenvalues, it can happen that the invariant subspace only exists after we enlarge the base field. We will say a representation is *absolutely irreducible* when this does not happen: even over bigger fields, there is no invariant subspace.

Well, look at $\bar{\rho}_{E,3}$. It may or may not be absolutely irreducible. If it is, Wiles calls upon a famous theorem of J. Tunnell, based on work of R.P. Langlands (see

[**Tun81**], [**Lan80**]) to show that it is modular, and hence, using the deformation theory, that the curve is modular.

If $\bar{\rho}_{E,3}$ is not absolutely irreducible, Wiles shows that there is another elliptic curve which has the same representation modulo 5 as our initial curve, but whose representation modulo 3 *is* absolutely irreducible. By the first case, it is modular. Hence, its representation modulo 5 is modular. But since this is the same as the representation modulo 5 attached to our original curve, we can apply the deformation theory for $p = 5$ to conclude that our original curve is modular.

If Wiles' strategy is successful, we get:

Theorem 2. *The Shimura-Taniyama-Weil conjecture holds for any semistable elliptic curve.*

And, since the Frey curve is semistable,

Corollary 1. *For any $n \geq 3$, there are no non-zero integer solutions to the equation $x^n + y^n = z^n$.*

Of course, this is just *one* corollary of the proof of the STW conjecture for semistable curves, and it is certain that there will be many others still. For example, as Serre pointed out in [**Ser87b**], one can apply Frey's ideas to many other diophantine equations that are just as hard to handle as Fermat's. These are equations that are closely related to the Fermat equation, of the form

$$x^p + y^p = Mz^p,$$

where p is a prime number and M is some integer. From Serre's argument and Wiles' result, one gets something like this:

Corollary 2. *Let p be a prime number, and let M be a power of one of the following primes:*

$$3, 5, 7, 11, 13, 17, 19, 23, 29, 53, 59.$$

Suppose that $p \geq 11$ and that p does not divide M. Then there are no nonzero integer solutions of the equation $x^p + y^p = Mz^p$.

The proof is precisely parallel to what we have done before: given a solution, construct a Frey curve, and consider the resulting modular form. Apply Ribet's theorem to lower its level, and then study the space of modular forms of that level to see if the form predicted by Ribet is there. If there is no such form, there can be no solution.

In fact, one can even get a general result, as Mazur pointed out:

Corollary 3. *Let M be a power of a prime number ℓ, and assume that ℓ is not of the form $2^n \pm 1$. Then there exists a constant C_ℓ such that the equation $x^p + y^p = Mz^p$ has no nonzero solutions for any $p \geq C_\ell$.*

However successful they may be in the end at proving the Shimura-Taniyama-Weil conjecture, Wiles' new ideas are certain to have enormous impact.

REFERENCES

[Ami75] Y. Amice, *Les nombres p-adiques*, Presses Universitaires de France, Paris, 1975.

[C$^+$] Henri Cohen et al., *GP-PARI*, a number-theoretic "calculator" and C library. Available by anonymous ftp from math.ucla.edu.

[Cas86] J. W. S. Cassels, *Local fields*, Cambridge University Press, Cambridge 1986.

[Cas91] J. W. S. Cassels, *Lectures on elliptic curves*, Cambridge University Press, Cambridge, 1991.

[Con] Ian Connell, *APECS: arithmetic of plane elliptic curves*, an add-on to *Maple*. Available by anonymous ftp from math.mcgill.edu.

[Cre92] J. E. Cremona, *Algorithms for modular elliptic curves*, Cambridge University Press, Cambridge, 1992.

[CS86] Gary Cornell and Joseph H. Silverman (eds.) *Arithmetic geometry*, Springer-Verlag, Berlin, Heidelberg, New York, 1986.

[Fla92] M. Flach, *A finiteness theorem for the symmetric square of an elliptic curve*, Invent. Math. 109 (1992), 307–327.

[Fre86] G. Frey, *Links between stable elliptic curves and certain diophantine equations*, Annales Univesitatis Saraviensis, Series math. 1 (1986), 1–40.

[Fre87a] G. Frey, *Links between elliptic curves and solutions of $A - B = C$*, J. Indian Math. Soc. 51 (1987), 117–145.

[Fre87b] G. Frey, *Links between solutions of $A - B = C$ and elliptic curves*, Number Theory, Ulm 1987 (H.P. Schlickewei and E. Wirsing, eds.) Lecture Notes in Mathematics, vol. 1380, Springer-Verlag, 1987.

[Gou93] Fernando Q. Gouvêa, *p-adic numbers: an introduction*, Springer-Verlag, Berlin, Heidelberg, New York, 1993.

[Hus87] Dale Husemöller, *Elliptic curves*, Springer-Verlag, Berlin, Heidelbert, New York, 1987.

[Kna92] Anthony W. Knapp, *Elliptic curves*, Princeton University Press, Princeton, 1992.

[Kob84] N. Koblitz, *p-adic numbers, p-adic analysis, and zeta-functions*, second ed., Springer-Verlag, Berlin, Heidelberg, New York, 1984.

[Kol91] V. Kolyvagin, *Euler systems*, The Grothendieck Festschrift, vol. 2, Birkhauser, 1991, pp. 435–483.

[Lan76] Serge Lang, *Introduction to modular forms*, Springer-Verlag, Berlin, Heidelberg, New York, 1976.

[Lan80] R.P. Langlands, *Base change for* GL(2), Ann. of Math. Stud., vol. 96, Princeton University Press, Princeton, NJ, 1980.

[Lan91] Serge Lang, *Number theory III*, Encyclopedia of Mathematical Sciences, vol. 60, Springer-Verlag, Berlin, Heidelberg, New York, 1991.

[Maz89] Barry Mazur, *Deforming Galois representations*, Galois Groups Over $\mathbb{Q}$ (Y. Ihara, K. A. Ribet, and J.-P.Serre, eds.) Springer-Verlag, 1989.

[Maz91] Barry Mazur, *Number theory as gadfly*, American Mathematical Monthly 98 (1991), 593–610.

[Maz93] Barry Mazur, *On the passage from local to global in number theory*, Bull. Amer. Math. Soc 29 (1993), 14–50.

[Miy89] Toshitsune Miyake, *Modular forms*, Springer-Verlag, 1989.

[Rib79] Paulo Ribenboim, *13 lectures on Fermat's Theorem*, Springer-Verlag, Berlin, Heidelberg, New York, 1979.

[Rib90] Kenneth A, Ribet, *On modular representations of* $\mathrm{Gal}(\overline{\mathbb{Q}}/\mathbb{Q})$ *arising from modular forms*, Invent. Math. 100 (1990), 431–476.

[Ser87a] Jean-Pierre Serre, *Lettre à J-F Mestre*, Current Trends in Arithmetical Algebraic Geometry (Kenneth A. Ribet, ed.), Contemporary Mathematics, vol. 67, American Mathematical Society, 1987.

[Ser87b] Jean-Pierre Serre, *Sur les représentations modulaires de degré 2 de* $\mathrm{Gal}(\overline{\mathbb{Q}}/\mathbb{Q})$, Duke Math. J. 54 (1987), 179–230.

[Shi71] G. Shimura, *Introduction to the arithmetic theory of automorphic forms*, Princeton University Press, 1971.

[Sil86] Joseph H. Silverman, *The arithmetic of elliptic curves*, Springer-Verlag, Berlin, Heidelberg, New York, 1986.

[Sil93] Joseph H. Silverman, *Taxicabs and sums of two cubes*, American Mathematical Monthly 100 (1993), no. 4, 331–340.

[ST92] Joseph H. Silverman and John Tate, *Rational points on elliptic curves*, Springer-Verlag, Berlin, Heidelberg, New York, 1992.

[SvM]   J. H. Silverman and P. van Mulbregt, *EllipticCurveCalc*, a *Mathematica* package. Available by anonymous ftp from gauss.math.brown.edu; contact jhs@gauss.math.brown.edu for information.
[Tun81] J. Tunnell, *Artin's conjecture for representations of octahedral type*, Bull. Amer. Math. Soc. (N. S.) **5** (1981), 173–175.
[Was82] Larry C. Washington, *Introduction to cyclotomic fields*, Springer-Verlag, Berlin, Heidelberg, New York, 1982.
[Wei83] André Weil, *Number theory: an approach through history, from Hammurapi to Legendre*, Birkhäuser, 1983.
[Z$^+$] H. G. Zimmer et al., *SIMATH*, a computer algebra system with main focus on algebraic number theory. Contact simath@math.uni-sb.de for more information.

Colby College
Department of Mathematics and Computer Science
Waterville, ME 04901
fqgouvea@colby.edu

Originally appeared as:
Gouvêa, Fernando Q. "A Marvelous Proof." *American Mathematical Monthly*. vol. 101, no. 3 (March 1994), pp. 203–222.

Second Helping

Fernando Q. Gouvêa

My article "A Marvelous Proof" was written in the heady days just after Wiles' announcement, at the Newton Institute in Cambridge, that he had proved the semi-stable case of the Modularity Conjecture (which I call "STW" in the paper) and therefore had also proved Fermat's Last Theorem. By the time my article was published, Wiles had announced that a gap had been found. The final triumph was still some six months in the future, which accounts for the tentative tone taken at the end of the article.

I was visiting Harvard in the Fall of 1994, and one day I walked into the department to find a couple of preprints in my mailbox: an article by Andrew Wiles and Richard Taylor supplying a crucial step, and the main article by Andrew Wiles containing the finished proof. The semistable case of the Taniyama Conjecture was now a theorem, and as a result Fermat's Last Theorem was at last a real theorem. The papers were published as [12] and [10].

"A Marvelous Proof" largely avoids getting into the technical details of Wiles' proof that the (appropriate) deformation ring $\mathbf{R}$ is indeed isomorphic to the (appropriate) Hecke algebra $\mathbf{T}$. Since that is where the trouble was, there is not a lot to add to the description in section 3.3. The final argument relies on constructing a large family of deformation rings $\mathbf{R}_\Sigma$ by varying the conditions (encoded in Σ) on the deformed representations. The relationship between these various rings and the corresponding Hecke algebras is what eventually gives the proof. This idea has since been extended and codified into a theory of "Taylor-Wiles systems," which has now become a standard tool in proving more general theorems of the "$\mathbf{R} = \mathbf{T}$" type.

Another fruitful idea in Wiles' proof is the one at the very end, where he managed to propagate the property of modularity from a "seed" representation to a whole family. This idea has been used over and over to prove any number of important results. The most important of these is probably the full proof of the Modularity Conjecture by Breuil, Conrad, Diamond, and Taylor in a series of papers culminating in [1].

More recently, Serre's conjecture "pour optimistes" was proved by using these techniques. The conjecture, stated in [8], says that (odd, absolutely irreducible) Galois representations into $\mathrm{GL}_2(\mathbb{F}_p)$ comes from a modular form of specified level and weight. The case in which the predicted modular form has level 1 was proved by Chandrashekhar Khare in [6]. Khare and Jean-Pierre Wintenberger then extended the result to forms of odd level, and showed that the full conjecture would follow from the general form of a lifting theorem of Mark Kisin. Kisin has recently proved that result, so that

Serre's Conjecture is now proved. (Visit the authors' web sites for preprints of their papers.) The Khare-Wintenberger-Kisin Theorem has important consequences, including the odd, two-dimensional case of Artin's Conjecture. It can be viewed as the strongest evidence so far for the general philosophy that all interesting arithmetical objects should be somehow connected to automorphic forms.

Several books and survey articles on Wiles' proof and the work that followed have been published. In particular, one might cite the proceedings of the 1995 conference at Boston University [2] and the books by van der Poorten [11], Hellegouarch [5]. The recent book on modular forms by Diamond-Shurman [3] has as its main goal to lead the reader to a full understanding of the modularity conjecture. Also worth consulting are the Séminaire Bourbaki "Exposés" on this work, [9, 7, 4].

References

[1] Christophe Breuil, Brian Conrad, Fred Diamond, and Richard Taylor. "On the Modularity of Elliptic Curves over $\mathbf{Q}$: Wild 3-adic Exercises." *J. Amer. Math. Soc.* **14** (2001), 843–939.

[2] Gary Cornell, Joseph H. Silverman, and Glenn Stevens. *Modular Forms and Fermat's Last Theorem.* Springer Verlag, New York, 1997.

[3] Fred Diamond and Jerry Shurman. *A First Course in Modular Forms.* Springer-Verlag, New York, 2005.

[4] Bas Edixhoven. "Rational Elliptic Curves are Modular (after Breuil, Conrad, Diamond and Taylor)." *Séminaire Bourbaki, 1999/2000. Astérisque,* **276** (2002), 161–188.

[5] Yves Hellegouarch. *Invitation to the Mathematics of Fermat-Wiles.* Translated from the second (2001) French edition by Leila Schneps. Academic Press, San Diego, CA, 2002.

[6] Chandrashekhar Khare. "Serre's Modularity Conjecture: the Level One Case." *Duke Math. J.*, **134** (2006), 557–589.

[7] Joseph Oesterlé. "Travaux de Wiles (et Taylor,...). II." *Séminaire Bourbaki, 1994/95,* Exposé 804. *Astérisque* **237** (1996), 333–355.

[8] Jean-Pierre Serre. "Sur les représentations modulaires de degré 2 de Gal($\overline{\mathbf{Q}}/\mathbf{Q}$)." *Duke Math. J.* **54** (1987), 179–230.

[9] Jean-Pierre Serre. "Travaux de Wiles (et Taylor,...). I." *Séminaire Bourbaki, 1994/95,* Exposé 803. *Astérisque* **237** (1996), 319–332.

[10] Richard Taylor and Andrew Wiles. "Ring-Theoretic Properties of Certain Hecke Algebras." *Annals of Mathematics,* 2nd Ser., **141** (3) (1995), 553–572.

[11] Alf van der Poorten. *Notes on Fermat's Last Theorem.* Wiley, New York, 1996.

[12] Andrew Wiles. "Modular Elliptic Curves and Fermat's Last Theorem." *Annals of Mathematics*, 2nd Ser., **141** (3) (1995), 443–551.

About the Editors

Ezra (Bud) Brown grew up in New Orleans and has degrees from Rice University and Louisana State University. His doctoral advisor was Gordon Pall, who encouraged him to "write your mathematics as if you wanted someone to read it." Since 1969 he has been in the Mathematics Department at Virginia Tech, where he is currently Alumni Distinguished Professor.

He spent one sabbatical year (and every summer since 1993) in Washington, D. C. and another in Munich. He is the author of some sixty papers, mostly in number theory and discrete mathematics. His translation of *Regiomontanus: his life and work*, Ernst Zinner's biography of that fifteenth century mathematician/astronomer, was published in 1990 by North-Holland.

He is a member of the Mathematical Association of America (MAA) and Pi Mu Epsilon national mathematics honor society. He received the Outstanding Teacher Award from the MD/DC/VA Section of the MAA, and he currently (2008) serves as that section's governor. He received the Carl Allendoerfer Award (2003) and three George Pólya Awards (2000, 2001, 2006) from the MAA for expository writing. He enjoys his family, singing, playing piano, and gardening, and he occasionally bakes biscuits for his students.

Arthur (Art) Benjamin earned his B.S. in Applied Mathematics from Carnegie Mellon and his Ph.D. in Mathematical Sciences from Johns Hopkins. Since 1989, he has taught at Harvey Mudd College, where he is Professor of Mathematics and past Chair. In 2000, he received the Haimo Award for Distinguished Teaching by the Mathematical Association of America. He served as the MAA's Pólya Lecturer from 2006 to 2008.

His research interests include combinatorics and number theory, with a special fondness for Fibonacci numbers. Many of these ideas appear in his book (co-authored with Jennifer Quinn), *Proofs That Really Count: The Art of Combinatorial Proof*, published by the MAA. In 2006, that book received the Beckenbach Book Prize by the MAA. From 2004 to 2008, profesors Benjamin and Quinn were the editors of the *Math Horizons* magazine, published by MAA.

Art is also a magician who performs his mixture of math and magic to audiences all over the world, including the Magic Castle in Hollywood. He has demonstrated and explained his calculating talents in his book *Secrets of Mental Math* and on numerous television and radio programs, including the Today Show, CNN, and National Public Radio. He has been featured in *Scientific American*, *Omni*, *Discover*, *People*, *Esquire*, *New York Times*, *Los Angeles Times*, and *Reader's Digest*. In 2005, *Reader's Digest* called him "America's Best Math Whiz."